OTTER SURVEY OF ENGLAND 1991

A REPORT ON THE DECLINE AND RECOVERY OF THE OTTER IN ENGLAND AND ON ITS DISTRIBUTION, STATUS AND CONSERVATION IN 1991–1994

R. Strachan and D.J. Jefferies

The Vincent Wildlife Trust
1996

The Vincent Wildlife Trust, 10 Lovat Lane, London EC3R 8DT

'Otter at dusk' – from a painting by Bruce Pearson.

CONTENTS

1. SUMMARY

The third Otter Survey of England was carried out from October 1991 to January 1994. A total of 3,188 full survey sites (each 600m long) was examined and signs of otters were found at 706 (22.15%) of them. In addition, 112 spot checks were made (ie. searches under bridges and surrounds) and 45 of these were found to be positive (ie. with signs of otters). Results are presented on the basis of percentage of full survey sites found positive within alternate 50km squares of the National Grid and within the ten Regions of the National Rivers Authority (NRA).

In the baseline survey of England in 1977-79 (Lenton *et al*, 1980) a total of 2,940 full survey sites was examined and signs of otters were found at 170 (5.78%) of these. In the second otter survey of 1984-86 (Strachan *et al*, 1990) a total of 3,188 sites was examined, including the 2,940 of 1977-79. 286 sites (8.97%) were found to be occupied. A comparison of the results for the 2,940 sites examined in all three surveys shows 170 (5.78%), 284 (9.66%) and 687 (23.37%) occupied in 1977-79, 1984-86 and 1991-94 respectively. There was a 67.13% increase in percentage of sites occupied between 1977-79 and 1984-86 and a 141.93% increase between 1984-86 and 1991-94. This is a 304.33% increase in 14 years. These three increases are statistically significant.

Thus, the recovery of the English otter population suggested by the 1984-86 survey has been confirmed and extended by that of 1991-94. Also, recovery has been confirmed in separate surveys of Wales (Andrews & Crawford, 1986; Andrews *et al*, 1993) and Scotland (Green & Green, 1987). Otters are now present in every one of the 32 50km squares surveyed in England and in every one of the ten NRA Regions. There were statistically significant increases in the numbers of occupied sites in eight out of the ten NRA Regions between 1984-86 and 1991-94. The remaining two Regions showed small increases. Recovery is from the west (ie. south-west England and the Welsh borders) towards the east and from the north towards the south. Population expansion and recolonisation is occurring both through breeding and by movement.

The two population strengthening programmes started in 1983 have released 55 captive-bred otters (the Otter Trust) to East Anglia and central-southern England and 25 rehabilitated otters (The Vincent Wildlife Trust (VWT)) to North Yorkshire up to the end of 1993. These two programmes have been successful, with frequent proven breeding in the wild. Some of the young otters seen must represent the third generation. A considerable area and number of rivers in North Yorkshire and East Anglia are now occupied by these stocks. Without them the East Anglian population would very probably have been extinct by 1986.

However, 78% of riparian sites in England are still without otters in 1994, and the populations of the Midlands, central-southern and south-eastern England are still very small and sparsely distributed. Calculation of the recovery curve (which fits the population changes of Scotland and Wales as well as England) shows that recovery to 75% site occupation over all of England (still far below the 1980-81 level for Ireland; Chapman & Chapman, 1982) is unlikely before the year 2025.

The causes of the sudden population crash of the otter in southern Britain, starting in 1957, and its recovery in the 1970s and 1980s are re-examined using the increased data from the three otter surveys and other information from the records of otter hunts. All of the available information supports a hypothesis that the immediate cause of the crash was a very high mortality of breeding-age adults following the introduction of the highly toxic and persistent organochlorine insecticides dieldrin and aldrin as seed dressings and sheep-dips in 1956. However, the otter population suffered severe persecution in the decades preceding dieldrin use and this had already reduced its numbers to a low level. This persecution was particularly severe in the east of England. The eventual crash would not have been so serious without this previous reduction in population density. The otter population of the west was not so severely affected by organochlorines, as the acreage treated with dieldrin was much less than in the east. Also, recovery started earlier in the west because of the early (1966) bans on the use of dieldrin in sheep-dips. Hence the observed recovery is progressing from the west towards the east, as would be expected. Only two major actions have been taken to bring about this recovery and these are (1) the bans on all agricultural uses of the organochlorine insecticides by 1981-82 and (2) legal protection of the otter in 1978 and 1982. These again correlate with the above hypothesis.

There has been no delay in recovery, which started within a short period of locally important bans on dieldrin use. Recovery of some western Regions, particularly in Wales with much suboptimal upland habitat, may be depressed for several years as much of the annual cub production may move out eastwards into neighbouring Regions and into England, while many prime sites and habitat remain empty or under-used there. However, eventually even suboptimal upland streams may be inhabited if recovery continues.

Further examination of the survey data collected shows that there is a significant positive relationship between the percentage site occupation of a Region and the number of spraints per occupied site in that Region (spraint density). That is, as one increases, so does the other. There are other significant negative relationships between percentage site occupation and spraint density and the altitude, ie. as altitude increases, then site occupation and spraint density both decrease. The only conclusion which can be drawn from these consistent patterns is that spraint density is indicative of otter density. Thus, spraint density can be used as a 'fine adjustment' for the results of otter spraint surveys. For example, spraint density, and so otter density, may increase in an area in which percentage site occupation remains the same.

The long-discussed problem of the relationship between the feral American mink and the otter is clarified. The mink was not a causal factor in the otter's decline, nor did its presence prevent the otter's recolonisation of mink-occupied areas. Indeed, otters appear to show strong antagonism towards mink and to cause a decrease in their numbers and site occupation (presumably by predation) everywhere that the two species occur together and where the otter density has shown a marked recovery (eg. in the south-west of England). Here there has been a 50% reduction in mink occupation since 1984-86.

The factors influencing the size of the British otter population and threats to its continuation are discussed, and information on conservation problems is updated. The high mortality on the road, the low age at death, affecting the rate of recruitment and replacement, and the potential problem of contagious lethal diseases contracted from captive mustelids appear to be the most serious of the threats. The report concludes with a discussion of past, present and the requirement for future conservation action for the otter.

2. INTRODUCTION

The otter *Lutra lutra* (L. 1758) in England is largely a secretive and nocturnal animal and is rarely seen. However, its presence can be readily determined by searching for the animal's spraints (faeces) and footprints or other signs.

The baseline survey of the distribution and density of occupied sites for the otter in England, which was based on a systematic search for the animal's field signs, was carried out over the years 1977-79. A total of 2,940 sites was examined and evidence of otters was located at only 170 (5.78%) of these (Lenton *et al*, 1980). This provided further evidence of the results of a widespread decline which started in 1957 and continued through the 1960s and 1970s (Chanin & Jefferies, 1978). The results of the baseline survey showed that the species was sparsely distributed or absent over most of lowland and central England, but still present at many of the sites in the south-west and on the Welsh borders. In northern England, East Anglia and the counties of the south coast the population appeared to be small and fragmented.

Using the same standardised technique, concurrent surveys were also carried out in Wales and Scotland (and slightly later in Ireland). These found otters present at 20% of sites in Wales (Crawford *et al*, 1979), 73% of sites in Scotland (Green & Green, 1980) and 92% of sites in Ireland (Chapman & Chapman, 1982).

After an interval of seven years the surveys for England, Scotland and Wales were repeated using exactly the same methods and visiting all of the same series of sites, keeping as close as possible to the same sequence and date as in the original survey. This was to facilitate a direct comparison without the complication of seasonal variations in weather, water levels and bankside vegetation cover.

The Otter Survey of England 1984-86 showed a marked (67%) overall improvement in the results, with 9.66% sites positive. However, at the Regional level, although seven of the ten areas showed some evidence of increase, one remained negative and two had continued to decline further (Strachan *et al*, 1990).

After an interval of another seven years, the surveys were again repeated to assess the trends in the changing distribution and status of the otter in each of the three countries. The results for England for the period 1991-94 are detailed in this report.

Now that the results of three England surveys are to hand (as well as one of Ireland, two of Scotland and three of Wales), the data can be further analysed as a whole in order to show the national rate of recovery and the shape of the recovery curve, which can be used for predictive purposes. The larger database also allows examination of the possibility of an interaction between otters and feral American mink in several areas and a further exploration of the causes of the 1957 crash in the British otter population. In addition, we have returned to an analysis of the value of spraint density in surveying otters.

The field surveying, 1991-94 data collection and Section 5 were completed by R. Strachan, and the data analysis and writing of the remainder of the report were carried out by D.J. Jefferies.

3. METHODS

3.1 History

Confirmation of the severe decline in the otter populations of England and Wales (Chanin & Jefferies, 1978) allowed a case to be made by the Nature Conservancy Council for legal protection for the species under the Conservation of Wild Creatures and Wild Plants Act 1975 (O'Connor *et al*, 1979). Inclusion on Schedule 1 of that Act provided the otter with legal protection in England and Wales from 1st January 1978. The Masters of Otter Hounds Association anticipated the scheduling of the otter and announced the suspension of otter hunting in England and Wales in 1977. This, of course, ended any further use of hunting records (the method used by Chanin and Jefferies, 1978) for future monitoring of the status and distribution of the species. However, it was essential to have some means of monitoring in order to see if the situation was worsening or improving and to determine whether conservation measures were having any effect. An alternative technique was necessary.

One of the authors (DJJ) was asked by the Joint Otter Group to design and co-ordinate a national survey programme for England, Scotland and Wales in 1977. The survey design needed to provide information on distribution and relative density of occupied sites in Regions and countries. Also, as they were to be used to monitor changes in status with time, surveys needed:

(1) to have a standard, objective and easily repeatable method so that they could be repeated at intervals for comparative purposes

(2) to be carried out over as short a period as possible so that results would be representative of the situation at one point in time (ie. two years rather than ten)

(3) to be carried out by as few experienced surveyors as possible in order to reduce any biases due to differing skills.

Initially, the use of couples of otter hounds for surveying purposes was suggested and offered, but it was considered that these might have caused too much disturbance to the remaining otter population and could have produced difficulties in interpretation. Thus, although there appeared to be potential problems (reviewed by Jefferies, 1986) associated with techniques reliant on finding spraints, since sightings and good footprints were so infrequent in England and Wales, spraints were the only relatively common signs of presence left for analysis. Consequently, a survey was designed based on spraint (and footprint) searches at a series of sites of standard length. The first survey was started in England, Scotland and Wales in 1977. The original survey design is included in Appendix A of Lenton *et al* (1980).

3.2 Survey method adopted as standard for Britain

Full survey sites:

These were originally selected (in 1977-79) at about 5km intervals along each waterway, coast or lake/reservoir shore, giving a mean frequency of six sites per 10km square. With the very large area and the expected large number of negatives increasing the time spent searching at each site, only half of England could be covered by one surveyor in two years, though Wales and Scotland could be covered completely. Consequently, only alternate 50km squares (north-west and south-east diagonal quarters of each 100km square of the National Grid) were surveyed in England, in order to give representative cover of all regions. A 600m walk along one bank (300m upstream and 300m downstream from a starting point such as a bridge) was mapped and examined for otter signs at each of these full survey sites. Otter presence, in terms of the percentage of sites found to be positive in each region or 50km square, could then be compared on a geographical or temporal basis. Owing to the impossibility of proving a negative by this means, the species could only be termed "sparse to absent" where no otter signs could be found.

Figure 1. The alternate 50km squares (north-west and south-east diagonals of each 100km square of the National Grid) which were surveyed in the otter surveys of England. The code letters used by the Ordnance Survey are shown for each 100km square. When reference to a particular survey square is made in the text, these code letters plus the diagonal reference is given, eg. SK n/w.

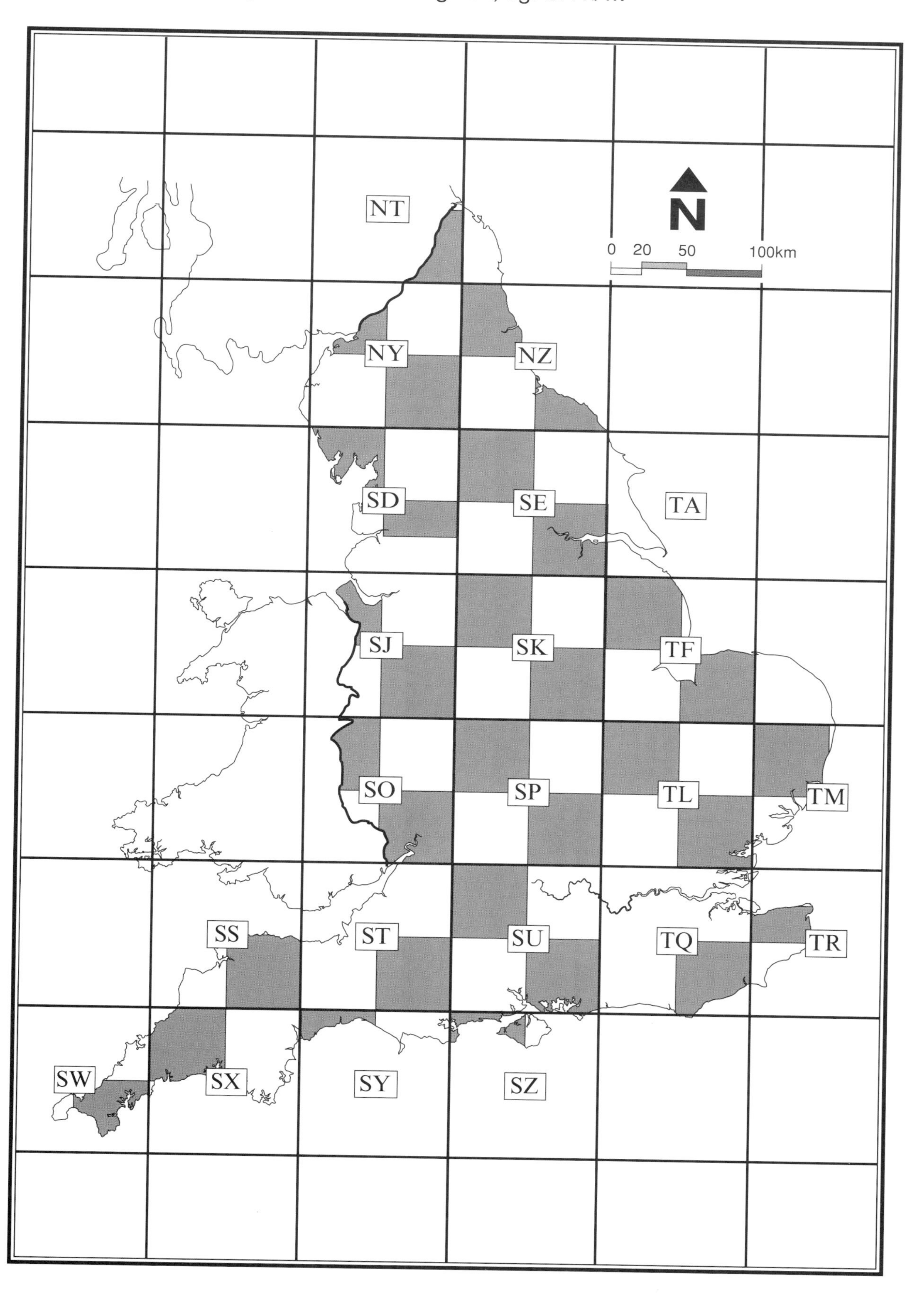

Number of sites and timing of survey:

2,940 sites were mapped and surveyed in 1977-79 to form the baseline from which to monitor change. In 1986, a further 248 sites were surveyed (making 3,188 in all) when four new 50km squares were added (72, 34, 30 and 112 sites in squares SK n/w, TR n/w, NZ s/e and TL s/e, respectively) to improve the information on distribution. These 3,188 previously selected sites were resurveyed during 1991-94 to provide the database analysed here. Thus, the total distance surveyed in this sample was 1,912.8km or 1,189 miles. Full survey sites were re-examined in the same chronological order and at the same time of year (usually within three weeks of the original date) as in the first two surveys. This was done in order to reduce any differences in the circumstances between surveys, ie. regarding water level, weather, vegetative cover and spraint density.

Spraint density:

It was decided that, as in 1984-86, the full 600m should be examined for spraint sites and all spraints counted even if spraints were found (and so the site declared positive) in the first few metres. This enabled a very useful analysis of spraint density to be made (see Section 6.1), which provided further information on otter presence at each positive site.

Data recorded:

The recording sheet used in the 1991-94 survey followed the same format as that used in the baseline survey (see Appendix 1 for an example). Type of waterway, current, width and depth, shore type and vegetation cover were recorded at each site. Adjacent land use, water use and any riparian engineering works were also noted together with a score for apparent disturbance (on a scale of 0 to 5, where 5 = highly disturbed). Any changes to the site or survey route were marked on the annotated sketch map. The presence of both otter and mink was recorded and a description made of each spraint site encountered. These data have yet to be analysed.

Spot checks:

Full surveys were supplemented by ‘spot checks’. These were made at selected bridges which provided suitable ledges or earth banks as potential sprainting places for otters. Here, only the immediate surrounds of the bridge were searched for 10m of both banks. The results of spot checks, which were not included in the statistical analysis of differences between surveys (see Section 4), provided further information on distribution, eg. on the limits of occupation along a waterway. Sometimes they provided the sole indication of otter presence. A total of 174 spot checks was made in each of the 1977-79 and 1984-86 surveys, but only 112 in 1991-94 (see Table 1). However, each of the positive spot check sites was revisited in the third survey. The results shown for spot checks in each of the NRA Regions (see Section 5) are those for the 112 spot checks repeated in all surveys.

Non-visited, assumed negative sites:

Non-visited and assumed negative sites were introduced in the baseline survey to make coverage consistent over the whole country, especially in those areas which included large urban conurbations such as Birmingham or Liverpool. Without this category a disproportionately high percentage of positive sites could result from survey of a 50km square consisting of mainly urban areas, if only fringe sites were visited and found to hold otters. 241 non-visited and assumed negative sites were incorporated in the 1977-79 and 1984-86 surveys. In 1991-94, only 85 sites were non-visited (SU s/e (11 sites), SX n/w (6), NZ n/w (9), SP n/w (32), SD s/e (20), SJ s/e (7)). 156 of the original 241 sites previously assumed negative were upgraded and visited in 1991-94, and some were found to have otter spraints.

Surveyor:

The 1977-79 survey of England was carried out by one full-time surveyor, because it was expected that a single surveyor could apply constant effort throughout, providing even coverage, particularly in those areas holding few or no otters. The 1984-86 survey was begun by one surveyor in October 1984 and completed by another of similar skills in

September 1985. The 1991-94 survey was again carried out by a single surveyor (RS), who had completed the second otter survey of England, thus providing a continuity of survey effort. This limits any biases in the results and permits a more reliable comparison with the 1984-86 survey findings.

3.3 Tests of the method adopted

At the time of the first survey, several tests of the method were made and reported. Thus, in England, Lenton *et al* (1980) surveyed a 6km test length of several rivers known to be inhabited by otters in 1978 and 1979. If these lengths are divided into 600m stretches (the length of a survey site), each starting at 50m intervals, then 109 different stretches could have been selected for examination within the 6km. On the River Wansbeck (Northumberland), for example, 16 spraint sites were found and 99 of the 109 stretches would have contained a sprainting site and only ten would not. Of these 99 positive sites, 33, 47, 16 and 3 would have contained one, two, three and four sprainting sites, respectively, and 91% of the possible 600m survey sites would have been positive. A similar survey of a 6km test length of a Shropshire river found 18 spraint sites and again 91% of the 109 possible survey sites would have been positive. This percentage is very large and indicates a high probability of determining occupation. On the other hand, on the River Wensum (Norfolk), where otter activity was very low in 1978, only three sprainting sites were found along the 6km test length and only 27 (or 25%) of the 109 possible survey sites would have been positive. Thus, where otter activity is low or otters may be transient and passing through, the chances of proving presence from one survey site in 6km are much reduced below that when they are in residence.

As the greatest distance between two sprainting sites was 1,100m on the Wansbeck and 950m on the Shropshire river, one could only increase the chances of finding a sprainting site on the test 6km of these occupied rivers to 100% by increasing the single survey site length from 600 to 1,100 or 950 metres. However, it should be remembered that in practice 600m survey sites are placed at 5km intervals along a river, so a 6km length measured from the start of any 600m survey site would contain almost two such survey sites (or 1,000m). This considerably increases the chances of finding signs when an otter is in residence in that 6km. Further, in the field survey itself, rivers have at least three or four and many over ten 600m survey sites in their length, so establishing that a river with resident otters is indeed occupied is virtually certain. Thus, nine out of ten sites may be found positive on rivers with a resident population like the Wansbeck (with 91% of possible survey sites positive) and one out of four where otter activity is low or itinerant as on the Wensum (with 25% of possible survey sites positive). Such rivers with an almost continuous series of occupied sites or, alternatively, just sporadic positive sites can be found on examination of the Regional survey maps in Section 5. This indicates that it should not be assumed that there is no otter activity or sprainting on the occasional stretches of river with negative survey sites in an otherwise positive river, but just that the spraint density was too low for the survey technique to pick them up every time. Even where all the sites along a river are found to be negative one cannot assume absence as lone individuals may not mark the banks traversed (see Section 3.4).

Green and Green (1980) tested the survey technique with regard to river occupation after surveying Scotland in 1977-79. They revisited and resurveyed 227 sites on 31 rivers after the survey had ended. Only eight of the 227 provided new records. Status on 29 rivers remained unchanged but presence was established on two rivers in Lothian which were previously found to be negative. These (rivers Tyne and South Esk) were the only two rivers to be found occupied in Lothian Region in the 1984-85 survey too, so the resurvey may have detected the first signs of re-occupation.

In conclusion, the overall pattern which emerges after plotting all the positive and negative sites on the maps of the rivers is a definite rather than a random one. Also, repeated surveying shows that not only do occupied rivers remain occupied and many apparently empty rivers stay so, but there are progressive changes in occupation between surveys. This indicates that the technique as designed in 1977 is basically reliable for monitoring the status and distribution of the otter in Britain. The correlation of the results from all eight surveys of England, Scotland and Wales (see Section 6.2) provides confirmation.

3.4 Factors likely to influence survey results and methods of obviating them

Habitat type:

The physical characteristics of some sites makes them easier to survey than others. This depends on factors such as the steepness of the banks, the depth of the water and the presence of boulders and trees. A spraint is more likely to persist and remain visible on a hard substrate than on a tuft of grass. Some sites may be more attractive to otters or may be sprainted upon more frequently (Macdonald *et al*, 1978; Macdonald & Mason, 1983).

Weather:

Sites in the south-west and north-west of England receive a higher rainfall than those of the east. As well as affecting the persistence of spraints, the timing of heavy rain before a survey can affect results in other ways. Spates and floods make sprainting sites inaccessible and will conceal or wash away signs. On the other hand, fresh sandy banks, ideal for detecting tracks, may be exposed when the flood water subsides. Low water and drought conditions cause dry banks with a low likelihood of tracks being made, though spraints persist longer. Recent snowfall may allow tracks to be found even in areas of very low otter density.

Bridges:

It had been known for a long time that certain bridges (those with ledges or banks underneath the arches) were frequently used as sprainting sites. Spraints also persist longer when under cover. Thus, otter evidence is more likely to be found where suitable bridges are present and they occur more frequently in some localities than in others. The only evidence which could be found of otter presence on the rivers Stour (Dorset) and Avon (Hampshire) in the 1977-79 survey was under such bridges (Lenton *et al*, 1980).

Season:

Dense bankside vegetation creates problems for surveying lowland areas in summer and leaf fall provides problems in wooded areas in the autumn. Spate conditions are more frequent in upland areas in winter, with melting snow. Bank maintenance and dredging of lowland waterways and dykes are regularly carried out in summer and autumn.

Otter density:

It is likely that the presence of other otters stimulates marking behaviour because of territorial needs, while their absence may depress it (Green *et al*, 1984). Indeed, there is evidence with other mustelids – weasels (Lockie, 1966), pine martens (Strachan *et al*, in press) and badgers (Skinner *et al*, 1991a, b) – that the territorial system may break down altogether when the population is at a very low density, and individuals may wander over large areas, leaving little sign. Thus, the absence of spraint does not necessarily mean no otters, only that they are likely to be "sparse to absent" (Lenton *et al*, 1980).

Sprainting activity cycles:

It has been found independently in several areas that sprainting activity may show cycles (see Section 6.1.1). However, the peak season may vary considerably, and indeed Marshall (1991), working in Devon, found no cycling of sprainting activity at all.

Methods of obviating some of these factors:

(1) As far as possible, upland regions were surveyed mainly in summer, when water levels were likely to be low. Lowland regions were visited in winter, when bankside vegetation was less dense.

(2) As noted in Section 3.2, all surveys subsequent to the baseline were to be carried

out as close as possible to the timing of the original survey (ie. within three weeks). This was for comparative reasons, ie. so that each site would be resurveyed under similar conditions to those experienced at the initial survey.

(3) With regard to sprainting cycles, spraints do not immediately disappear after production, neither is there a period of no spraints. Mason and Macdonald (1986) found that marked spraints lasted from 3 to 15 weeks. Thus, there is always a carry-over through the period of low sprainting activity. Again, resurvey at the same time of year should ensure that results are comparable between surveys.

3.5 Particular factors influencing the 1991-94 survey results

Lowland England had been experiencing a three-year drought in the late 1980s and this was still in evidence at the time of the start of the second re-survey in October, 1991. There were notably low water flows and indeed complete drying-out of the upper reaches of some rivers owing to failed subterranean aquifers and springs. These conditions could have influenced the survey results for some of the chalk streams in Wessex, Thames and Southern Regions in particular, where the effects of the drought (drying-out) were most marked at the time of survey. Towards the end of the survey in winter 1993-94, the weather conditions were reversed and there was frequent and prolonged rainfall causing severe flooding throughout the country.

Dry banks reduce the chances of good footprints and complete drying-out would mean that the otters would temporarily vacate that stream or river to fish elsewhere, with consequent absence of spraints. Spate and flood conditions not only make sprainting sites inaccessible to the surveyor, but also remove any otter sign that had been present. Any extreme weather conditions which might have influenced the chances of locating otters at any particular site were noted on the recording sheet and have been reported in the paragraph about that river (see Section 5).

3.6 Conclusions

Notwithstanding the many potential problems inherent in a spraint survey (Jefferies, 1986), the survey technique used has been found to produce very useful and repeatable results in practice, with the results of each survey fitting into definite regional and overall British trends. In fact, far from giving just a broad picture, the maps of positive and negative sites have given much detail of the fragmentation of the population in the south and the build-up of the populations derived from released stock. Also, repeat surveys have confirmed many of the areas where otters are still apparently absent.

Results would appear to suggest that the early criticisms of the method (Kruuk & Conroy, 1987) are unfounded and the effects of suggested problems minimal in practice.

4. OVERALL RESULTS FOR ENGLAND

4.1 Results for 1991-94

The results for the third Otter Survey of England 1991-94 are summarised for each of the 50km squares surveyed in Table 1 and for each of the National Rivers Authority Regions in Table 2.

3,188 full surveys were carried out and signs of otters were located at 706 of these, which is an overall success rate of 22.15% positive. In addition, 112 spot checks were carried out at suitable bridges and 46 of these were found to be positive. Some of these spot checks provided information on otters which was not obtained from the full surveys. For example, on the lower Salisbury Avon in Hampshire and on the River Deben in Suffolk, the presence of otters was only confirmed by a bridge spot check when adjacent full surveys failed to show any signs.

When the first survey was carried out in 1977-79, only 2,940 sites were surveyed (see Section 3.2). Survey of these same 2,940 sites in 1991-94 showed that 687 were positive for otters. Calculation shows this to be a slightly higher overall success rate at 23.37%.

The percentage site occupation in each of the 32 50km squares surveyed in 1991-94 is shown 'rounded up' in Figure 2, and the distribution of the occupied 10km squares is shown as a 'dot map' in Figure 4. These Figures include data from the full 3,188 sites surveyed.

In Figure 6 all the presently occupied rivers and watersheds over the whole of England have been mapped using a combination of data from the present survey and from regional and county otter surveyors in areas (50km squares) not covered by the national survey (detailed in Section 5). Those rivers considered to have been colonised or reinforced by released otters and/or their progeny are distinguished from those occupied by wild stock. Thus, Figure 6 can also be used to indicate the extent of the success of the two release programmes (see Section 7.2.6).

4.2 Comparison with the results of the surveys of 1977-79 and 1984-86

The results for the baseline (1977-79) and second (1984-86) English otter surveys are listed for each of the 50km squares surveyed in Table 1 and for each of the National Rivers Authority Regions in Table 2. The percentage site occupation in each of the 50km squares surveyed in 1977-79 and 1984-86 is shown 'rounded up' in Figures 3a and b, the distribution of the occupied 10km squares is shown as 'dot maps' in Figures 5a and b and the occupied rivers and watersheds at those dates are shown in Figures 7a and b. These allow comparison with the results of the present survey.

3,188 full surveys were carried out in 1984-86 while only 2,940 sites were surveyed in 1977-79. The increased number was completed in order to improve the information on distribution (see Section 3.2). However, when comparing the overall success rate with that in the baseline, one can only make a comparison between those sites which were examined in both surveys. Thus, in 1977-79 the 2,940 sites given a full survey were found to be 5.78% positive (170 sites with signs). On resurveying these same 2,940 sites in 1984-86 the percentage which was positive had increased to 9.66% (284 sites with signs), which is 1.67 times the initial figure. Occupation of these same 2,940 sites had risen to 23.37% (687 sites with signs) in 1991-94, which is 2.42 times the figure for 1984-86 and 4.04 times the figure for 1977-79, some 14 years before the latest survey.

There are now (1991-94) signs of otters in every one of the 32 50km squares surveyed in England (see Figure 2). This is a marked progressive improvement over the situation in 1984-86 when 9 out of 32 (28%) and in 1977-79 when 11 out of 28 (39%) squares were apparently empty of otters (see Figures 3a and b).

4.3 Results for the National Rivers Authority Regions

The individual survey results for each of the ten NRA Regions in England are detailed and discussed in Section 5. A summary of the results for each Region is given in Table 2 in terms of the numbers and percentages of sites found to be occupied in each of the three surveys. The percentage changes in occupation between 1977-79 and 1984-86, 1984-86 and 1991-94 and 1977-79 and 1991-94 are shown in Table 3. The level of statistical significance of these changes is provided in Appendix 2. The increase in the number of occupied sites between 1984-86 and 1991-94 was found to be 'highly significant' in seven Regions (and overall). Of the remaining three Regions, the difference in occupation was 'probably significant' in Wessex and showed a small but non-significant upward change in Thames and Southern Regions.

4.4 Conclusions

The results show that otters are now present in all surveyed 50km squares and all ten NRA Regions. The overall number of occupied sites in England increased significantly (by a factor of x1.67) between 1977-79 and 1984-86 and then more than doubled (from 284 to 687; factor x2.42) between 1984-86 and 1991-94. This increasing rate of upward change in occupation is to be expected, but gains are likely to decrease between future seven-year surveys because of the sigmoid form of the calculated recovery curve (see Section 6.2).

The increasing otter populations of Wales and western and northern England have spread eastwards and southwards and now link with the augmented eastern population. However, otters are still very few indeed and the population is discontinuous across the Midlands and into the southern and south-eastern parts of the country.

Although this increase and the present overall percentage site occupation of 23.37% are very encouraging for conservationists, it is nevertheless true that over 76% of waterway sites in England are still without otters, whereas only 8% are in Ireland (Chapman & Chapman, 1982).

Due to the reduced scale of the Regional maps, the number of full survey sites illustrated may not agree with the figures given in the text. This is because in some instances overlapping and the close proximity of some sites may not allow them to be shown separately.

Figure 2. The percentage site occupation by otters in each of the 32 50km squares surveyed in 1991-94. These results have been 'rounded up' to the nearest whole number. The individual numbers of occupied sites and sites surveyed in each square are provided in Table 1.

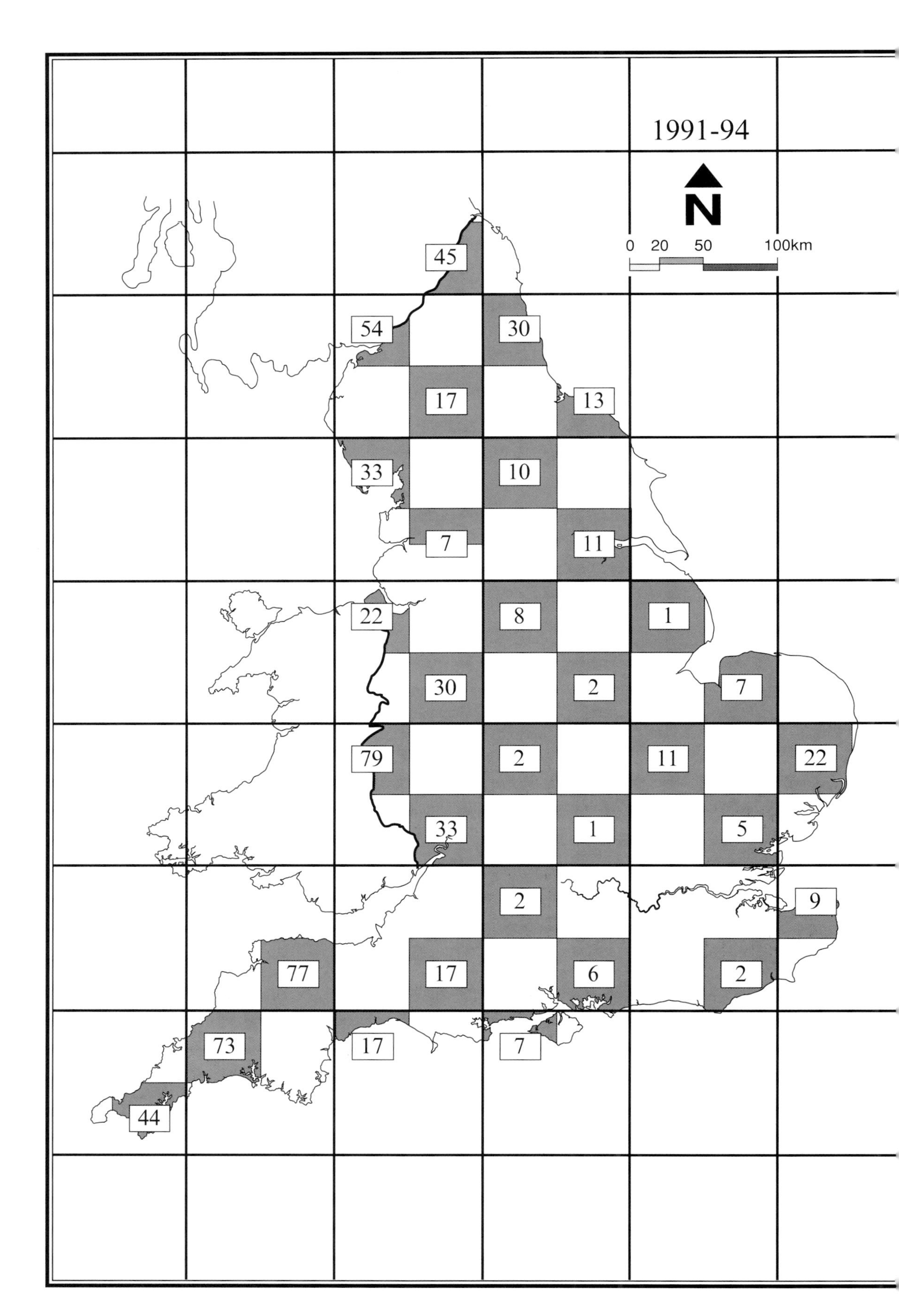

Figures 3a & b. The percentage site occupation by otters in each of the 50km squares surveyed in (a) 1977-79 and (b) 1984-86. These results have been ‘rounded up’ to the nearest whole number. The individual numbers of occupied sites and sites surveyed in each square are provided in Table 1. Data from Lenton *et al* (1980) and Strachan *et al* (1990).

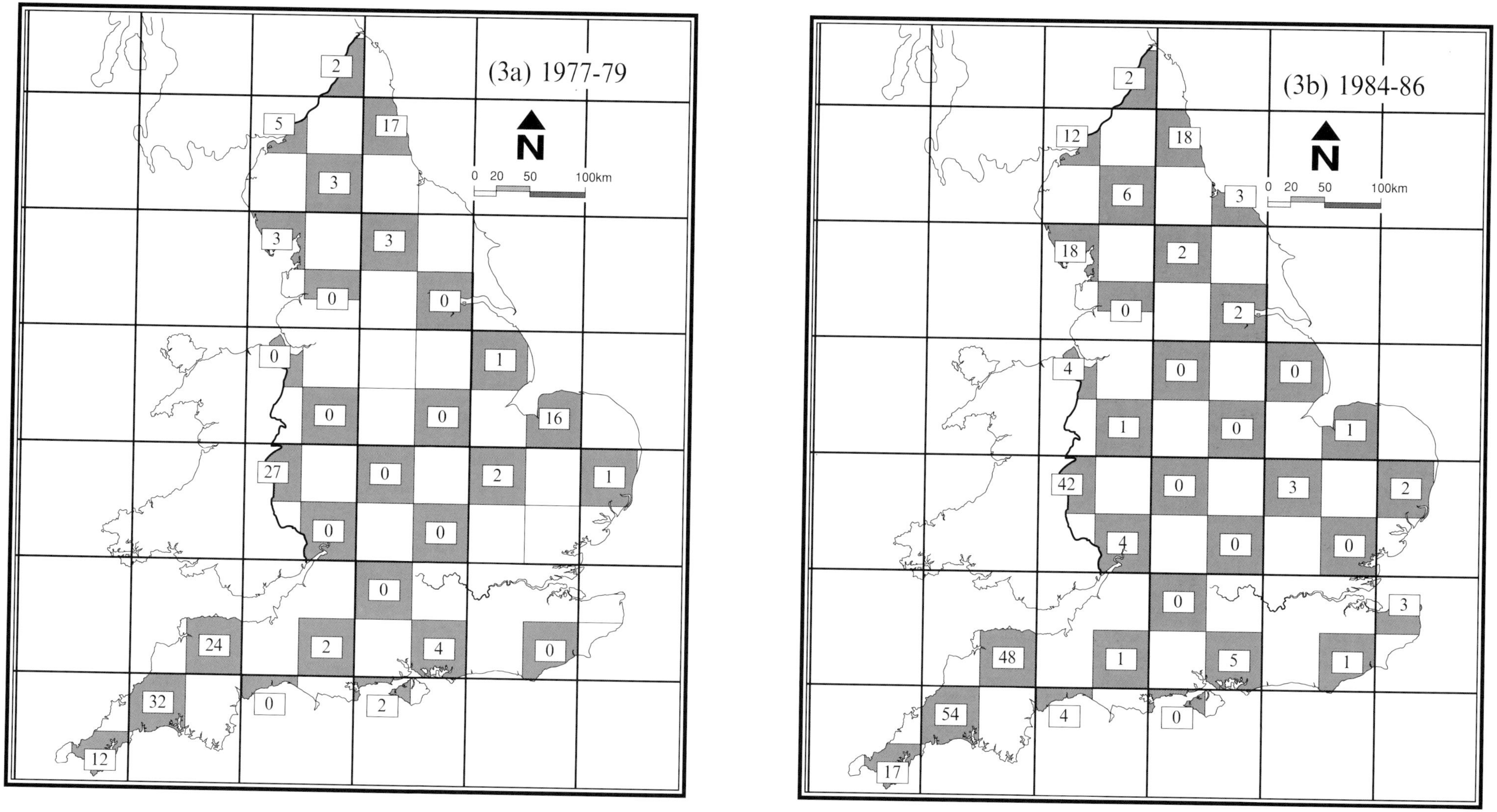

(3a) 1977-79

(3b) 1984-86

Figure 4. Otter distribution in 1991-94 based on occupied 10km squares. Survey areas are shaded and Water Authority boundaries shown.

Key ● Positive full survey result in 10km square
o Positive spot check only

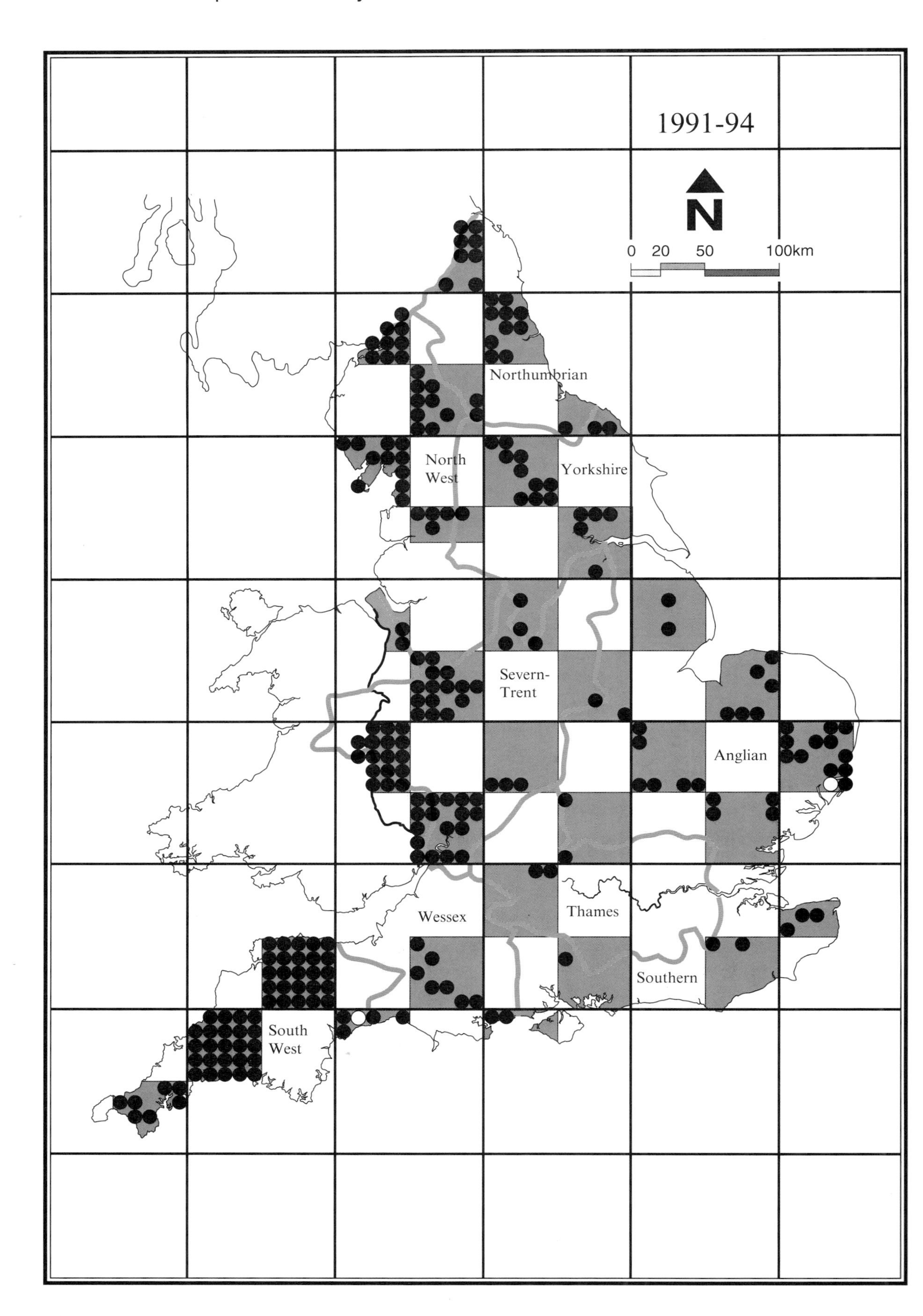

Figures 5a & b. Otter distribution in (a) 1977-79 and (b) 1984-86 based on occupied 10km squares. Survey areas are shaded and Water Authority boundaries shown. Data from Lenton *et al* (1980) and Strachan *et al* (1990).

Key ● Positive full survey result in 10km square
o Positive spot check only

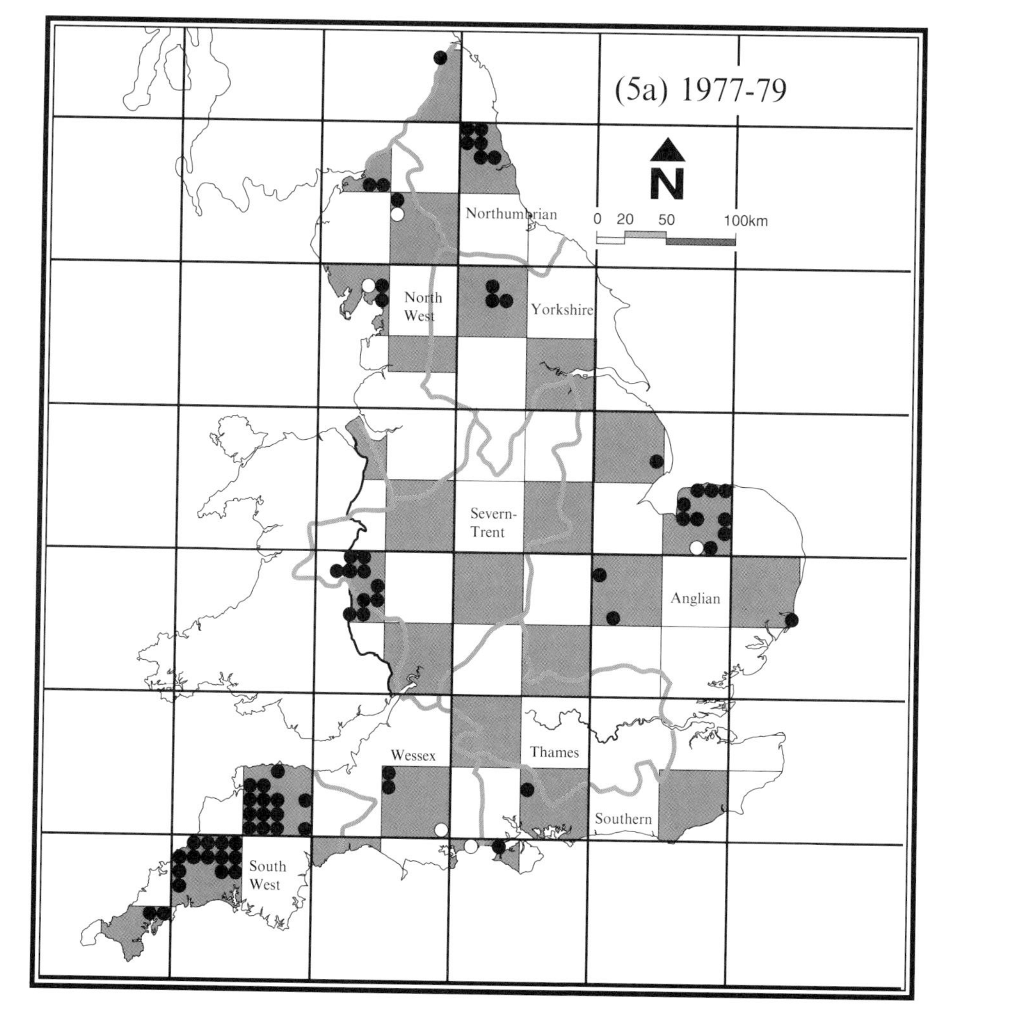

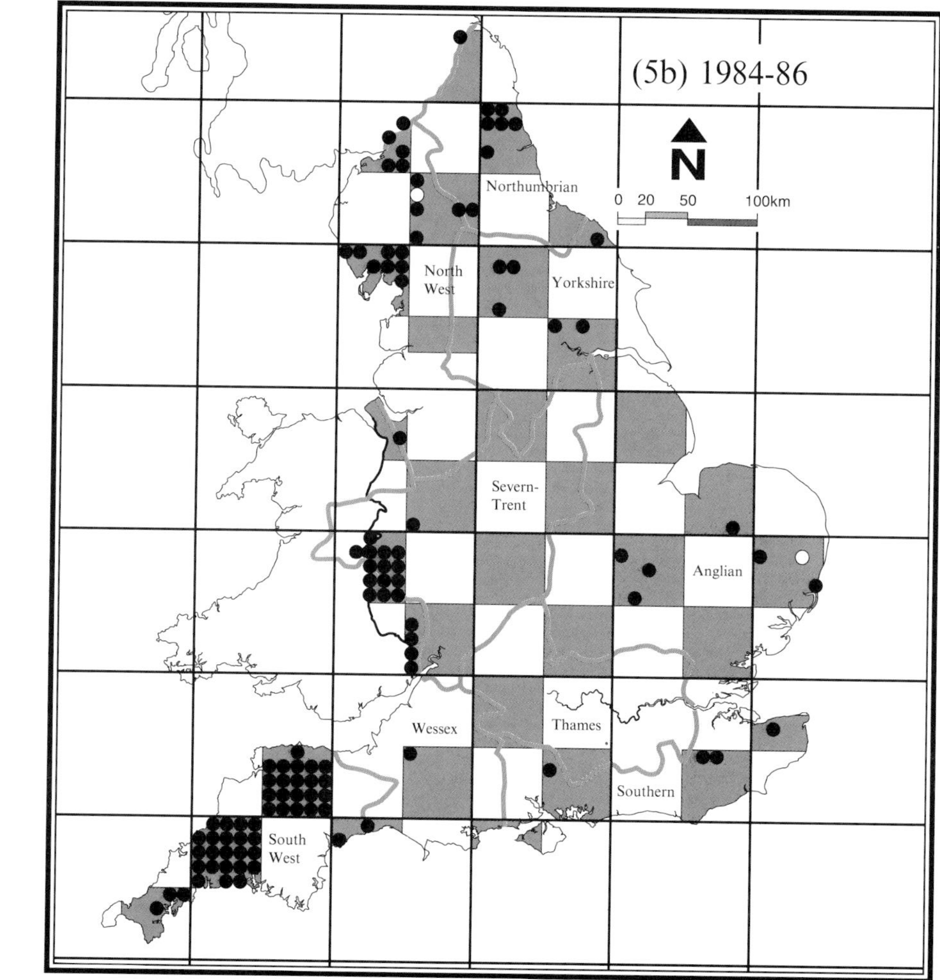

Figure 6. The rivers and watersheds of England known to be occupied by otters in 1991-94. This map has been prepared using a combination of data from the present survey and that provided by regional and county otter surveyors in areas (50km squares) not covered by the national survey. These surveys and surveyors are listed in Section 5 under the appropriate Region. Those rivers considered to have been colonised or reinforced by released otters and/or their progeny are distinguished from those occupied by wild stock.

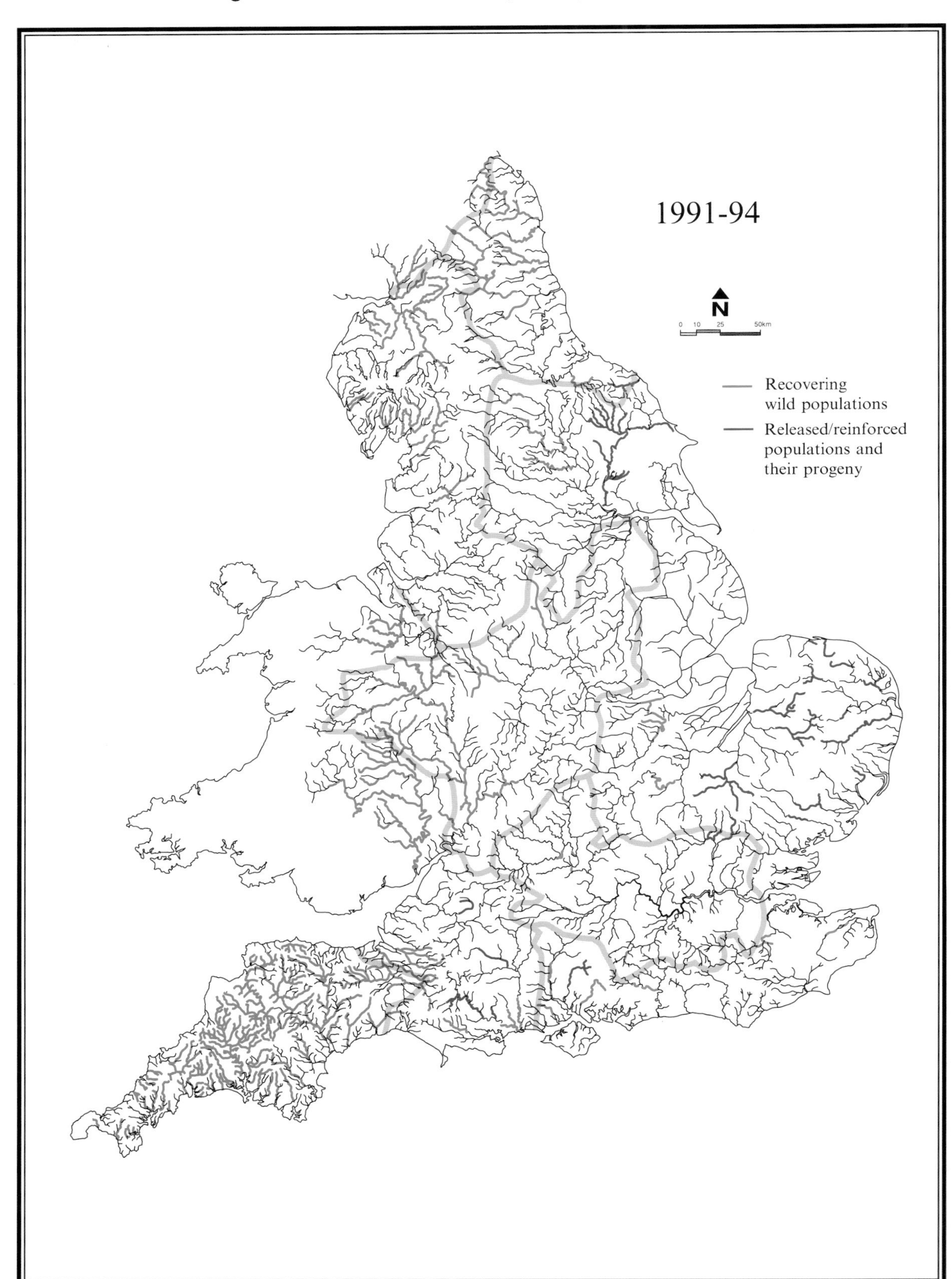

Figures 7a & b. The rivers and watersheds of England known to be occupied by otters in (a) 1977-79 and (b) 1984-86. These maps have been prepared using a combination of data from the national surveys and that provided by any available local information in areas not covered by surveyed 50km squares. Those rivers considered to have been colonised or reinforced by released otters and/or their progeny by 1984-86 are distinguished from those occupied by wild stock in that map.

(7a) 1977-79

N

0 10 25 50km

— Surviving populations

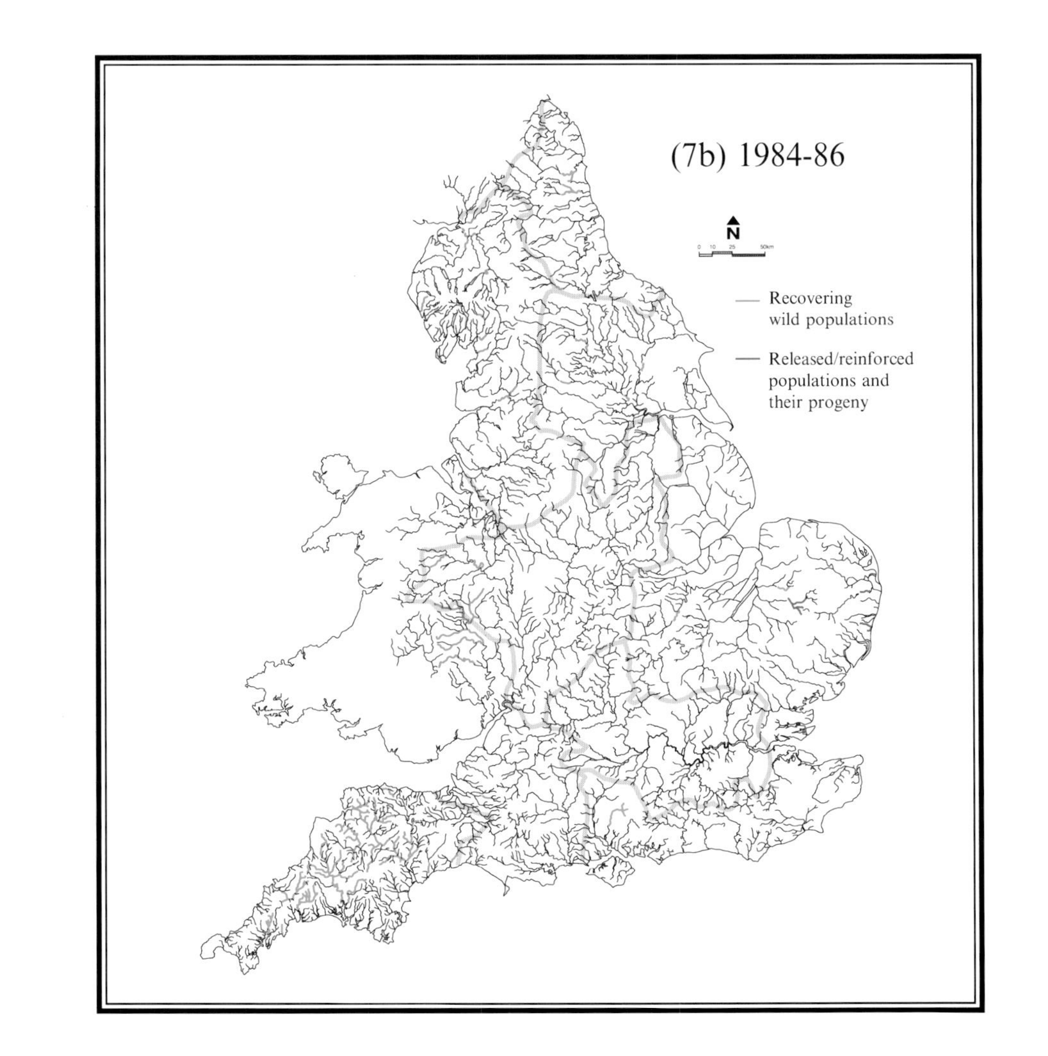

Figure 8. Mink distribution in 1991-94 based on occupied 10km squares. Data are from this survey. Survey areas are shaded and Water Authority boundaries shown.

Key ● Surveyor's records
o Other records

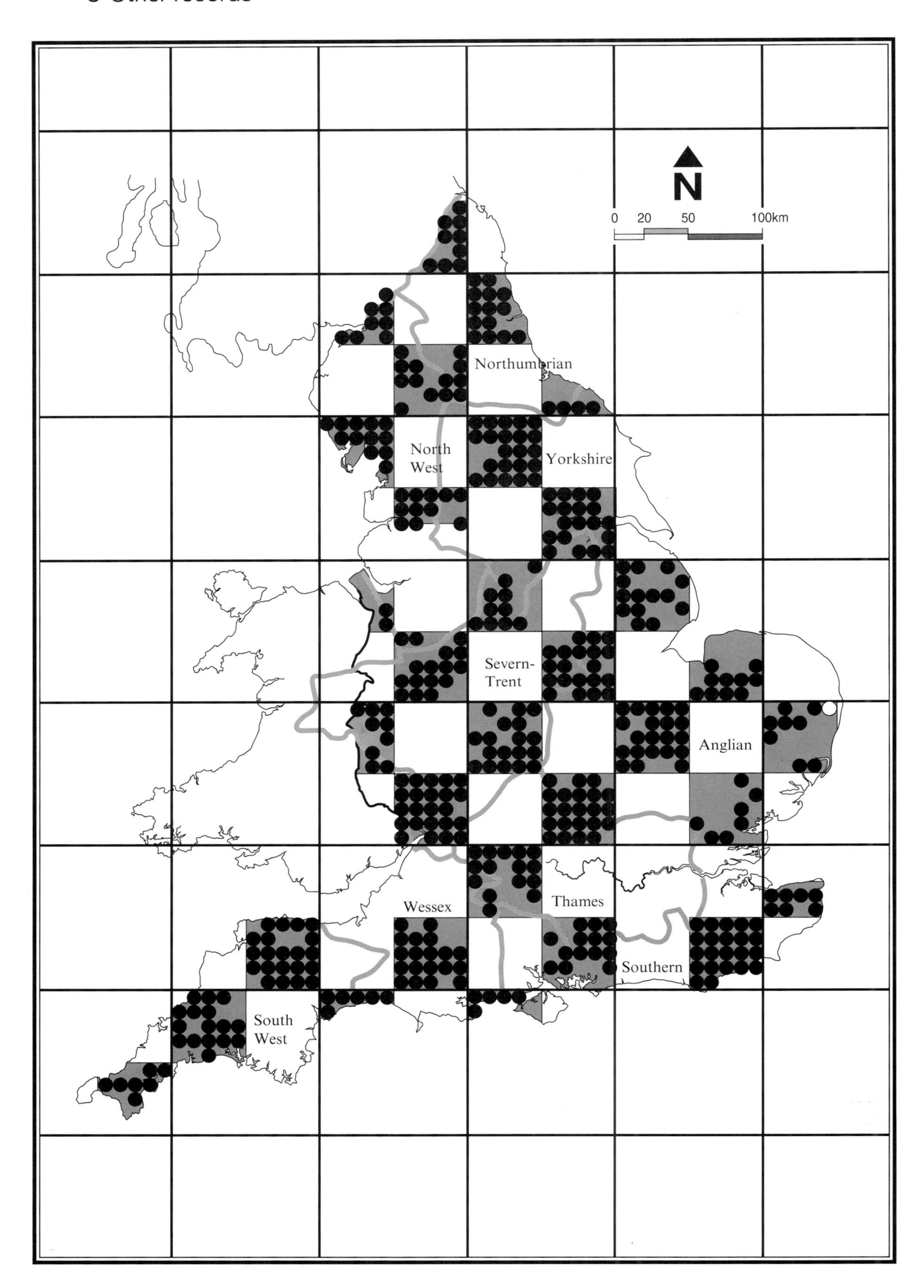

4.5 Distribution of mink

The Otter Survey of England has also acted as an American mink survey since its inception in 1977. The presence of mink signs was recorded on each survey site form at the time of each search for otter spraints. 10km square 'dot maps' of mink distribution have been prepared for each survey report. That for 1991-94 is shown in Figure 8. There has been a continuation of the movement to occupy midland and eastern England since 1984-86. (An intermediate stage in the distribution can be seen in the results of the water vole and mink survey of 1989-90; Figure 5 in Strachan & Jefferies, 1993.) However, the increasing otter population of the western counties of England has brought about a reduction in the density of mink in those areas. This antagonistic relationship is discussed in Section 6.3. Otters are still too few in the Midlands and east of England to have an effect on the increasing mink population there.

Table 1. A summary of the survey results for each 50km square (listed in chronological order of survey) for each of the three surveys of England.

* The four extra 50km squares surveyed in 1984-86 and 1991-94, making 3,188 survey sites instead of the 2,940 surveyed in 1977-79. The comparable results for these 2,940 sites when surveyed in 1984-86 and 1991-94 are also shown in the totals.

** The total number of survey sites for these squares includes 85 non-visited sites (assumed not to be able to support otters or it was impossible to gain access).

This Table contains corrections of some of the figures from the 1984-86 survey report. The number of spot checks made in 1991-94 was lower than that in the previous two surveys, though all the previously positive spot check sites were revisited.

50km Square	Date Surveyed in 1991-94 (month/year)	Results for 1991-94			Results for 1984-86			Results for 1977-79		
		Full survey sites Positives/ No. surveyed	Full surveys Percentage positive	Spot checks Positives/ No. surveyed	Full survey sites Positives/ No. surveyed	Full surveys Percentage positive	Spot checks Positives/ No. surveyed	Full survey sites Positives/ No. surveyed	Full surveys Percentage positive	Spot checks Positives/ No. surveyed
ST s/e	10/91	16/92	17.39%	1/3	1/92	1.09%	0/3	2/92	2.17%	0/3
SY n/w	11/91	8/46	17.39%	1/4	2/46	4.35%	0/4	0/46	0.00%	0/4
SU s/e	11/91	6/100**	6.00%	0/4	5/100	5.00%	0/4	4/100	4.00%	0/4
SZ n/w	12/91	3/42	7.14%	1/1	0/42	0.00%	0/1	1/42	2.38%	1/1
SW s/e	1/92	26/59	44.07%	2/2	10/59	16.95%	0/2	7/59	11.86%	0/2
SU n/w	2/92	2/93	2.15%	0/4	0/93	0.00%	0/4	0/93	0.00%	0/4
SP s/e	3/92	2/147	1.36%	0/6	0/147	0.00%	0/6	0/147	0.00%	0/6
TF s/e	4/92	7/98	7.14%	3/21	1/98	1.02%	1/22	16/98	16.33%	5/22
SX n/w	5/92	95/130**	73.08%	–	70/130	53.85%	0/6	41/130	31.54%	0/6
NY s/e	6/92	19/111	17.12%	4/9	7/111	6.31%	2/9	3/111	2.70%	0/9
NZ n/w	7/92	23/76**	30.26%	1/2	14/76	18.42%	0/2	13/76	17.11%	0/2
NT s/e	7/92	19/42	45.24%	1/1	1/42	2.38%	0/1	1/42	2.38%	0/1

Table 1 continued

50km Square	Date Surveyed in 1991-94 (month/year)	Results for 1991-94			Results for 1984-86			Results for 1977-79		
		Full survey sites Positives/ No. surveyed	Full surveys Percentage positive	Spot checks Positives/ No. surveyed	Full survey sites Positives/ No. surveyed	Full surveys Percentage positive	Spot checks Positives/ No. surveyed	Full survey sites Positives/ No. surveyed	Full surveys Percentage positive	Spot checks Positives/ No. surveyed
SE n/w	8/92	13/129	10.08%	–	3/129	2.33%	0/7	4/129	3.10%	0/7
SE s/e	9/92	13/114	11.40%	2/4	2/114	1.75%	0/7	0/114	0.00%	0/7
TM n/w	10/92	27/121	22.31%	10/14	3/121	2.48%	2/8	1/121	0.83%	0/8
SK n/w*	11/92	6/72	8.33%	–	0/72	0.00%	–	–	–	–
SK s/e	12/92	3/138	2.17%	0/5	0/138	0.00%	0/5	0/138	0.00%	0/5
TF n/w	1/93	2/142	1.41%	–	0/142	0.00%	0/8	1/142	0.70%	0/8
TL n/w	2/93	15/132	11.36%	–	4/132	3.03%	0/14	2/132	1.52%	0/14
SO s/e	3/93	47/143	32.87%	–	6/143	4.20%	0/10	0/143	0.00%	0/10
SP n/w	4/93	4/160**	2.50%	–	0/160	0.00%	0/1	0/160	0.00%	0/1
SS s/e	5/93	140/182	76.92%	–	87/182	47.80%	0/2	43/182	23.63%	0/2
NY n/w	6/93	31/57	54.39%	–	7/57	12.28%	–	3/57	5.26%	–
SD n/w	6/93	36/108	33.33%	5/10	19/108	17.59%	1/18	3/108	2.78%	1/18
SD s/e	7/93	6/82**	7.32%	–	0/82	0.00%	0/5	0/82	0.00%	0/5
SO n/w	8/93	72/91	79.12%	8/12	38/91	41.76%	6/12	25/91	27.47%	3/12
SJ s/e	9/93	44/148**	29.73%	6/7	1/148	0.68%	0/7	0/148	0.00%	0/7
SJ n/w	10/93	5/23	21.74%	–	1/23	4.35%	–	0/23	0.00%	–
TQ s/e	11/93	3/134	2.24%	1/3	2/134	1.49%	0/6	0/134	0.00%	0/6
TR n/w*	12/93	3/34	8.82%	–	1/34	2.94%	–	–	–	–
NZ s/e*	12/93	4/30	13.33%	–	1/30	3.33%	–	–	–	–
TL s/e*	1/94	6/112	5.36%	–	0/112	0.00%	–	–	–	–
Totals		**687/2,940** **706/3,188**	**23.37%** **22.15%**	**46/112**	**284/2,940** **286/3,188**	**9.66%** **8.97%**	**12/174**	**170/2,940**	**5.78%**	**10/174**

Table 2. Comparison of the results obtained in each of the ten Regions of the National Rivers Authority. The numbers of the full survey sites found to be positive in each Region are shown, together with the percentages of the total sites surveyed. At the time of the 1984-86 survey an additional 248 sites were examined besides the 2,940 examined in 1977-79. This augmented total (3,188) was surveyed in 1991-94 also. The Regions affected are marked (*). Percentages have been calculated for these Regions for both 1977-79 and 1984-86 numbers of survey sites.

Region	Number and percentage of full survey sites found to be positive					
	1977-79		1984-86		1991-94	
North West	9/324	(2.78%)	31/324 31/333*	(9.57%) (9.31%)	93/324 93/333	(28.70%) (27.93%)
Northumbrian	14/168	(8.33%)	17/168 17/174*	(10.12%) (9.77%)	45/168 46/174	(26.79%) (26.44%)
Yorkshire	4/226	(1.77%)	5/226 6/270*	(2.21%) (2.22%)	25/226 28/270	(11.06%) (10.37%)
Severn-Trent	13/567	(2.29%)	22/567 22/610*	(3.88%) (3.61%)	120/567 126/610	(21.16%) (20.66%)
Anglian	20/623	(3.21%)	8/623 8/725*	(1.28%) (1.10%)	52/623 58/725	(8.35%) (8.00%)
Thames	0/170	(0.00%)	0/170 0/180*	(0.00%) (0.00%)	4/170 4/180	(2.35%) (2.22%)
Wessex	2/154	(1.30%)	1/154	(0.65%)	29/154	(18.83%)
South West	91/386	(23.58%)	169/386	(43.78%)	259/386	(67.10%)
Southern	5/241	(2.07%)	7/241 8/275*	(2.90%) (2.91%)	9/241 12/275	(3.73%) (4.36%)
Welsh	12/81	(14.81%)	24/81	(29.63%)	51/81	(62.96%)
Totals	**170/2,940**	**(5.78%)**	**284/2,940** **286/3,188***	**(9.66%)** **(8.97%)**	**687/2,940** **706/3,188**	**(23.37%)** **(22.15%)**

Table 3. Comparison of the results in terms of differences and percentage differences between the surveys of 1984-86 & 1977-79, 1991-94 & 1984-86 and 1991-94 & 1977-79. The figures shown are the percentage site occupation at the second survey minus the percentage site occupation at the first survey, providing the difference between the two survey results. This difference is then calculated as a percentage (shown in brackets) of the percentage site occupation at the first survey, to indicate the extent of the increase (or decrease) in site occupation between the two surveys. Two answers are available for some of the comparisons between the 1991-94 and 1984-86 surveys because these surveys were both done with the augmented number of survey sites (3,188 rather than 2,940). Survey sites were not augmented in all Regions, depending on the coverage already obtained in the 1977-79 survey.

Region	No. of sites used in comparison	Comparison 1984-86 to 1977-79	Comparison 1991-94 to 1984-86	Comparison 1991-94 to 1977-79
North West	324 333	9.57 – 2.78 = +6.79 [+244.24%]	28.70 – 9.57 = +19.13 [+199.90%] 27.93 – 9.31 = +18.62 [+200.00%]	28.70 – 2.78 = +25.92 [+932.37%]
Northumbrian	168 174	10.12 – 8.33 = +1.79 [+21.49%]	26.79 – 10.12 = +16.67 [+164.72%] 26.44 – 9.77 = +16.67 [+170.62%]	26.79 – 8.33 = +18.46 [+221.61%]
Yorkshire	226 270	2.21 – 1.77 = +0.44 [+24.86%]	11.06 – 2.21 = +8.85 [+400.45%] 10.37 – 2.22 = +8.15 [+367.12%]	11.06 – 1.77 = +9.29 [+524.86%]
Severn-Trent	567 610	3.88 – 2.29 = +1.59 [+69.43%]	21.16 – 3.88 = +17.28 [+445.36%] 20.66 – 3.61 = +17.05 [+472.30%]	21.16 – 2.29 = +18.87 [+824.02%]
Anglian	623 725	1.28 – 3.21 = –1.93 [60% decrease]	8.35 – 1.28 = +7.07 [+552.34%] 8.00 – 1.10 = +6.90 [+627.27%]	8.35 – 3.21 = +5.14 [+160.12%]
Thames	170 180	0.00 – 0.00 = 0.00 [No change]	2.35 – 0.00 = +2.35 ∝ 2.22 – 0.00 = +2.22 ∝	2.35 – 0.00 = +2.35 ∝
Wessex	154	0.65 – 1.30 = –0.65 [50% decrease]	18.83 – 0.65 = +18.18 [+2,796.92%]	18.83 – 1.30 = +17.53 [+1,348.46%]
South West	386	43.78 – 23.58 = +20.20 [+85.67%]	67.10 – 43.78 = +23.32 [+53.27%]	67.10 – 23.58 = +43.52 [+184.56%]
Southern	241 275	2.90 – 2.07 = +0.83 [+40.10%]	3.73 – 2.90 = +0.83 [+28.62%] 4.36 – 2.91 = +1.45 [+49.83%]	3.73 – 2.07 = +1.66 [+80.19%]
Welsh	81	29.63 – 14.81 = +14.82 [+100.07%]	62.96 – 29.63 = +33.33 [+112.49%]	62.96 – 14.81 = +48.15 [+325.12%]
Totals	**2,940** **3,188**	**9.66 – 5.78 = +3.88** **[+67.13%]**	**23.37 – 9.66 = +13.71** **[+141.93%]** **22.15 – 8.97 = +13.18** **[+146.93%]**	**23.37 – 5.78 = +17.59** **[+304.33%]**

5. RESULTS FOR NATIONAL RIVERS AUTHORITY REGIONS

5.1 NORTH WEST REGION

Full surveys were carried out in the following 50km squares – NY n/w (57 sites), NY s/e (56), SD n/w (108), SD s/e* (69), SJ n/w (14), SJ s/e (20), SK n/w (9).

* SD s/e includes 20 sites non-visited and assumed negative.

Description of Region

The river catchments of the Region rise in the mountains and hills of the Lake District and north Pennines, where some peaks reach a height of over 1,000m, and flow towards the western coast. Each hydrometric area differs slightly in its landform, with some like the Eden, Lune and Ribble flowing through long, broad valleys which were carved out by glaciers, while others like the Mersey in the southern part of the Region are more low-lying and run a meandering course. Many of the northern rivers such as the Derwent, Leven, Crake and Eamont flow through large deepwater lakes that give the name Lake District to the area. The geology is varied but primarily dominated by old igneous granites, metamorphosed shales (slate) and areas of limestone pavement outcrop.

Land use of the Region is predominantly upland sheep grazing, with some cattle pasture and arable farming in the valley bottoms and low-lying coastal plains. Some of the higher ground has been planted up with extensive conifer plantations.

The human population for the Region is largely concentrated in the southern half, with the large conurbations of Liverpool (469,600 inhabitants), Manchester (445,900), Stockport (290,900), Bolton (263,000), Salford (235,600), Oldham (219,500), Rochdale (206,800), Blackburn (134,400), Birkenhead (99,000) and Burnley (85,400) centred on the Mersey catchment. In the northern part of the Region the populated areas are concentrated around the coast or principal river mouths and include Blackpool (143,800), Lancaster (131,100), Preston (128,100), Carlisle (102,000) and Barrow (72,400). The population in this area is greatly swollen in the summer months with holiday-makers, being centred on the coast, lakes and mountains with increased recreation on the rivers.

Pollution problems

Industrial spillages and effluent problems have largely been confined to the Mersey catchment and lower Ribble, although most estuaries show heavy metals in their sediments. Elsewhere, pollution from agricultural sources, notably slurry run-off, sheep-dip and agrochemical input, has affected most rivers at some time.

Upland catchments may also be prone to increasing acidification from acid rain and conifer plantations, although these effects may be ameliorated where limestone occurs.

The majority of the rivers in the Region meet the NRA River Quality Objectives and are classed 1A, 1B or 2 (see Appendix 3).

Fisheries on principal rivers

The Border Esk, Eden, Lune and Ribble support the principal salmon and sea trout fisheries together with good populations of brown trout. The last two salmonids can also be found on most of the other rivers in the Region and NRA sampling of fry and parr throughout the Region has shown a healthy recruitment on all systems in their upper reaches, meeting biomass targets.

European Community (EC) designated salmonid fisheries should exceed a sampling biomass of $15g/m^2$ while designated cyprinid fisheries should exceed a sampling biomass of $20g/m^2$.

Large populations of coarse fish are restricted to the lower reaches of these predominantly spate rivers or in some lakes, although most salmonid rivers also support reasonable densities of eel, minnow, stone loach and bullhead.

Figure 9. North West Region

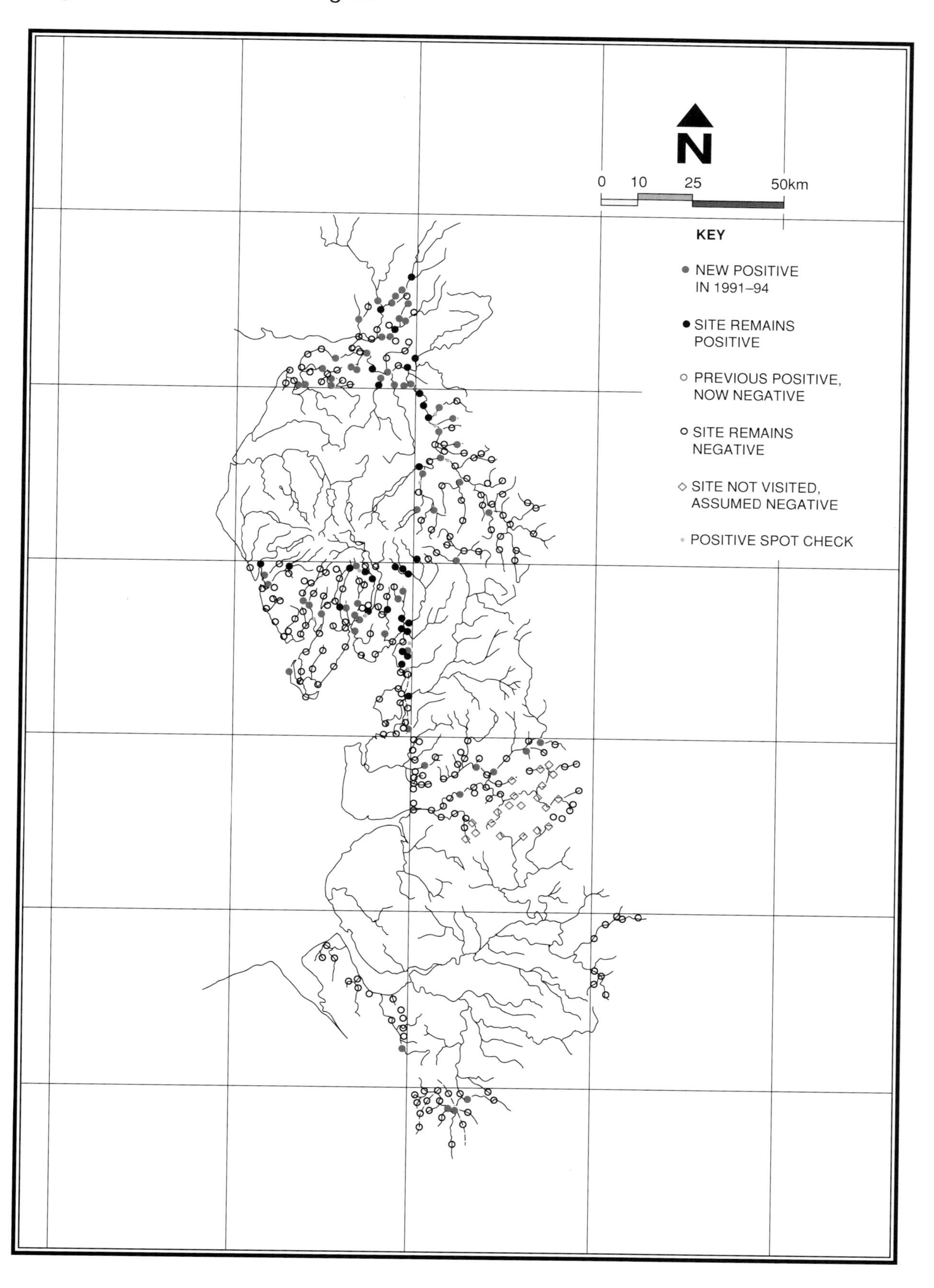

Otter conservation effort

Suitable otter habitat occurs along the lower and middle reaches of most of the rivers, but this becomes sparse where there is intensive farming, around most urban centres and in the higher fells.

Conservation effort in the Region has concentrated on the establishment of 'otter havens' and cultivating sympathetic riparian habitat management with landowners. In particular, The Vincent Wildlife Trust's Otter Haven Project played a vital role in the 1980s and was succeeded by the Royal Society for Nature Conservation's (RSNC) Otters and Rivers Project in conjunction with the Cumbria Wildlife Trust. An otter project officer has been employed from 1990 to 1994 to carry out work principally on the River Eden catchment.

Results for 1991-94

Total full survey sites	333
Positive full survey sites	93
Positive spot checks	9

Comparison of results for the same survey sites 1977-79 and 1984-86

	1977-79	1984-86	1991-94
Full survey	9/324 (2.8%)	31/324 (9.6%)	93/324 (28.7%)
Spot check	2/19	3/19	9/19

Summary of results for each river system (see Figure 9)

Esk (including Lyne and Liddel Water)

12/19 full survey sites were positive, compared to 3/19 in 1986 and all negative in the 1979 baseline survey. This represents a continued range expansion and consolidation of otters on this border river system. Records and reports have been numerous over the last five years.

Sark

1/2 full survey sites was positive on this small river that forms the national boundary between Scotland and England. Both sites were negative in previous surveys.

Solway estuary

1/5 full survey sites was positive on these coastal grazing marshes, which provide very little cover and very few potential spraint sites. (Otter presence was determined at one site as a clear track in the silt under a small tractor bridge and might easily have been over-looked if the tides had been higher.) Recent reports suggest that otters may be resident on the estuary but roam over a large area.

Waver and Wampool

5/15 full survey sites were positive, compared to none in both previous surveys. This is an encouraging range expansion into the low-lying area below the Solway despite the apparent lack of suitable cover and potential lying-up places.

Lower Eden (including Irthing, Petteril and Caldew)

12/16 full survey sites were positive, compared to four in 1986 and three in 1979, showing that the otter population has further consolidated its range over the interim period. Otter haven work carried out on the Eden catchment by the Cumbria Wildlife Trust is thought to have aided this recovery. Records and reports of otters have been numerous.

Middle Eden

14/48 full survey sites and four spot checks were positive on the main river and its tributaries (the Eamont, Lowther, Lyvennet, Croglin Water and Raven Beck). This represents a considerable expansion since the two previous surveys, which revealed three positive full survey sites and one positive spot check in 1978 and four positive full survey sites and two positive spot checks in 1985.

It is possible that the population on the Eden has been reinforced by otters moving down from Scotland as signs have increased on the lower part of the system, particularly on some minor tributaries in the last three years, although the upper reaches have yet to show regular use (M. Twiss, pers. comm.).

Ravenglass Esk, Mite and Irt

4/9 full survey sites were positive, compared to two in 1986 and all negative in the 1979 baseline survey. These west Cumbrian rivers continue to support a small otter population with most of the recent reports coming from the Irt.

Annas and Whicham Brook

(0/7) The seven full survey sites were negative, as in the previous surveys.

Duddon (including River Lickle and Kirby Pool)

3/15 full survey sites and one spot check were positive, compared to all negative in the previous surveys. Signs were sparse, suggesting that, as yet, only small numbers of otters may be using this excellent looking wooded river valley. National Park rangers received reports of otters in the vicinity of Ulpha bridge in 1991 and 1992.

Walney Island

1/3 full survey sites was positive, based on a single fresh spraint beside one of the natterjack toad pools of North Walney nature reserve. No recent reports were received from the area and all sites showed no signs in the previous surveys. This was probably a transient animal, although some excellent cover can be found in this part of the island.

Furness peninsula

(0/8) The eight full survey sites were negative, as in the previous surveys.

Crake and Coniston Water

2/7 full survey sites were positive, the same sites where evidence was found in the 1986 survey. Signs were sparse, suggesting irregular use by otters (perhaps alternating with Rusland Pool – see next entry).

Rusland Pool and Colton Beck

3/5 full survey sites showed sparse evidence of otters (a single spraint at two and only tracks at the other). All sites were negative in previous surveys.

Leven and Windermere (including Cunsey Beck)

5/10 full survey sites and two spot checks were positive, compared to three full survey sites in 1986 and only a single spot check in 1979.

Recent local surveys have found regular spraints on Cunsey Beck and the lower Leven, and there have been reports of otters from the north Windermere tributaries Brathay, Rothay, Troutbeck and Langdale Beck, suggesting an expansion in the range of the otter population on this system.

Cartmel peninsula

1/7 full survey sites was positive, based on a single spraint on Skelwith Pool. The whole area showed no signs in both previous surveys.

Winster

2/6 full survey sites showed otter signs, a slight improvement from a single positive site in 1986.

Kent (including Gilpin, Pool, Gowan)

9/12 full survey sites were positive, a slight improvement from eight in 1986 and one in 1979. This system continues to support a healthy otter population and there have been numerous records throughout most of its length over the interim period between national surveys (M. Twiss, pers. comm.).

1/2 full survey sites was positive on the upper tributaries (Sprint and Bannisdale Beck) as a single old spraint, suggesting irregular use. This site was positive in 1985 but not in 1978.

Leighton Beck (Leighton Moss) and Keer

4/5 full survey sites and two spot checks were positive. Again a slight improvement on the 1986 survey, with a regular breeding population centred on Leighton Moss nature reserve (where almost daily sightings are recorded).

Lower Lune and coastal sites (including Conder and Cocker)

2/9 full survey sites showed evidence of otters, one on the River Lune and the other on the Conder, the latter being a single spraint but representing a range expansion to the south, possibly via the Lune estuary. The same Lune site was positive in the 1986 survey.

Upper Lune

1/6 full survey sites was positive on the upper reaches of this system, which provides patchy habitat for otters. All sites were negative in 1978 and 1985.

Wyre

1/9 full survey sites was positive on the upper reaches of this system. (None of the main river came into the survey area.) A single fresh spraint provided evidence of otters on the River Brock, which offers good bankside woodland cover. Both previous surveys were negative.

Ribble (including Darwen, Calder, Hodder and Stockbeck)

5/37 full survey sites showed the presence of otters, including three on the main river, one on the lower Hodder and one on the Swanside Beck, a minor tributary. The evidence was low at each positive site, consisting of one or two spraints, and the distribution of the positive sites was fragmented, with negative sites in between them. This suggests that the population of otters on this system is currently small and 'fragile'.

Recent reports are few and largely unconfirmed.

Lancaster Canal

(0/8) The eight full survey sites were negative, as in both previous surveys in this largely undisturbed habitat which offers few good sprainting places that are easy to check.

Mersey

(0/9) The nine full survey sites were negative along the upper reaches of the Goyt and Etherow rivers. Potentially good habitat can be found in the Goyt valley above Whaley Bridge and an unconfirmed report was noted from this area.

Weaver

3/14 full survey sites were positive along these upper parts of the catchment, compared to none in previous surveys. The evidence was sparse and scattered, being found at single sites on the River Weaver, Barnett Brook and Checkley Brook. Some good habitat for otters remains along these watercourses as well as around several meres in the form of reed swamp/carr, but these proved difficult to survey adequately.

Gowy and lower Mersey tributaries (Birkenhead)

1/14 full survey sites was positive in this largely urban area. A set of clear tracks was the only evidence found and probably represents a transient animal. All sites were found to be negative in both previous surveys.

Shropshire Union Canal

(0/6) The six full survey sites were negative, as in both previous surveys along this canal that offers unlikely otter habitat.

Complementary information from additional surveys on rivers in the adjacent 50km squares:

NY n/e - Tributaries and adjacent rivers were investigated by M. Twiss as part of the River Eden Otter Project over the same period as the Otter Survey of England 1991-94. Otters were found to be making regular use of the following watercourses – the Black Lyne, White Lyne, King Water, River Irthing and River Gelt. In particular, a chance visit to the middle reaches of the River Irthing during June 1993 revealed fresh tracks of a female with cubs, confirming breeding for the site (R. Strachan, pers. obs.).

NY s/w - The Eden tributaries Petteril and Caldew were surveyed by the River Eden Otter Project using the same method as the national survey (ie. systematic survey) and were found to be used by otters along their middle and lower reaches. The west Cumbrian systems of the rivers Ellen, Derwent, Cocker and Ehen were surveyed by the Cumbria Wildlife Trust during 1992 without any great success, suggesting an absence of otters from a large part of this area, which includes Derwent Water, Bassenthwaite Lake, Buttermere and Crummock Water, Loweswater and Ennerdale Water.

SD n/e - Information about otters from the principal rivers of this 50km square has been gleaned from fisheries officers of the NRA, as no systematic surveys for otters had been carried out in recent years on either the Lune or the Ribble. The reports suggest that otters are now using the majority of the River Lune but are rarely encountered on the Ribble (J. Martin, pers. comm.).

Summary

93/333 (27.93%) full survey sites were positive in this Region, with the most significant gains from those systems in the northern part, suggesting an expansion in range by otters out of Scotland and into England, principally via the rivers Esk and Eden. Elsewhere, the species appears to have consolidated its range. Much of west Cumbria and the systems in the southern part of the Region remain unoccupied.

The weather conditions for surveying were generally favourable in this Region of high rainfall, but flooding and spate conditions could have made some of the results unreliable on some rivers. However, since bad weather affected both previous surveys in a similar way, the results should still be directly comparable.

5.2 NORTHUMBRIAN REGION

Full surveys were carried out in the following 50km squares – NT s/e (42 sites), NZ n/w* (76), NZ s/e (5), NY s/e (51).

* NZ n/w includes nine sites non-visited and assumed negative.

Description of Region

The rivers of this Region rise in the Cheviot hills and northern Pennines, where they reach a height of 800m above sea-level, and flow eastwards towards the coast, where the topography is more gentle.

Land use is predominantly sheep grazing on the higher ground, with arable farmland in the valley bottoms and on the coastal plain. Extensive blocks of conifer plantation occur in some parts of the Region.

Opencast mining is widespread and heavy industry is now largely confined to mouth of the rivers Tees and Tyne.

The human population is scattered and rural, apart from the urban and industrial centres of Sunderland (296,100), Newcastle upon Tyne (279,600), North Shields (192,900), Stockton-on-Tees (176,800), South Shields (155,700), Middlesbrough (143,200), Darlington (99,700), Durham (85,800), Blyth (78,500) and Berwick (26,800).

Pollution problems

Most rivers have minor pollution problems stemming from farm run-off (sheep-dip and slurry) or storm-water overflow from sewage treatment plants. Acidification problems have been noted from the uppermost reaches of the Till, Coquet, Tyne, Wear and Tees. Serious industrial incidents in the Region have been few over the last seven years and confined largely to the lower reaches of those rivers which support urban and industrial centres, such as the Tyne, Wear and Tees.

The majority of the rivers in the Region meet the NRA River Quality Objectives and are classed 1A, 1B or 2 (see Appendix 3). Only the lower parts of those rivers mentioned above fail to achieve these objectives.

Fisheries on principal rivers

The rivers of this Region are, for the most part, game fisheries but mixed populations of coarse fish occur along the slower and deeper stretches. Migratory salmonids (salmon and sea trout) are a regular feature on the Tweed (Till), Coquet, Tyne and Tees, and NRA sampling of fry and parr throughout the Region has shown a healthy recruitment on all systems in their upper reaches, meeting biomass targets. In addition to the surveyed migratory salmonids, brown trout, numerous small coarse fish (minnow, bullhead and stone loach) and eel are recorded throughout every river system in the Region at the electro-fishing sampling points.

Otter conservation effort

Otter havens and artificial holts were created in the 1980s by The Vincent Wildlife Trust on 'key' rivers in the Region. This work was continued in recent years, with havens established along the Tyne, Wansbeck and Coquet systems by the Otters and Rivers Project, together with local periodic surveying of otter distribution by the Northumberland (NWT) and Durham (DWT) Wildlife Trusts.

An Otter Project Officer has been employed by the NWT to encourage the species back along the River Till (Tweed tributary) and has fostered sympathetic riparian habitat management by landowners (D. Glen and A. Bielinski, NWT, pers. comm.). The RSNC's Five Rivers Project targeted key rivers in Northumberland, Cumbria and Yorkshire to promote a natural corridor for the dispersal of otters southwards into Yorkshire. Otter forums with conservation organisations, statutory bodies and landowners were set up to facilitate the project (H. Smith, L. Collins and I. Drury, RSNC, pers. comm.).

Figure 10. Northumbrian Region

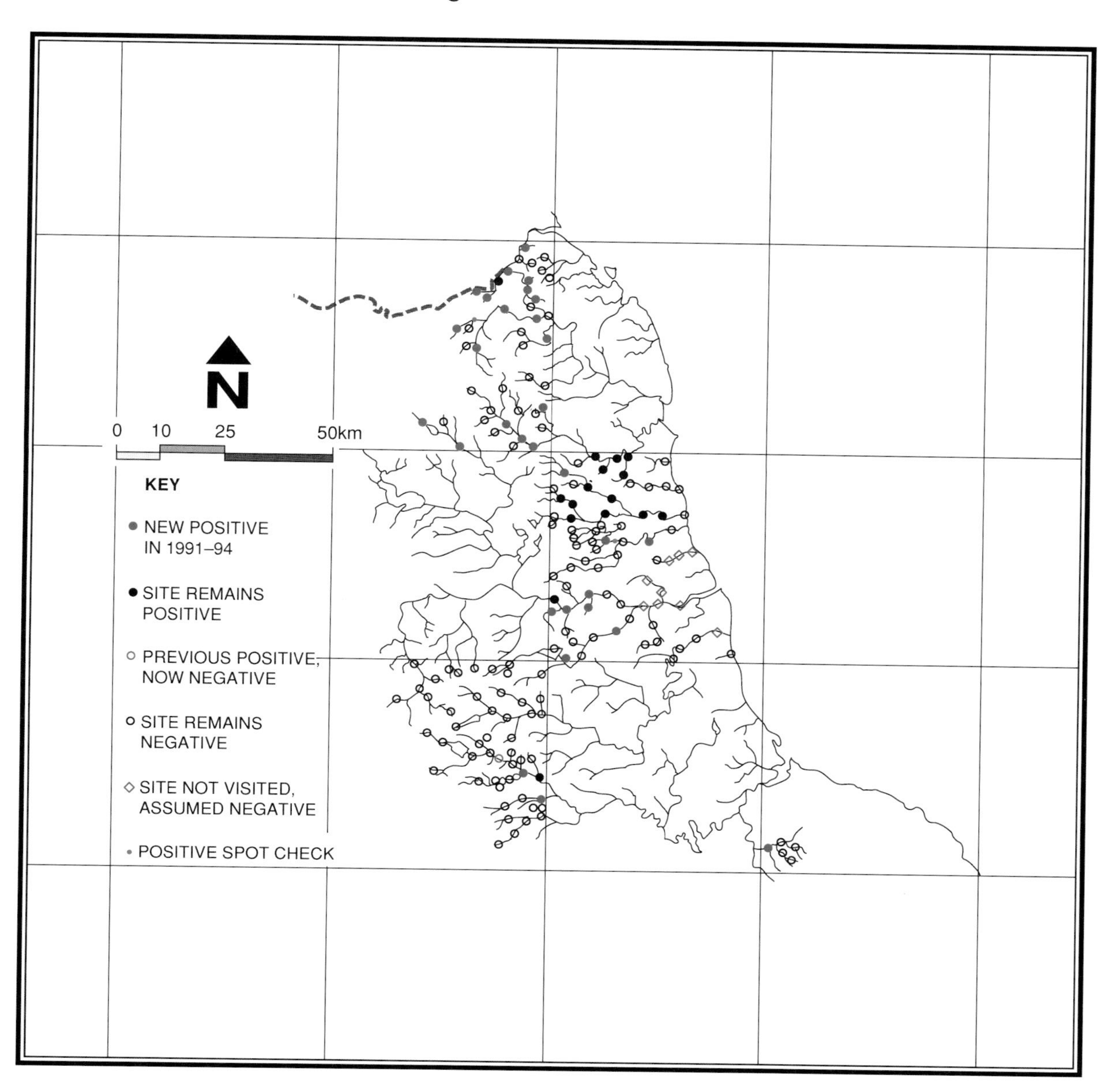

Research into the factors limiting otter distribution on the River Tyne catchment is presently being carried out (T. Thom, University of Durham).

Results for 1991-94

Total full survey sites	174
Positive full survey sites	46
Positive spot checks	2

Comparison of results for the same survey sites 1977-79 and 1984-86

	1977-79	1984-86	1991-94
Full survey	14/168 (8.3%)	17/168 (10.1%)	45/168 (26.8%)
Spot check	0/3	0/3	2/3

Summary of results for each river system (see Figure 10)

Tweed (including Till, Glen, Bowmont, Wooler Water and Breamish)

13/23 full survey sites and one spot check were positive on this system, compared to just a single positive site in both 1978 and 1985.

This represents a considerable improvement for the system and probably reflects an expansion in the otter population on the Tweed catchment in the Scottish Borders. Otter evidence was strong (up to 27 spraints/600m) where bankside cover was dense and offered good tree-root holts. Elsewhere, such as along the middle reaches of the Till, virtually no riparian cover suitable for otters was present and no evidence could be located, although otters must travel through it. The North Northumberland Otter Project is presently trying to address this problem by creating havens among 'oxbow lakes' on the system.

North Low

(0/3) The three full survey sites remained negative on these small coastal streams which provide some suitable cover but appear to offer poor water quality.

Coquet

4/13 full survey sites were positive along the upper reaches of this system, suggesting a range expansion in recent years. No evidence was located in either of the previous surveys in 1978 and 1985. Sparse cover is present in upper Coquetdale but disturbance is low particularly on Ministry of Defence (MOD) land.

5/6 full survey sites were positive on the middle/lower part of this system. This represents no change since 1985. Otters have been continually reported from this part of the river since the last survey, including an otter road death from the A1, south of the Paxtondean Burn bridge during 1991.

Lyne and Chevington Burn

(0/5) The five full survey sites were negative, as in both previous surveys. The systems are small and showed mine waste pollution.

Wansbeck

9/14 full survey sites were positive. This is an increase of one site since 1985. Evidence remains good on this system and reports of otters have been consistent over recent years, although the number of sites which have shown spraints in local surveys have fluctuated, causing some concern (A. Bielinski, pers. comm.).

Blyth

2/18 full survey sites and one spot check were positive, whereas no sites were positive in 1985, though the same two sites were positive during the baseline survey of 1978. Otters are believed to have only recently (ie. within the six months previous to survey) returned to this system, possibly from the expanding North Tyne otter population.

Seaton Burn

(0/4) The four full survey sites were negative, as in the previous surveys. The system flows through a largely built-up area and shows poor water quality.

North Tyne

2/3 full survey sites were positive on the upper reaches of the Rede, confirming the presence of otters on this system, which largely falls outside the survey squares. The otter population on the North Tyne has expanded its range considerably since the last survey and there have been many reports of sightings, including a female with cubs.

Tyne

(0/9) The nine full survey sites were negative on the upper reaches of the South Tyne and Allen. A few of the sites offered good riparian habitat, but most lacked suitable cover being shallow, boulder-strewn streams subject to spate conditions.

Lower Tyne including River Derwent

7/28 full survey sites were positive, representing the continued range expansion of otters down the Tyne, with spraint located only 5km from the start of the built-up area of Newcastle. Seven lower sites on the Tyne were assumed negative, as in the baseline survey.

No sites showed otter evidence in 1978, but a single positive site was found during the 1985 survey. Recent reports have been fairly numerous, although largely unconfirmed, mainly from the River Derwent. But the most notable record was during 1986, when a single animal was seen by several observers at Shibdon Pond, Gateshead (K. Bowie, pers. comm.).

Wear

(0/10) The ten full survey sites were negative along the upper reaches of this system. Reports were received of otters occurring further downstream (outside the survey square), with a good sighting near Wolsingham during 1991 and a series of tracks and fresh spraint/sign-heaps found along the length of the River Browney in March, 1988 (T. Coult, confirmed R. Strachan).

Lower Wear

(0/5) The five full survey sites were negative on this lower part of the system, as in previous surveys.

Tees

3/28 full survey sites were positive in the upper catchment of this system. This compares with two positive sites in 1985 and none in 1978, suggesting that a small relict population still survives on the Tees but its situation is still very vulnerable.

Reports of otters were received from the middle reaches of the Tees, where breeding was suspected in 1990. There have been no records since that time.

Lower Tees

1/5 full survey sites was positive on the River Leven, a tributary of the lower Tees. These sites were first surveyed in December 1986, when no otter evidence could be located. It is possible that otters could have spread to this river from the headwaters of the River Esk to the south.

Complementary information from additional surveys on rivers in the adjacent 50km squares:

NU s/w - Concurrent systematic surveys of a series of sites in this square were carried out by the Northumberland Wildlife Trust using a team of volunteers. The Aln was found to be used by otters throughout its length.

NY n/e - Concurrent systematic surveys of a series of sites in this square were carried out by the Northumberland Wildlife Trust using a team of volunteers. The North Tyne, including the Rede, was found to support otters throughout its length. The River South Tyne, including the Allen, however, showed only sparse signs at its lower reaches. The South Tyne catchment is currently under detailed investigation to establish the reasons for its apparent lack of otters (University of Durham).

NZ s/w - Concurrent systematic surveys of a series of sites in this square were carried out by the Durham Wildlife Trust using a team of volunteers. Their results were very disappointing, showing an apparent absence of otters on the River Wear and a very small population of otters remaining on the middle reaches of the Tees.

Summary

46/174 (26.44%) full survey sites were positive, with the biggest gains in the north of the Region and most notably on the Tweed, Till, Coquet and North Tyne. Evidence was found that the species was once more using the River Blyth, and a notable range extension was most evident along the lower River Tyne, down to the outskirts of the City of Newcastle.

In the south of the Region, otters were largely absent. However, the fisheries data, water quality data and suitability of the riparian habitat suggest that both the Wear and Tees should be able to support a thriving population of otters.

The weather conditions at the time of survey for this Region were generally favourable, with a period of heavy rainfall influencing the results at only a few sites.

5.3 YORKSHIRE REGION

Full surveys were carried out in the following 50km squares – NY s/e (4 sites), NZ s/e (24), SD s/e (13), SE n/w (129), SE s/e (80), SK n/w (20).

Description of Region

The principal rivers of this Region rise either in the Pennines, where the hills reach an altitude of 700m, or in the North Yorkshire Moors, among lower hills reaching 250m above sea-level.

Apart from the River Esk, which covers a relatively short route west to east along the northern flanks of the North Yorkshire Moors, all the other rivers meander their way through low-lying land to join together in the Humber estuary. These lower reaches provide a fertile plain for farming, while the land bordering the upper reaches is used for sheep grazing.

Large conifer plantations occur on the North Yorkshire Moors, which may influence the water quality of the River Derwent.

The human population is concentrated in the south of the Region on the Aire, Calder, Don and Rother, in the industrial centres of Leeds (709,600), Sheffield (528,300), Bradford (464,100), Huddersfield (315,300), Doncaster (291,600), Rotherham (251,800), Barnsley (220,900) and Halifax (195,900). Along the Ouse system and Humber estuary the main centres are Hull (247,000), Harrogate (147,000) and York (100,600). Elsewhere there are many smaller market towns.

Pollution problems

Frequent minor pollution problems have been recorded on most rivers, stemming largely from farm run-off (sheep-dip and slurry) or storm-water overflow from sewage treatment plants. Acidification problems may occur at some headwater sites, but their effects may be ameliorated by the limestone geology.

Serious industrial incidents of pollution in the Region have been numerous only on the Aire, Calder, Don and Rother, which support the larger urban and industrial centres.

Apart from those mentioned above (classed as 3 or 4), the majority of the rivers in the Region meet the NRA River Quality Objectives and are classed 1A, 1B or 2 (see Appendix 3).

Fisheries on principal rivers

The River Esk is the only game fishery in this Region. Although brown trout occur on most of the other systems, these are predominantly mixed fisheries supporting good populations of coarse fish, especially throughout the middle and lower reaches of each system, where electro-sampling has shown that they meet NRA biomass targets. Eels are recorded as numerous throughout.

Otter conservation effort

This Region had received very little direct otter conservation effort until the late 1980s, when the RSNC's Five Rivers Project was set up to promote a natural corridor for the dispersal of otters southwards into Yorkshire from Northumberland via Cumbria. Otter forums of conservation organisations, statutory bodies and landowners were set up to facilitate the project (H. Smith, L. Collins, I. Drury, RSNC/Yorkshire Otter Forum, pers. comm.).

Following a detailed survey of the North Yorkshire Moors National Park rivers, a number of sites was selected to receive otters for release from 1990 onwards. This restocking programme was co-ordinated by The Vincent Wildlife Trust, and rehabilitated orphaned and injured wild otters were introduced into the River Derwent system (J. Green and G. Woodroffe, pers. comm.).

Results for 1991-94

Total full survey sites	270
Positive full survey sites	28
Positive spot checks	2

Comparison of results for the same survey sites 1977-79 and 1984-86

	1977-79	**1984-86**	**1991-94**
Full survey	4/226 (1.8%)	5/226 (2.2%)	25/226 (11.1%)
Spot check	0/4	0/4	2/4

Summary of results for each river system (see Figure 11)

Esk (including coastal denes and becks)

3/24 full survey sites were positive on the middle and lower reaches of the River Esk. This compares with a single positive site in 1986, but in both surveys the weather conditions were difficult, with the river in spate.

During 1993, the small number of otters using this system were reinforced by the release of rehabilitated animals. Surveys in early 1994 showed that the majority of the catchment was being used, from the headwaters to Whitby (G. Woodroffe, pers. comm.).

Swale

(0/4) The four full survey sites on the uppermost reaches of the system were negative, as in both previous surveys. There were abundant potential spraint sites.

Swale (including Cod Beck, River Wiske and Arkle Beck)

3/43 full survey sites were positive, compared to none in both previous surveys.

A small amount of moderately fresh spraint was located at two consecutive sites along the upper reaches of the Swale and probably represented activity from a single animal. At the confluence with the River Ure a single site revealed fresh tracks of an otter but no spraint could be found.

Ure

5/40 full survey sites showed the presence of otters.

In both previous surveys a similar localised distribution was recorded, involving four consecutive sites in 1978 and only two sites in 1985, suggesting that a small isolated population continues to occupy this system between Ripon and Leyburn.

A number of recent reports of otter sightings and field signs refer to this stretch of river.

Nidd

5/27 full survey sites showed evidence of otter presence. All sites in both previous surveys were negative.

Three consecutive sites along the lower Nidd (including one under the A1 trunk road bridge) showed the presence of otters as fresh tracks only and could easily have been overlooked if the water level had been high at the time of survey.

Further upstream spraint was found along the well-wooded riverbank of the Nidd Gorge.

Wharfe

(0/19) The 19 full survey sites showed no evidence of otters despite recent reports. A single site was recorded as positive in 1985, although all were found negative in the baseline survey.

Figure 11. Yorkshire Region

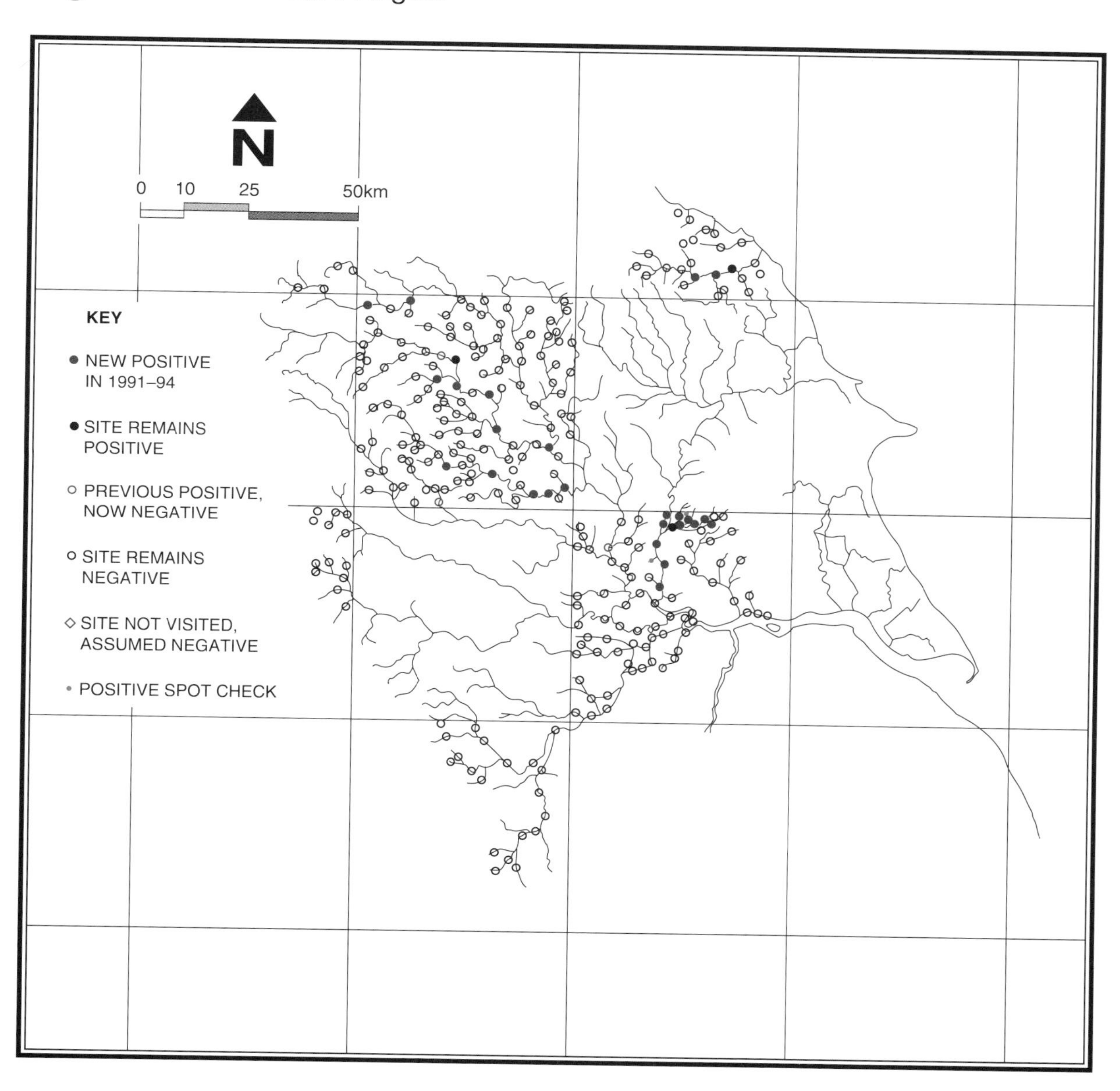

Apparently suitable habitat occurs throughout this system, especially along its middle reaches, which were outside the survey square.

All sites remain negative on the lower reaches of this system.

Kyle

(0/3) The three full survey sites were negative. This minor tributary of the River Ouse also failed to show any signs of otters in both previous surveys.

Ouse (including Stillingfleet Beck, Skipwith Beck and Warping Drain)

(0/17) The 17 full survey sites failed to reveal any signs of otters, with conditions made difficult by heavy rainfall and high water levels concealing possible spraint sites.

A single site was positive in 1985, probably from a transient animal, and since that time there have been occasional unconfirmed sightings of single otters from this stretch of the river.

Aire (including Selby Dam and canals)

(0/10) The ten full survey sites were negative, as in both previous surveys. These sites lack good cover and suffer from disturbance and poor water quality.

Don (including Rother, Went and canals)

(0/40) The 40 full survey sites remained negative. These poor quality sites offer limited potential for otters.

The sites on the Don and Rother rivers have poor riparian habitat and water quality. As noted in 1986, there appears to have been little improvement since the baseline survey.

Humber

(0/4) The four full survey sites remained negative on the lower reaches of this system.

Foulness

(0/10) The ten full survey sites were negative, as in both previous surveys and no recent reports were received.

Derwent (including Pocklington Canal and Blackfoss Beck)

12/16 full survey sites and two spot checks were positive on this system, which provides some excellent habitat. It is likely that all the evidence encountered was from otters released to this system (see Section 7.2.5.1 and Appendix 13). One site was positive in 1985, but none in 1978.

Aire and Calder

(0/11) The 11 full survey sites were negative, as in both previous surveys. There were no recent reports from these systems, which offer sparse suitable habitat and poor water quality.

Leeds and Liverpool Canal

(0/2) The two full survey sites, which were heavily disturbed, showed no signs of otters.

Complementary information from additional surveys on rivers in the adjacent 50km squares:

SE s/w - No systematic surveys are available for this square, but the rivers Wharfe, Calder and Aire had been investigated as part of a general collation of information on otters in North Yorkshire (L. Collins, pers. comm.). Only the Wharfe had otters reported from it in recent years and then only irregularly, suggesting transient animals rather than a resident population.

SE n/e - Surveys of the River Derwent and rivers of the North Yorkshire Moors National Park had been carried out annually since 1985 (G. Woodroffe and L. Winter, pers. comm.). The overall conclusions for the period 1985-90 were that these rivers held a small and fragmented otter population which was vulnerable (Woodroffe, 1994). Since 1990, the area has received introduced otters which have successfully established themselves as a breeding population and now occupy the majority of the Derwent catchment.

Summary

28/270 (10.37%) full survey sites were positive, showing a slight expansion in the range of otters using the Swale, Ure and Nidd. The very much improved situation on the River Derwent, and to a lesser extent on the River Esk, was wholly due to a restocking experiment at selected receiving sites on both rivers. By the end of January 1994 (the end of the survey), 25 rehabilitated otters had been released into the systems (see Section 7.2). Weather conditions were generally favourable during the survey period, but some sites on the rivers Swale, Ure and Wharfe were affected by heavy rainfall and subsequent spate.

5.4 SEVERN-TRENT REGION

Full surveys were carried out in the following 50km squares – SE s/e (24 sites), SK n/w (43), SK s/e (97), SJ s/e* (126), SP n/w* (158), SO n/w (61), SO s/e (101).

* SJ s/e and SP n/w include seven and 32 sites non-visited and assumed negative respectively.

Description of Region

This Region comprises two adjacent hydrometric areas, the River Severn and its tributaries, which rise in the Welsh hills at about 600m above sea-level, and the River Trent and its tributaries, which rise in the Peak District of Derbyshire and Staffordshire among hills of around 500m altitude. Both catchments then run a meandering natural course through low-lying land, the Severn in a southerly direction to its estuary and the Bristol Channel, the Trent initially south but then in a predominantly north-easterly direction to the Humber estuary.

Low-lying land offers good mixed and arable farming, while higher ground is used for grazing.

The large urban and industrial centres of Birmingham (993,700), Coventry (306,200), Wolverhampton (294,400), Walsall (262,300) and Solihull (205,000) influence tributaries of both catchments. Other urban centres in the Trent hydrometric area are Nottingham (273,500), Stoke (246,800), Derby (215,400), Newark (63,000) and Scunthorpe (60,500). In the Severn hydrometric area the other urban centres are Shrewsbury (90,900), Gloucester (90,500), Bromsgrove (89,600), Tewkesbury (87,200) and Worcester (81,200). This large human population and the infrastructure of roads, development and recreational use must make an impact on the habitat for otters as well as increasing disturbance.

Pollution problems

The upper reaches of the Severn catchment meet River Quality Objectives but show signs of acidification. Middle and lower reaches of the system, however, show signs of water quality deterioration from a number of agricultural and industrial sources. In particular the River Avon, below Coventry to Leamington and between Stratford-upon-Avon and its confluence with the River Arrow, was recorded as poor quality from a biological survey carried out in 1990 (Biological data band C or National Water Council Class 3; see Appendix 3).

The upper reaches of the Trent and its tributaries, Penk, Sow and Tame fail to comply with River Quality Objectives, showing poor water quality as Class 3 or even grossly polluted Class 4 near industrial and urban centres such as Birmingham.

The middle reaches of the Trent system benefit from the clean water arriving from the Dove and Derwent but along its lower reaches, below the confluences of the Soar and Maun, the water quality deteriorates from Class 2 to 3.

Fisheries on principal rivers

Both systems are primarily cyprinid fisheries, with non-migratory salmonids present throughout but dominating a few upper tributaries such as the Derwent, Wye, Dove, Churnet on the Trent and the Teme and upper tributaries on the Severn. Overall, electro-fishing surveys meet designated fisheries targets and show a good distribution of mixed coarse fish. Eels are widespread, but concern about recruitment has led to several restockings of elvers in the upper Trent (A. Crawford, pers. comm.).

Otter conservation effort

Throughout the 1980s, The Vincent Wildlife Trust's Otter Haven Project was instrumental in preparing the ground for the otter to recover its former range along the Severn catchment. This has involved practical conservation work by way of sympathetic riparian management, creation of secure breeding holts and monitoring of the changing distribution and status of the otter in the Region.

From 1990, the RSNC's Otters and Rivers Project, in conjunction with the Worcestershire Wildlife Trust, has continued the work on the middle/lower Severn, River Teme and River Avon. Upgrading and strengthening of habitat to consolidate the small population of otters over the 'advancing front' forms the main focus of activity and effort. The project has concentrated effort into the construction of permanent pipe-and-chamber artificial holts along the lower and middle reaches of the River Avon, much of which is navigable and largely devoid of natural holt sites. A number of the holts constructed on the Severn and lower Avon are showing positive evidence of otter usage (P. Hoban, pers. comm.).

Results for 1991-94

Total full survey sites	610
Positive full survey sites	126
Positive spot checks	9

Comparison of results for the same survey sites 1977-79 and 1984-86

	1977-79	1984-86	1991-94
Full survey	13/567 (2.3%)	22/567 (3.9%)	120/567 (21.2%)
Spot check	0/16	1/16	9/16

Summary of results for each river system (see Figure 12)

Trent (including River Torne and drains)

1/24 full survey sites was found to show evidence of otters as a single spraint on a wooded stretch of the River Torne. A recent report concerns a probable otter sighting on the lower part of the River Trent, suggesting that transient animals are using this system.

Both previous surveys failed to reveal the presence of otters at these mostly canalised and heavily maintained sites.

Trent (including Devon, Smite, Soar and Wreake)

2/97 full survey sites were found to be positive during difficult conditions of high water and extensive flooding. Otter signs were found at two adjacent sites on the Wreake (both had a single, fairly old spraint), suggesting a transient animal rather than a resident population. Much of this part of the Trent system lacks suitable riparian habitat for otters and shows a high level of disturbance.

Soar

(0/4) The four full survey sites remained negative on the upper reaches of this river.

Trent – main river

2/8 full survey sites showed otter evidence along this upper part of the Trent that provides some suitable looking habitat (although the water quality and fish stocks are poor around Stoke-on-Trent and Stone). All previous surveys failed to show the presence of otters, and it is believed that the animals have only recently crossed the watershed between the Severn and the Trent (within the last 2-3 years).

Trent – Sow and tributaries including Clanford Brook

4/15 full survey sites and one spot check were positive. All sites were negative during the 1986 and 1979 surveys, but a catchment survey by P. Howell during May 1990 revealed

Figure 12. Severn-Trent Region

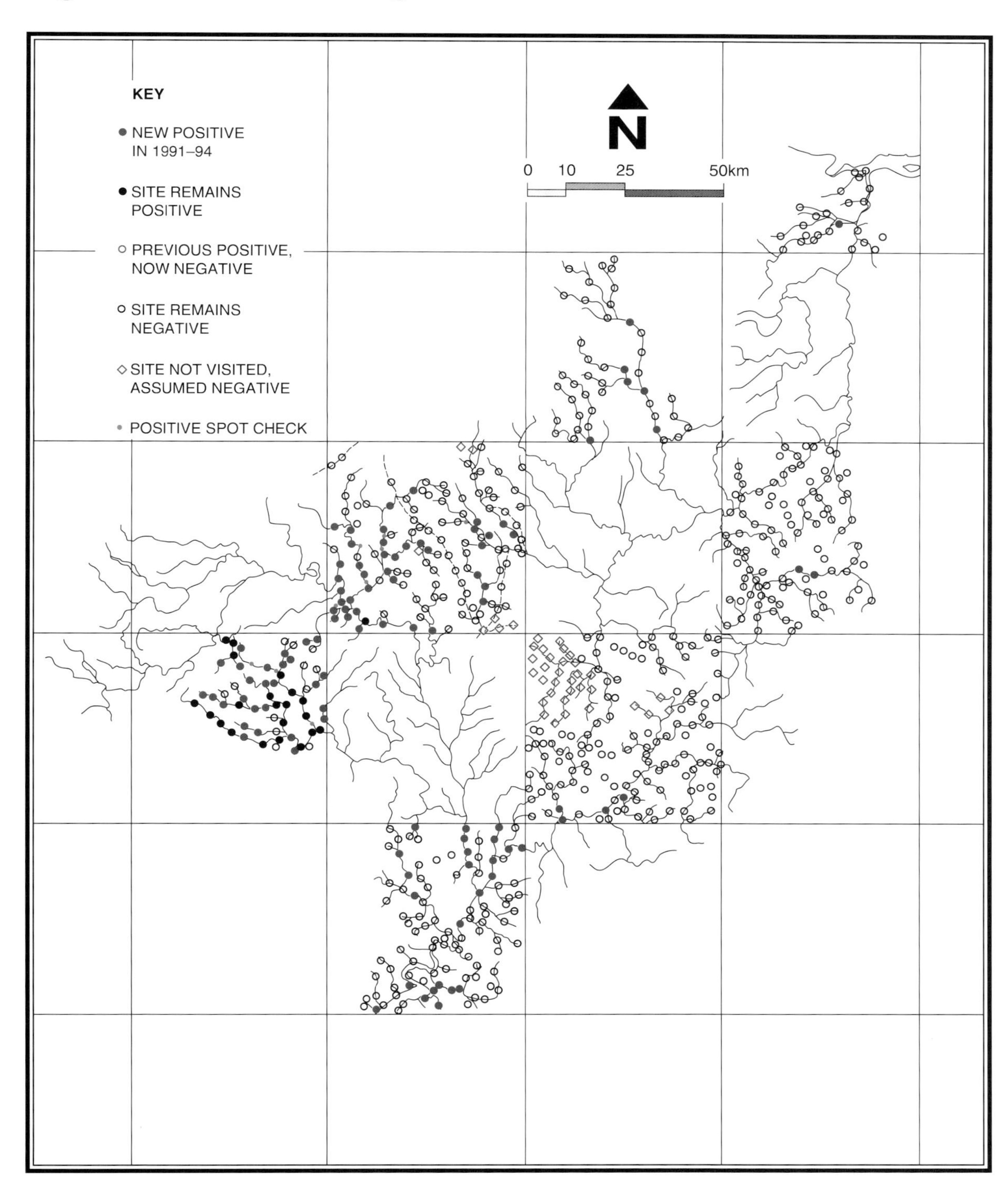

signs at a single site on the Clanford Brook. However, other reports from the system are largely unconfirmed.

Penk

3/10 full survey sites showed otter evidence on this system, which suffers from poor water quality but offers some suitable riparian habitat. All previous surveys were negative and there have been no recent reports.

Blithe

(0/3) The three full survey sites on the uppermost reaches of the river were negative, as in the previous surveys.

Tame and Anker

(0/25) The 25 full survey sites were negative, as in the previous surveys. The Tame and Anker are tributaries of the Trent system which run through industrial Birmingham and Coventry and show poor water quality and sparse cover for otters.

Dove

1/13 full survey sites was positive, a single spraint being located at the confluence of the Dove and Manifold. As with the evidence on the Derwent system, sparse signs and lack of recent reports suggest the recent arrival of otters to this upper part of the Dove. (These may be transient animals, possibly expanding into the Trent system from the Severn system.)

Tean

(0/1) The single full survey site remained negative on this small upper tributary of the Dove.

Derwent

5/31 full survey sites were positive on the River Derwent and its main tributary the Wye. Excellent habitat for otters remains on the system, but evidence of otters was sparse, suggesting very few animals. The lack of recent confirmed reports suggests that otters arrived on the system only shortly before the survey.

Severn (Camlad and upper Cound Brook)

7/8 full survey sites were found to be positive on these upper/middle tributaries. This compares with three positive sites on the Camlad in 1986 and a single positive site in 1979. These tributaries provide some excellent habitat for otters, and breeding was confirmed for the Camlad (cub tracks). Recent sightings were also reported.

Severn – lower Cound Brook and minor tributaries

8/9 full survey sites and one spot check were positive. All sites were negative in both previous surveys.

Teme (Clun, Onny and Corve)

41/52 full survey sites and three spot checks were positive. This compares with 18 positive full survey sites and one positive spot check in 1986, and 12 positive full survey sites and no positive spot checks in 1979. This represents a consolidation of otter activity on the main rivers and most of the minor tributaries, except at their uppermost reaches.

Records and reports have been consistent during the period between the last two surveys.

Severn – main river

5/5 full survey sites were positive, compared to a single positive site on the main river near Cressage in 1986. This was the prelude to a steady recolonisation process that has taken place over the last seven years.

Records and reports have been consistent during the period between the last two surveys.

Worfe

1/7 full survey sites showed evidence of otters, compared to none in previous surveys. Reports suggest a periodic occupancy of this river.

Rodden

5/12 full survey sites and two spot checks were positive. Otters have been regularly reported on this river since 1987. All sites failed to reveal otter signs in both previous surveys.

Tern

9/22 full survey sites and two spot checks were positive, compared to all sites being negative in 1986 and 1979. As with the Rodden, there have been regular reports of otters since 1987.

Meese

4/10 full survey sites showed evidence of otters, compared to none in the previous surveys. Frequent signs have been noted along the length of the river to Aqualate Mere during the last seven years (A.K. Crawford, pers. comm.).

Severn (below Worcester and including lower Avon, Leigh Brook, Frome, Cam and estuarine sites)

24/104 full survey sites were positive, representing a dramatic range expansion covering over 70km of waterway down the River Severn and its major tributaries. Both previous surveys in 1979 and 1986 failed to locate any evidence of otters.

The main river revealed the presence of otters between Worcester and Tewkesbury, together with signs along the lower Avon between Tewkesbury and Evesham. Elsewhere, sparse signs were found on the Leadon above Dymock, on the Frome below Stroud and along some of the small streams that flow into the lower Severn estuary (such as the Cam near Slimbridge).

Recent reports suggest that the otter occupancy of the Severn system below Worcester may only have occurred in the last year or so.

Avon

4/64 full survey sites revealed evidence of otters, representing a dramatic expansion in range along the Avon from its confluence with the Severn at Tewkesbury. Both previous surveys failed to locate any otter signs in this 50km square. Signs were found between Bidford-on-Avon and Hampton Lucy (including Stratford-upon-Avon) and also on its tributary the Arrow below Alcester. There had been no recent reports of otters this far up on the system, the last confirmed record being from the 1960s.

Canals

(0/57) The 57 full survey sites were investigated on the following canals – Trent and Mersey, Staffs and Worcester, Shropshire Union, Stratford-upon-Avon, Worcester and Birmingham, Grand Union, Birmingham and Fazeley, Coventry, Oxford, Ashby-de-la-Zouch. As in the previous surveys no evidence of otters could be found. All sites were generally poor for otters, offering largely unsuitable riparian habitat and few potential spraint sites, and were likely to be disturbed.

Complementary information from additional surveys on rivers in the adjacent 50km squares:

SO n/e - A full systematic survey of this square was carried out by P. Howell during October 1991, and smaller scale surveys were carried out in 1992 and 1993 by P. Hoban. The 1991 survey showed otter evidence throughout the Teme (15/36 full survey sites were positive) down to its confluence with the Severn, but sparse evidence was found on the River Severn itself (8/50). These positive sites were found 20km upstream of Worcester and just below the confluence with the Teme. By 1993 there had been a slight improvement.

SP s/w - The Stratford Avon catchment was surveyed by G. Scholey during September – November 1993, using the same methods as those of the Otter Survey of England. All 21 sites failed to reveal any evidence of otters despite recent reports and evidence being found in adjacent areas as described above. The rivers Stour and Isbourne are possible colonisation routes which would allow otters to cross watersheds between the Severn catchment and the Upper Thames.

No information is available for the two alternate 50km squares SK s/w and SK n/e which comprise most of the Trent catchment. However, parts of the Trent and lower Dove were visited for a local water vole and mink study in 1993, during which fresh spraints were located at a small number of sites (C. Strachan, pers. comm.).

Summary

126/610 (20.66%) full survey sites were positive. This showed a dramatic improvement throughout the Severn catchment, and also a range expansion into the upper Trent and its tributaries, Dove and Derwent. The otters probably crossed the watershed between the two catchments during their investigatory searching for unoccupied water.

The otters of the upper Severn have consolidated their range, as can be seen from the many spraints at numerous sites, together with the presence of cub tracks, indicating successful breeding. On the lower Severn otter signs have been located on the River Avon, suggesting an expansion eastward along this system into the heartland of England.

The weather conditions for surveying were generally very favourable without any period of prolonged flooding.

5.5 ANGLIAN REGION

Full surveys were carried out in the following 50km squares – SE s/e (10 sites), SK s/e (41), SP s/e (79), TF n/w (142), TF s/e (98), TL n/w (132), TL s/e (102), TM n/w (121).

Description of Region

The river systems of this Region drain the lowlands of eastern England with land rarely exceeding 200m above sea-level (in the Chiltern Hills), and over an extensive part of the area the topography is flat. These are the former flood plains of many of the principal rivers which now provide fertile cultivated land on peaty soils, particularly around the Wash.

Flooded peat diggings on the Yare and Bure systems have given rise to the Broads of Norfolk. Farming is predominantly intensively arable with the main crops being cereals (wheat and barley), sugar beet and oil seed rape. Other root vegetables and brassicas are grown on the fen soils. Elsewhere flood plains and coastal marshes are used for grazing.

The human population is largely rural, being centred on market towns and cities, notably Northampton (182,100), Milton Keynes (177,600), Peterborough (152,900), Chelmsford (151,700), Colchester (150,200), Bedford (136,800), King's Lynn (134,000), Norwich (117,300), Ipswich (114,900), Cambridge (97,800), Lincoln (80,400) and Corby (51,000).

Pollution problems

As would be expected, the rivers of the Region are prone to a large number of pollution incidents from agricultural sources. These are widespread throughout the Region and include fertiliser and pesticide spillages/run-off together with farm waste run-off from intensive livestock farms (cattle, pigs, chickens, geese and turkeys). Nitrate levels in the watercourses are monitored and appear to have increased over the last ten years.

Overall, most rivers meet their designated NRA Water Quality Objectives, although a number of rivers have deteriorated from Class 2 to Class 3 (see Appendix 3).

Industrial pollution incidents are largely confined to the River Nene and Great Ouse catchments, but were noted as occurring around some of the Region's larger towns.

Fisheries on principal rivers

All rivers in the Region are classed as cyprinid fisheries, although brown trout occur throughout and a number of the larger reservoirs are stocked as trout fisheries (rainbow and brown trout). Commercial eel netting occurs in a number of the river mouths, especially around the Wash.

Most rivers support a good population of mixed coarse fish and meet biomass targets, although many sites are regularly restocked by angling clubs.

Otter conservation effort

Limited direct otter conservation effort has been carried out in this Region, with otter havens and sympathetic riparian land management occurring on a few rivers including the rivers Black Bourn, Wensum, Bure, Thet, Minsmere, Glaven and Wissey.

Distribution surveys have also been limited, but a number of County Wildlife Trusts are now addressing the need for such information as follows:

River Great Ouse: by the Wildlife Trust for Bedfordshire, Cambridgeshire, Northamptonshire and Peterborough (BCNP) (J. Green).

River Nene: by the Wildlife Trust above (J. Green and A. Colsten).

River Witham: by the Lincolnshire Wildlife Trust (G. Smith).

Rivers Glaven, Stiffkey, Nar, Wissey, Wensum and Bure: by the Norfolk Naturalists' Trust volunteers.

The Otter Trust has also carried out periodic surveys in the Region as a prelude to any otter releases, or subsequent work to assess their success. Release of captive-bred otters by the Otter Trust has concentrated on a number of key rivers which had either 'lost' their otters or were regarded as only supporting a small and probably non-viable population and therefore requiring reinforcement. Up to January 1994, 37 animals have been released at ten sites which cover the rivers Waveney, Yare, Bure, Glaven, Wissey, Black Bourn, Thet, Minsmere, Deben and Tiffey.

Results for 1991-94

Total full survey sites	725
Positive full survey sites	58
Positive spot checks	13

Comparison of results for the same survey sites 1977-79 and 1984-86

	1977-79	1984-86	1991-94
Full survey	20/623 (3.2%)	8/623 (1.3%)	52/623 (8.3%)
Spot check	5/37	3/37	13/37

Summary of results for each river system (see Figure 13)

Ancholme and South Humberside drains

(0/23) The 23 full survey sites were negative, as in both previous surveys. These waterways are canalised, regularly dredged and have heavily maintained banks in their middle and lower reaches, but run a more natural course at the uppermost reaches.

No recent reports were noted from the area, though an otter was drowned in a fyke net set in Target Lakes, South Humberside, in 1988 (Jefferies, 1990b).

Great Eau, Louth Canal and fen drains

(0/41) The 41 full survey sites were found to lack otter signs, as in both previous surveys, although weather conditions were poor with high water masking any suitable silt banks or ledges for tracks or spraints.

An otter found still alive in a fyke net set at Ingoldmells, near Skegness, in November 1989 was the only confirmed record of any otters being present on this catchment (Jefferies, 1990b).

Steeping

(0/13) The 13 full survey sites remained negative in the area where a single spraint revealed the presence of otters in the baseline survey. Part of the system offers some promising habitat, but the river is small and shallow with a low fish biomass.

Upper Witham

(0/16) The 16 full survey sites were negative, but the water level was high at the time of survey. Some good habitat is present and recent reports suggest that otters do use this part of the system from time to time.

Witham (including Bain, Slea and Brant)

2/75 full survey sites were positive. The two sites were situated along the Bain and widely spaced between a series of negative sites. This suggests a highly fragmented population, which probably consists of a few transient animals using the system, which offers suitable riparian habitat in places. A survey of this system by the Lincolnshire Wildlife Trust in early 1994 was able to confirm the continued presence of otters along the River Bain (G. Smith, pers. comm.).

All sites were negative in both previous surveys.

Figure 13. Anglian Region

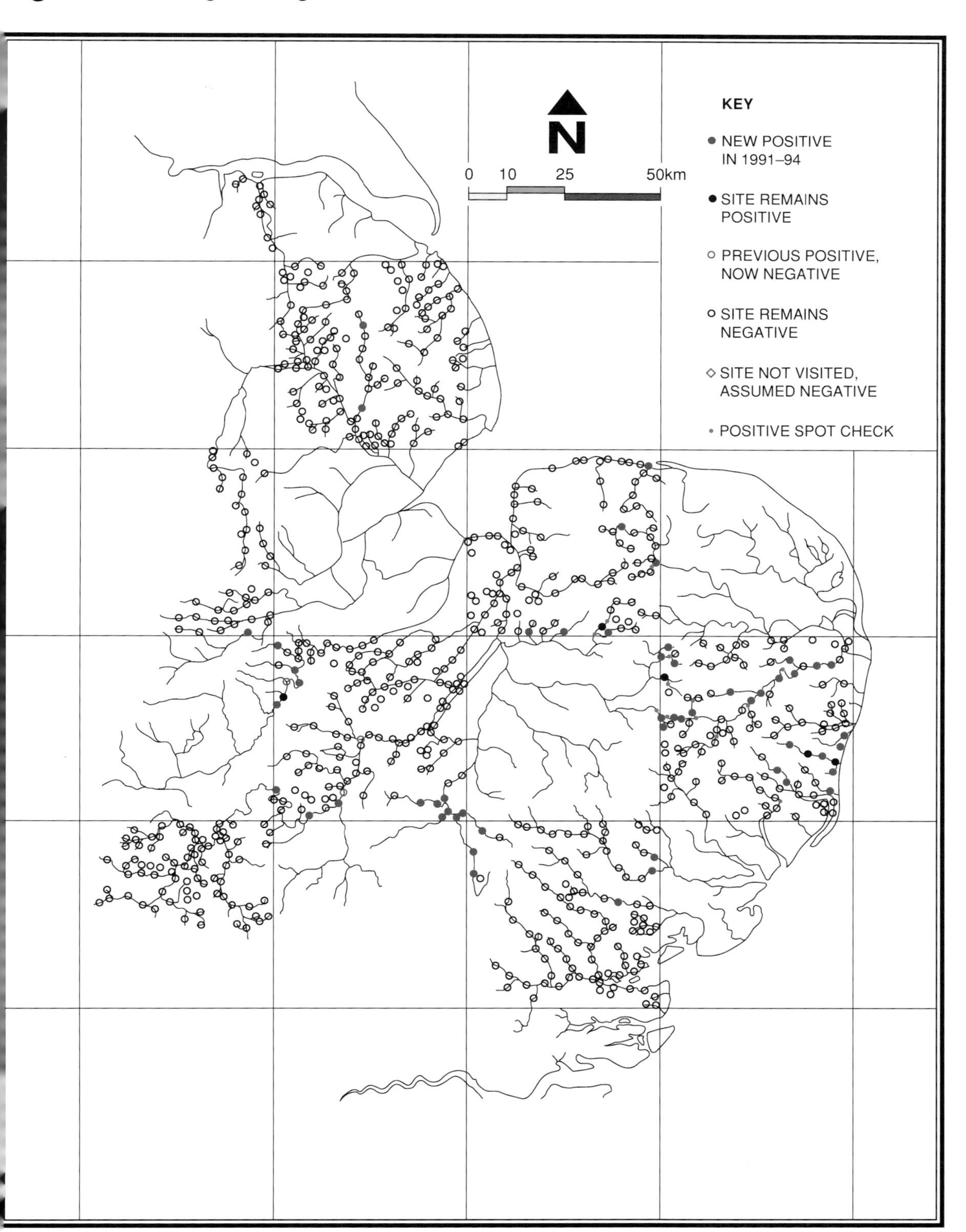

Welland (including Chater, Gwash and Glen)

1/24 full survey sites was positive, located on the main river. As in 1986, there were a number of unconfirmed reports of transient otters from some tributary sites and Rutland Water.

Nene

5/28 full survey sites were positive along the middle reaches of this system, compared to two in 1986 and none in 1979. This part of the river provides some suitable riparian habitat for otters, together with a series of disused and flooded gravel pits which have low disturbance and are well stocked with fish. Independent surveys on this river during 1993 confirm that a small population of otters is extant in the vicinity of the Titchmarsh Nature Reserve (J. Green - Wildlife Trust for BCNP, pers. comm.). At the uppermost reaches of the River Nene a single site revealed fresh otter evidence from a probable transient animal during the winter of 1993/94 (R.E. Stebbings, pers. comm.).

River Great Ouse (upper reaches)

(0/78) The 78 full survey sites were negative, as in 1985 and 1978 on the upper reaches of this system, which includes the tributaries River Ouzel and Tove, as well as the Great Ouse. Suitable riparian habitat occurs on parts of the Great Ouse and Tove, but the water quality is poor (Class 2/3) throughout much of its length and disturbance is locally high.

River Great Ouse and fen drains

3/94 full survey sites were found to be positive along the middle reaches of this system, suggesting that a small population of otters retains a fragile hold in this area. Two sites showed otter evidence in 1986, while only one was positive in 1979.

Recent records and reports are largely unconfirmed but refer to wide-ranging and transient animals, with no reference to any breeding. The series of flooded gravel pits along the watercourse offer suitable lying-up places on small, undisturbed and well vegetated islands.

Lower River Great Ouse and drains

(0/29) The 29 full survey sites were negative, as in the previous surveys. The area is intensively farmed and generally there is little or no cover except for a fringe of *Phragmites* reed or, more rarely, backwater pools with reedbed.

Wissey

4/14 full survey sites were positive, together with two positive spot checks on this system that supports some excellent habitat for otters. Sparse signs were encountered along the upper part of the river, but very fresh signs were located on a tributary and around the pools of Foulden Common.

Outside the 50km survey square the Wissey flows through the protected area of the MOD land at Stanford, where otter activity has continued to be recorded since the last survey (J. Goldsmith, pers. comm.). Captive-bred otters were released to the Wissey by the Otter Trust in 1992 and 1993.

In 1985, the River Wissey provided the only evidence of otters for this 50km square (as one positive full survey site and one positive spot check).

The baseline survey of 1978 found otter signs at three full survey sites and two spot checks on the upper reaches of the river.

Nar

(0/12) The 12 full survey sites showed no signs of otters as in 1985 and 1978, despite rumoured sightings. The river offers some good riparian habitat along its middle and upper reaches.

Babingley River and coast

(0/10) The ten full survey sites were negative, as in 1985, in this area of coastal Norfolk that overlooks the Wash and includes sea defence dykes, grazing marshes and arable farmland. Little cover is offered, except among a few areas of reedbed and pools. The baseline survey of 1978 found five positive full survey sites and one positive spot check, suggesting that the Babingley River held resident otters at that time. However, it should be noted that an incidental visit to this system by the surveyor early in 1994 again located fresh otter spraint and tracks, the first confirmed evidence in 16 years.

Burn and coast

(0/10) The ten full survey sites on this part of coastal Norfolk were negative. The area includes sea defence dykes and grazing marshes, with locally good expanses of reedbed. All sites were also negative for otters in 1985, but three full survey sites proved positive in the baseline survey of 1978.

Stiffkey

1/8 full survey sites showed signs of otters, compared to none in 1985 but two positive full survey sites during 1978. Evidence was located at the sluice with the Stiffkey/Blakeney saltmarshes and suggests an animal from the Cley marshes/River Glaven system where captive-bred otters were released in 1987, 1991 and 1992 (warden, pers. comm.). Work on wild otters in this area carried out by V. Weir in the 1970s has shown that the Stiffkey river formed part of the home range of the Cley marshes/River Glaven animals (Weir, 1984).

Suitable habitat for otters remains on this system.

Wensum

2/15 full survey sites and one spot check were positive on this system, which showed no evidence in 1985 but revealed four positive full survey sites and two positive spot checks in 1978. Good riparian habitat for otters can be found along most of the upper and middle reaches.

Cam

10/23 full survey sites were positive along the upper reaches of this system, among excellent riparian habitat. No evidence of otters was found in either of the two previous surveys.

Recent reports suggest that the occurrence of resident otters on this system may have only happened since 1992 and may represent dispersal from release sites on adjacent rivers (Stort, Thet or Black Bourn).

A subsequent survey by the Wildlife Trust for BCNP in 1993 was able to corroborate these findings.

Thet

4/5 full survey sites and two spot checks were positive on the uppermost reaches of this system, which fall within the survey square (ie. near Attleborough and Old Buckenham and the tributary through Quidenham). This compares with a single positive full survey site in 1985 and none during the baseline survey. The Thet was the river chosen for the second otter release group in July 1984 (see Appendix 13 and Section 7.2).

Outside of the survey square, a search at a number of bridges down to the confluence with the Little Ouse at Thetford revealed the presence of otters throughout the system.

Little Ouse

2/3 full survey sites and two spot checks were positive on the uppermost reaches of this system, which fall within the survey square (ie. Hinderclay/Thelnethan Fen and a minor tributary). This compares to a single positive spot check in 1985 and none in the baseline survey.

The Little Ouse has its confluence with the River Black Bourn some 10km downstream of the survey square and its confluence with the Thet some 15km downstream. The Black Bourn was the river chosen for the first otter release group (see Section 7.2) in July 1983 (Jefferies *et al*, 1986; Wayre, 1989), and otters have been in evidence ever since that date, with the latest report during the survey period being a sighting of a family group (three cubs) near the release site in October, 1992 (J. Wilson, pers. comm.). A search at a number of bridges on the Little Ouse and Black Bourn, away from the survey square, revealed the presence of otters throughout both, downstream to Brandon. Signs were particularly abundant near the confluence of the two rivers, with up to 17 spraints under one bridge and sign-heaps (see photograph in Strachan *et al*, 1990) under another.

An otter was watched swimming and fishing at dusk on 2nd November, 1992 near the Black Bourn release site (R. Strachan).

Waveney

11/38 full survey sites and five spot checks were positive. This compares with a single positive spot check in 1985 and none in 1978.

Very fresh evidence was located at the source of the Waveney on Redgrave Fen which, together with recent signs on adjacent tributaries, suggests that otters may have crossed the watershed between this system and the Little Ouse. Elsewhere on the Waveney, otter evidence was encountered on the middle reaches, with two centres of activity – upstream of Bungay and downstream of Beccles.

The Waveney was the river chosen for the third otter release group in October, 1984 (see Section 7.2). Further otters were released in March 1992.

Chet

(0/4) The four full survey sites were negative, as in both previous surveys. Survey conditions were made difficult by heavy rainfall and high water level which may have removed signs.

Tas

(0/11) The 11 full survey sites were negative, as in both previous surveys. As on the Chet, survey conditions were unfavourable for finding all but the most recent otter evidence.

Latymere Hundred River

(0/3) The three full survey sites were negative, as in 1985 and 1978. Better habitat occurs downstream, outside of the survey square.

Blyth (including coastal streams between Wangford and Benacre)

(0/12) The 12 full survey sites were negative, as in both previous surveys. Suitable otter habitat remains, but the area was frustrating to survey owing to heavy rainfall and high water level which may have removed signs. An unconfirmed otter sighting was reported near Halesworth earlier in 1992.

Dunwich River, Minsmere River and Hundred River

8/10 full survey sites were positive on these coastal rivers and marshes. In 1985, two sites were found positive along the Minsmere River, while the baseline survey found a single positive site on the Hundred River. Minsmere was the site of the fourth otter release group in July 1985 (see Section 7.2) and breeding has been recorded several times since. This survey located fresh tracks of a female with two small cubs (about three months old) among the dykes of the Sizewell Belts, and it is believed that two other females reared litters of two and three cubs on the Royal Society for the Protection of Birds reserve at Minsmere during 1991 (Reserve warden).

It appears that the Minsmere release has successfully provided a breeding nucleus from which otters have populated the Dunwich and Hundred rivers, the Alde and probably the Blyth.

Alde (including coastal marshes)

2/14 full survey sites were positive. All failed to reveal otter signs in both previous surveys. The evidence was located at two adjacent sites on the middle reaches of the river. Elsewhere, heavy rainfall and high water conditions may have removed any field signs.

Deben

(0/11) The 11 full survey sites were negative, but one spot check revealed a set of fresh otter tracks. All sites in both previous surveys were negative. Heavy rainfall and subsequent flooding may have influenced the chances of finding evidence.

Gipping

(0/9) The nine full survey sites were negative, as in 1985 and 1978. Some good potential habitat remains and good fish stocks were noted, but no otters have been reported from this system in recent years.

Stour

2/22 full survey sites were found to be positive on this system that failed to show any evidence during 1986, when the sites were added to the survey. Good riparian habitat remains, but sites were investigated under conditions of high water level.

Local reports suggest that otter signs were first noted during 1993, the animals possibly arriving by way of the Suffolk Black Bourn (ie. descendants of released stock).

Colne

1/20 full survey sites was positive, suggesting a transient animal, although otters had been rumoured to be on this system since 1992 (Colchester Natural History Society). All sites were negative in 1986.

Blackwater and Chelmer

(0/50) The 50 full survey sites were negative, as in the previous survey. The catchment offers some good potential riparian habitat for otters, especially along the middle and lower reaches and among the marshes of the Blackwater Estuary, a traditional haunt of the otter. No recent reports were received.

Complementary information from additional surveys on rivers in the adjacent 50km squares:

SP n/e - During 1994 a systematic survey was carried out on the River Nene catchment by the Wildlife Trust for BCNP, and 1/50 full survey sites was found to be positive (A. Colston, pers. comm.). The presence of otters was confirmed downstream of Northampton by a single spraint found during an otter survey training day in the same year (R. Strachan, pers. obs.).

TL n/e - The Cam was surveyed by the Wildlife Trust during 1993 and found to support otters upstream of Cambridge but not below. The Black Bourn and Thet continue to be monitored by the Otter Trust following the release of otters in 1983 and 1984 and were still showing signs of the animals in 1994.

Summary

58/725 (8.00%) full survey sites were positive, demonstrating continued successful breeding and range expansion, largely derived from otter releases.

The 1984-86 survey revealed only 8/725 (1.10%) full survey sites with any evidence of otters, suggesting a low point in the population, compared with 20/623 (3.21%) in 1977-79 (see Figure 42). The Waveney, Thet, Little Ouse and Wissey all showed a sustained improvement in otters signs, as did the Minsmere and Alde populations following the Otter Trust releases.

The River Cam above Cambridge showed good signs of resident animals for the first time, while the Ouse, Nene, Welland and Bain all revealed signs from a small, fragmented population or transient animals. Two positive full survey sites were located on the upper Wensum and a single positive site represented the otter population using the Norfolk coast. In addition, the presence of transient animals was also identified on the rivers Stour and Colne, possibly from a dispersal of animals across the watersheds between the Black Bourn or Cam and these rivers in the south of the Region.

Weather conditions were generally good throughout the survey period.

5.6 THAMES REGION

Full surveys were carried out in the following 50km squares – SO s/e (3 sites), SP n/w (2), SP s/e (68), SU n/w (76), SU s/e (21), TL s/e (10).

Description of Region

The Thames hydrometric area runs west to east in a low-lying basin between the Chilterns, Cotswolds and the North Downs. The geology of the Region is mainly chalk, gravels and clay, allowing for good mixed cattle, sheep and arable farming.

Apart from the urban conurbation of Greater London (6,731,000) on the lower Thames, the Region is largely rural with the main centres being Swindon (150,800), Reading (132,400), Oxford (115,400) and Newbury (108,700).

Pollution problems

The Thames catchment suffers from an increasing number of pollution incidents from agricultural sources, organic waste, silage liquor run-off, agrochemical fertiliser and herbicide/pesticide input. Industrial and domestic pollution is also widespread, but nevertheless the NRA River Quality Objectives are met over most of the catchment. The upper Thames rivers show a water quality of Classes 1A, 1B and 2, whereas lower reaches are Class 3 (see Appendix 3). A report on river quality (NRA, 1991) has found that a number of sampling sites have deteriorated when compared to the mid-1980s survey. However, this trend is likely to be reversed owing to capital investment and sewage treatment works upgrading.

A number of specific pollution incidents have been recorded as causing major fish kills. In the most serious of these a ton of fish was killed on the Blackwater near Aldershot in 1992 as a result of cyanide pollution.

Fisheries on principal rivers

Throughout most of the Region a good healthy diverse fish fauna is recorded, although at many sites it fails to meet the EC designated cyprinid and salmonid fisheries targets of $20g/m^2$ and $15g/m^2$ respectively.

Elvers were released into the Thames catchment in 1993 to help boost populations in tributaries where eels were caught during electro-fishing sampling. These areas were the Windrush and Coln in the upper Thames area, the Lee catchment in the north-east Thames area and the upper Wey in the south-east Thames area. 180,000 elvers were released (60,000 in each area) where the three Thames Region Otter Habitat Projects were in progress at that time.

Otter conservation effort

Historically, only a limited amount of direct conservation effort has been targeted in the Region, but survey work and particular projects involving sympathetic riparian habitat management and creation of otter havens were initiated in 1990 on the upper Thames (T. Sykes), Lea Valley system (M. Findlay) and River Wey (G. Scholey). The Kennet Otter Habitat Project commenced in 1994 (M. Satinet) and the Cherwell Otter Habitat Project in 1995.

Restoration work by NRA (Thames Region) has implemented some noteworthy schemes, such as the re-creation of an extensive area of water meadows along the River Windrush and River Evenlode (A. Driver, pers. comm.).

In addition to the above, the Otter Trust carried out two releases on the River Lea and the River Stort, and these were monitored by the Hertfordshire Wildlife Trust (M. Findlay).

Figure 14. Thames Region

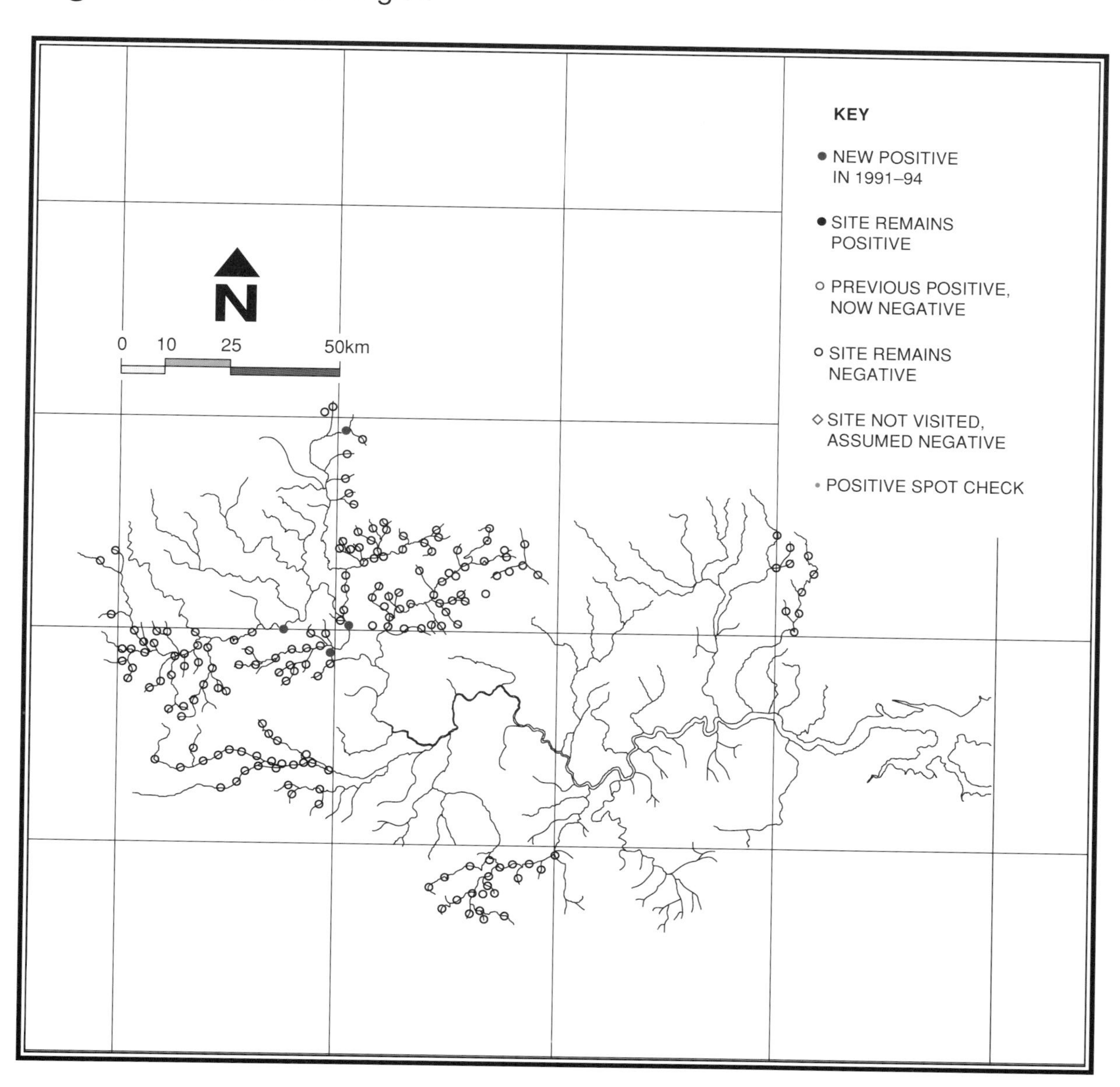

Results for 1991-94

Total full survey sites	180
Positive full survey sites	4
Positive spot checks	0

Comparison of results for the same survey sites 1977-79 and 1984-86

	1977-79	1984-86	1991-94
Full survey	0/170 (0.0%)	0/170 (0.0%)	4/170 (2.3%)
Spot check	0/6	0/6	0/6

Summary of results for each river system (see Figure 14)

Wey

(0/18) The 18 full survey sites on the upper catchment of this system (North and South Wey) were negative. They were surveyed during a spell of heavy rain, which may have affected the results. There were a few unconfirmed reports of probable transient otters in recent years, but the area had been intensively searched in a survey by G. Scholey (pers. comm.).

See also the note on complementary information below.

Kennet

(0/22) The 22 full survey sites on the upper and middle parts of the Kennet and its tributaries Lambourne and Enborne were negative. Very low water conditions affected these sites and seven were completely dry! Despite this, some good habitat for otters occurs, although there have been no recent reports of the animals. In addition to the above, five full survey sites were examined along the Kennet and Avon Canal, and as before all were negative.

Thames (upper)

2/51 full survey sites were positive on the upper Thames and its tributaries (including the Ock, Cole, Ray, Isis and confluence sites on the Churn and Coln). Old signs were found on the Thames below Abingdon, but fresher evidence was located further upstream above the confluence with the Windrush. No otter signs were located in either of the previous surveys, although there had been unconfirmed reports. Good riparian habitat remains along the upper catchment of this system.

Thames (middle)

2/74 full survey sites were positive on this middle part of the Thames (including the rivers Thame, Cherwell and Ray). One was on the Thames itself near Oxford and the other along the upper reaches of the River Cherwell, suggesting a small fragmented population perhaps occupying the rivers of the adjacent square SP s/w (Cherwell, Evenlode and Windrush). Good riparian habitat can be found along this part of the system which revealed no otter signs in the previous surveys.

Two road casualty otters have been found from this system in recent years (1986 and 1991), both identified as Asian short-clawed otters (Jefferies, 1990a, 1992a). No discernible difference could be detected in the spraints encountered here from those of *Lutra lutra* and unfortunately no tracks were seen.

Stort and Roding

(0/10) The ten full survey sites failed to reveal any signs of otters, as was the case in 1986. Suitable riparian habitat for the animals was lacking at the sites searched. However, three otters of captive-bred origin were released on the Stort by the Otter Trust during December 1991 and are thought to be still in the area, although an otter subsequently killed on the road at Bishop's Stortford a few months later (1992) may have come from this group.

Complementary information from additional surveys on rivers in the adjacent 50km squares:

TQ n/w, TQ s/w and SU s/e were systematically surveyed by the Wey Otter Project on the River Wey only. No evidence of otters was located (G. Scholey, pers. comm.).

However, an incidental visit to this system by R. Strachan in early January 1994 located fresh otter spraint among the water meadows just below Guildford (TQ 0051), confirming the recent reports of an otter in this area. Subsequent survey by G. Roberts located more evidence along the North Wey, suggesting that the transient otter or otters had perhaps crossed the watershed from the River Rother (Arun system) in the Southern Region.

SU n/e - Only one confirmed record was received from this 50km square and this concerned a dead otter found on the westbound hard shoulder of the M4 motorway, between Reading and Maidenhead (SU 854753). It was discovered on 30th July 1993 and the corpse was collected by G. Scholey of the NRA and sent to the Ministry of Agriculture Fisheries and Food (MAFF) for a post mortem examination. It was a healthy male animal when killed by a motor vehicle. It appears that the animal was attempting to cross the motorway in preference to using Twyford Brook culvert, to gain access to or from The Cut, another watercourse that runs within 100m of the motorway. It is well known that otters frequently use very tiny tributaries and eventually ditches in order to cross from one river system to the next (Jefferies, 1989c).

Although there were no previously confirmed records of otters in the area, a number of unconfirmed reports from both the Lodden and Kennet suggest that otters may have been present over a number of years.

TL s/w - Two rivers in this square were examined for otters, the Lea and Stort. Both of these systems were chosen by the Otter Trust for otter releases in October and December 1991. The success of the project has been monitored by the Hertfordshire Wildlife Trust, which confirms the continued presence of otters at the former but not the latter (M. Findlay, pers. comm.).

SP s/w - 83 sites were systematically surveyed on the upper Thames in a period from February 1992 to November 1993, using the methods of the national surveys. Two sites were found to show the presence of otters, one on the River Thames at Duxford (spraint) and a second on the Cherwell near Banbury (tracks), suggesting that both were transient animals.

Summary

4/180 (2.22%) full survey sites were positive, revealing the presence of otters in the Thames Region for the first time during the national survey (although confirmed reports had suggested that possible transient animals had been around in the catchment for the last seven years). Evidence was located at three sites in the middle and upper reaches of the River Thames near Oxford and at a single site on the upper Cherwell.

The adjacent position of the Upper Thames tributaries in the Cotswolds to the southern tributaries of the Severn and Avon is expected to provide a natural recolonisation route, with the species crossing the Cotswolds watersheds.

A period of sustained winter flooding may have influenced the success of finding otter spraints on many of the upper Thames rivers.

5.7 WESSEX REGION

Full surveys were carried out in the following 50km squares – SS s/e (17 sites), ST s/e (92), SU n/w (17), SY n/w (14), SZ n/w (14).

Description of Region

The catchment area of this Region encompasses the low-lying land of the Somerset Levels, together with rivers that originate in the Wiltshire Downs, the Dorset hills and the Brendon Hills in Somerset. The highest land is no more than 300m above sea-level.

The land provides rich mixed farming of cattle, sheep and arable, while the human population is largely rural. The main cities and towns are Bristol (377,700), Bournemouth (154,800), Yeovil (141,700), Poole (130,700), Bridgwater (96,400), Salisbury (96,200), Taunton (94,100) and Bath (84,800). Coastal sites experience a large seasonal increase in population from holiday-makers.

Pollution problems

The water quality over much of the Region meets its NRA River Quality Objectives and many of the principal systems are Classes 1A, 1B or 2; only a few upper tributaries of the Stour, Bristol Avon and Parrett show a water quality of Class 3 or 4 (see Appendix 3).

Pollution from agricultural sources (silage liquor and slurry run-off, fertiliser and herbicide input) was noted as being widespread. Industrial pollution incidents were recorded as uncommon over the last seven years but have still resulted in some major fish kills.

Fisheries on principal rivers

The Tone, Parrett, Yeo and Brue systems support good mixed coarse fish populations meeting biomass targets, with some large eel populations. Salmonids are heavily influenced by stocking and generally fall below targets.

Eels also feature as a major component of the biomass for the River Axe and Cheddar Yeo, where biomass is recorded up to $65g/m^2$.

The small coastal systems of Washford, Avill, Aller, Char, Brit, Frome and Piddle all show good populations of trout and eels.

The Dorset Stour and Salisbury Avon meet the target for cyprinid fisheries but show poor densities of salmonids. Trout and salmon are recorded as rare for the Stour, although good on some upper tributaries of the Avon.

Otter conservation effort

Suitable otter habitat occurs along the lower and middle reaches of most rivers, but this becomes sparse where there is intensive farming and around most urban centres. Conservation effort in the Region has concentrated on the establishment of 'otter havens' and cultivating sympathetic riparian habitat management with landowners. In addition, the changing distribution of the otter in the Region has been monitored by regular survey work (P. Howell, J. Williams).

The Otter Trust has carried out a number of releases in the Region "in advance of the natural expansion". These have included the River Wylye, the River Stour and Bristol Avon.

Results for 1991-94

Total full survey sites	154
Positive full survey sites	29
Positive spot checks	2

Figure 15. Wessex Region

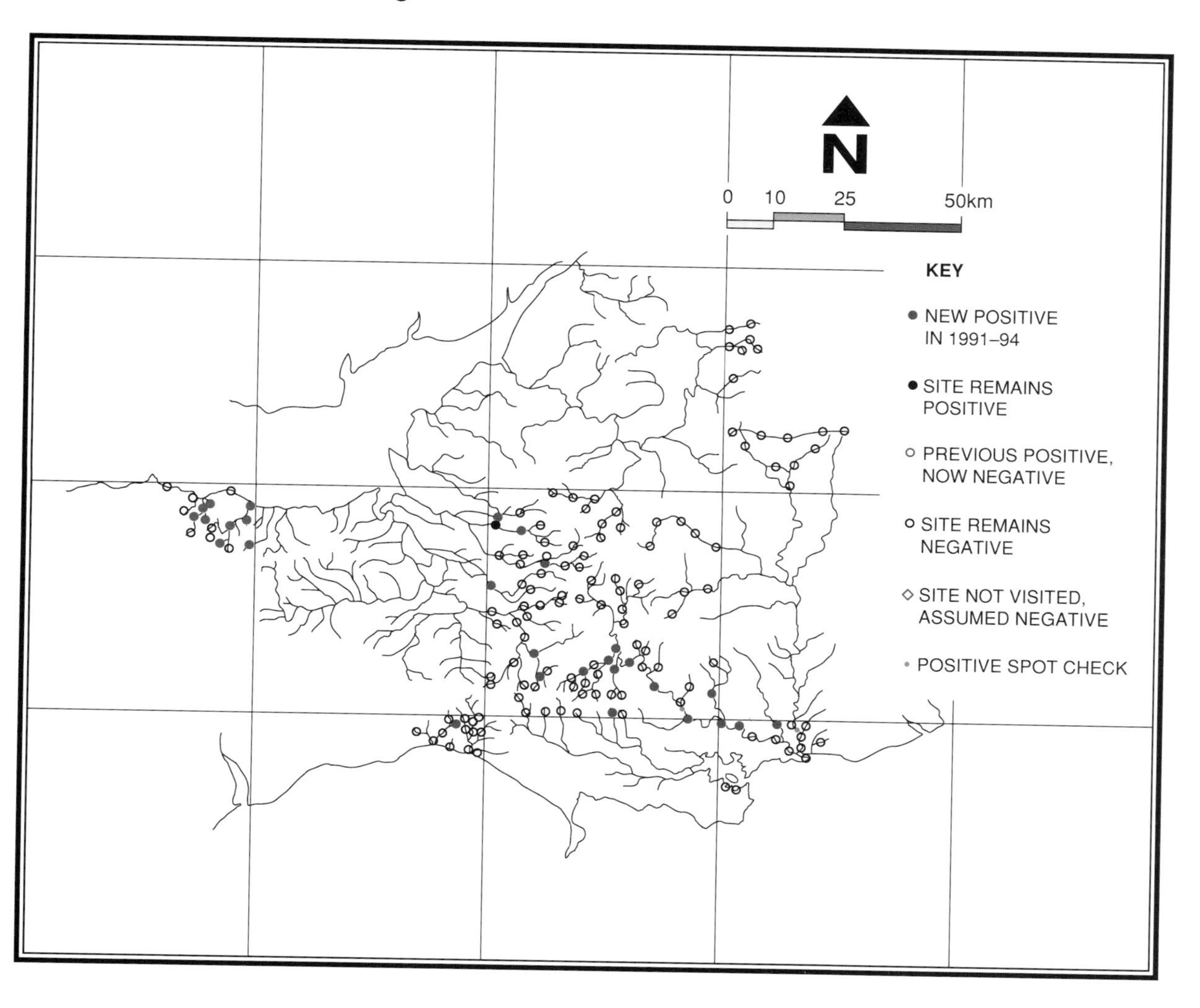

Comparison of results for the same survey sites 1977-79 and 1984-86

	1977-79	1984-86	1991-94
Full survey	2/154 (1.3%)	1/154 (0.6%)	29/154 (18.8%)
Spot check	2/8	0/8	2/8

Summary of results for each river system (see Figure 15)

Bristol Avon

(0/15) The 15 full survey sites on the uppermost part of this system (upper Avon, Frome and Mells rivers) were negative, as in the previous surveys. This was despite some suitable looking habitat, although some of the higher sites showed extremely low water conditions after the summer drought in 1991. During July 1990, two captive-bred otters were released by the Otter Trust on a tributary of the middle reaches of the system and signs were still being found close to the release site in 1992 (A. Moorhouse, pers. comm.).

Brue

4/15 full survey sites were positive on this system (Brue, Hartlake, Whitelake and Sheppey), compared with two in 1977 and a single positive site in 1984 (both times on the Hartlake only). This shows an encouraging expansion of the population known to be from the lower and middle reaches of the Brue.

Parrett

3/23 full survey sites were positive, compared to all sites being negative in both previous surveys. A set of fresh tracks gave the only evidence on the Cary, but upstream of Yeovil good signs were located at two neighbouring sites on the Wriggle (a possible recent arrival from the adjacent Stour catchment via the Lydden).

Coastal streams (including Washford, Avill, Aller, Horner and Hawkcombe)

9/15 full survey sites showed evidence of otters, compared to all sites being negative in both the previous surveys. This is a considerable improvement, suggesting the arrival of otters crossing the Exe headwaters. Recent reports suggest that they may only have occurred on these systems since 1990.

These small systems rising on the edge of Exmoor run through steep, wooded valleys offering good cover for otters.

Char and Brit

1/14 full survey sites was positive on these south coastal systems, which also include the Winniford, Simene and a single site on the lower Bride. Both previous surveys failed to find otters present on these systems, and it seems likely that the single spraint located on the Char was from a transient animal. Recent reports, however, suggest that otters are regularly using the Bride (P. Howell, pers. comm.).

Piddle and Frome

1/7 full survey sites was positive. It was located on the Devil's brook, a tributary of the Piddle. No otter signs were found in either of the previous surveys.

Studland Bay

(0/2) The two full survey sites were negative in this difficult area to survey, which has been known to hold otters in the past.

Stour

11/39 full survey sites and one spot check were positive. Evidence was located on the middle reaches of the Stour and its tributaries Allen, Divelish and Lydden. Negative sites were found between the positives, suggesting a small but wide-ranging population is present. The baseline survey found a single positive spot check on this system, but all sites were negative during 1984. The lower Stour was subsequently shown to hold otters in the interim period, and in 1989 and 1991 this population was reinforced by two releases on the middle Stour by the Otter Trust (P. Howell, pers. comm.).

Three positive full survey sites were located on the lower/middle Stour near Wimborne Minster and on the Moors River adjacent to Hurn Airport. Some very suitable habitat for otters remains despite urban encroachment.

Salisbury Avon (including Nadder, Wylye and Kennet, and Avon Canal)

(0/24) The 24 full survey sites examined were all negative, as in 1985 and 1978. However, a single spot check on the lower reaches of the main river revealed a set of clear otter tracks. Good riparian habitat remains comparable to that found along the Stour, suggesting that the system should be able to support a healthy otter population.

A male and female otter were released by the Otter Trust on the Wylye in December 1989 (in advance of the natural expansion of the otter population further west). The fate of these animals is unknown, but breeding by otters was confirmed along the lower reaches of the Avon during early 1992 (P. Howell and G. Roberts, pers. comm.).

Complementary information from additional surveys on rivers in the adjacent 50km squares:

ST s/w - The rivers Parrett and Tone are regularly monitored for otters and over the last five years a notable expansion in range has been seen, so that otters are now recorded throughout the systems (P. Howell and J. Williams, pers. comm.).

ST n/e - The Bristol Avon was surveyed between 1990 and 1993 and found to support otters only along the By Brook. This river was used as a release site by the Otter Trust in 1990 (P. Howell and A. Moorhouse, pers. comm.).

SU s/w - The Salisbury Avon was systematically surveyed in 1991 and 1992 with little success, otters apparently being confined to the lower reaches (P. Howell, pers. comm.).

Summary

29/154 (18.83%) full survey sites were positive, representing a considerable expansion in range through the Region's rivers. The evidence suggests a natural expansion eastward through the Somerset Levels into Dorset from the Devon rivers.

Previous surveys revealed sparse signs only on the Brue system (1/154 in 1984-86 and 2/154 in 1977-79), suggesting that a small, fragmented population had survived there. Restocking with captive-bred otters by the Otter Trust had been targeted ahead of the 'natural' range of the otter population at that time. The rivers Stour, Salisbury Avon and Bristol Avon were used for the small-scale release of groups of two or three animals per river. This appears to have resulted in an increased rate of recovery of otters back into this Region.

Heavy rainfall during the survey period may have led to under-recording of otters on some rivers.

5.8 SOUTH WEST REGION

Full surveys were carried out in the following 50km squares – SS s/e (165 sites), SW s/e (59), SX n/w* (130), SY n/w (32).

* SX n/w includes six sites non-visited and assumed negative.

Description of Region

This westernmost part of England has land rising to 600m in a chain of moorlands – Exmoor, Dartmoor and Bodmin. The rivers of the Region all run natural meandering courses to their various estuaries along the highly indented coastline. The geology is varied but is largely dominated by old sandstone and shale with igneous outcrops. This influences the acidity of many of the upper reaches of the watercourses, causing them to be oligotrophic, supporting poor faunal diversity.

The human population density is moderately low, being concentrated around the coast or along river mouths, with Plymouth (258,100), Torbay (115,700), Exeter (99,600), Newquay (85,700), Falmouth (85,000) and Penzance (59,300) being the largest towns. The popularity of this area with holiday-makers seasonally swells the population further, especially among the smaller market towns and villages, increasing the recreational use on the rivers. There is even a newly created long-distance footpath called the Tarka Trail that highlights the otter habitat of the Region.

Low-intensity mixed sheep and dairy farming form the principal land use, although there is some arable production on low-lying ground.

Pollution problems

Most rivers in the Region meet their NRA River Quality Objectives as Classes 1A, 1B or 2 (see Appendix 3). However, many are recorded as suffering episodic pollution incidents from farm run-off (slurry, silage liquor or sheep-dip) which result in localised fish kills.

A number of rivers remain grossly polluted through ground water seepage and mine-waste contamination. These include the Par, Fal, Carnon, St Austell, Clyst, Cranny, Rookery Brook, Knowlwater and Red River.

A major release of acid mine water, rich in toxic metals (arsenic, cadmium, mercury, tin, iron, zinc, copper and aluminium), occurred on the River Carnon on 14th January 1992 when a plug failed in the Wheal Jane Tin Mine. The mine ceased operating in 1991 and since its closure the 15 levels have flooded. This serious pollution from the mine continued for several months and is expected to remain a problem in the estuary for years. At the time of survey, contaminated water and sediment could be seen as far down as the Mylor creek and in the Carrick Roads towards the town of Falmouth. The implications for otters on this system are serious, with a reduction in food availability (by fish kills) and also the bio-accumulation of toxic metals resulting in lethal/sublethal effects for the animals at the top of the food chain.

Fisheries on principal rivers

The rivers of the Region are classed as game fisheries, supporting residential and migratory salmonids together with locally good populations of eels and mixed coarse fish.

The Fowey, Camel, Tamar, Taw, Torridge and Exe systems all support good salmonid fisheries which meet biomass targets. Only the Fal falls below standard, but it is now showing signs of improvement.

Otter conservation effort

The Vincent Wildlife Trust's Otter Haven Project has been instrumental in providing practical conservation work by way of sympathetic riparian management, creation of secure breeding holts and monitoring of the changing distribution and status of the otter in the Region from 1978 to 1989 (H. Marshall, pers. comm.).

Figure 16. South West Region

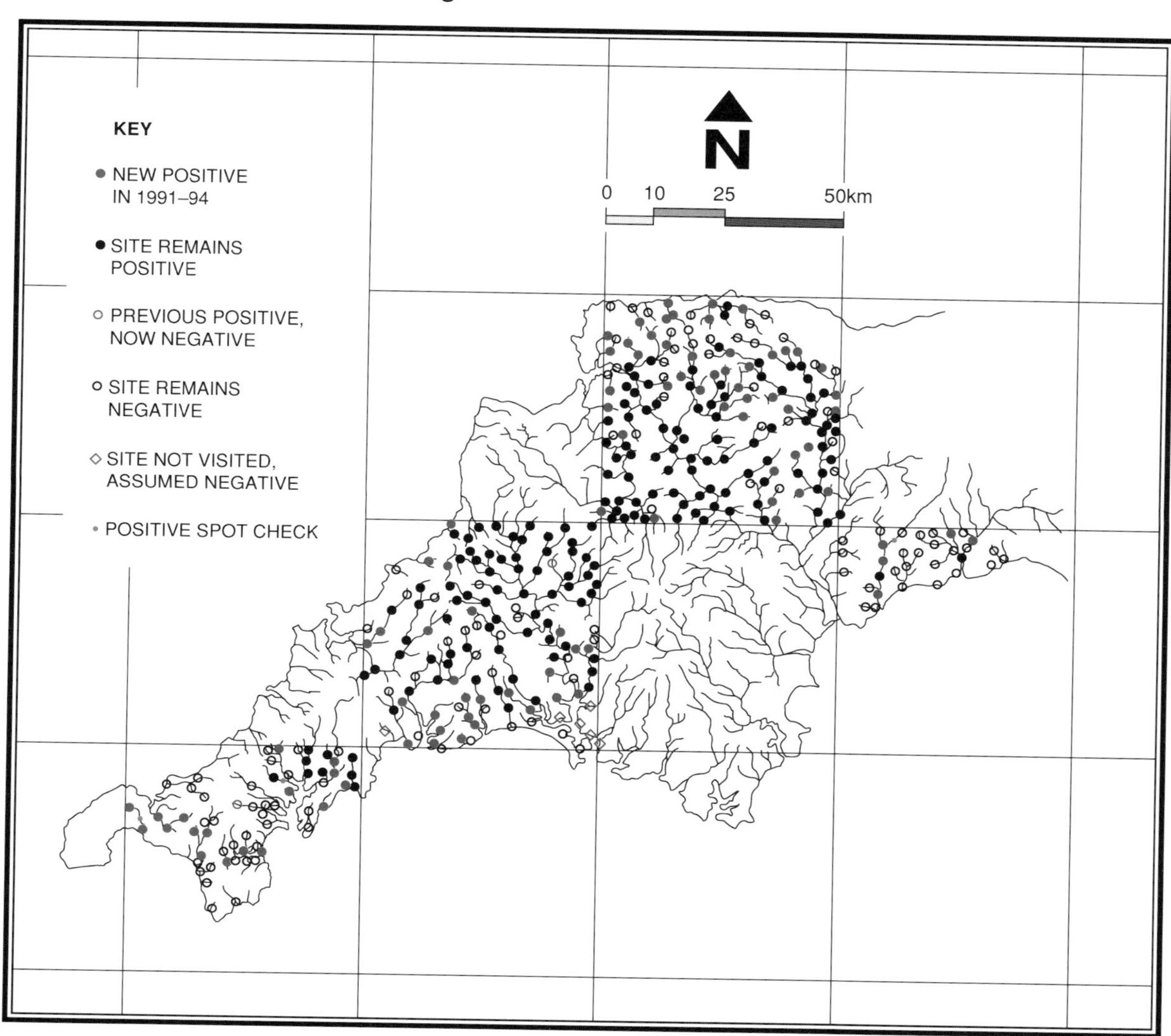

Since that time the RSNC's Otters and Rivers Project, in conjunction with Devon County Council's Tarka Project, has continued to foster sympathetic management for otters (M. Lane, pers. comm.).

Results for 1991-94

Total full survey sites	386
Positive full survey sites	259
Positive spot checks	3

Comparison of results for the same survey sites 1977-79 and 1984-86

	1977-79	1984-86	1991-94
Full survey	91/386 (23.6%)	169/386 (43.8%)	259/386 (67.1%)
Spot check	0/6	0/6	3/6

Summary of results for each river system (see Figure 16)

North Devon coast (Sterridge, Umber, Heddon, West Lyn and East Lyn)

9/15 full survey sites were positive, representing a considerable improvement since the previous surveys, when two were positive in 1986 and one in 1979 from the East Lyn.

Spraints examined from the Sterridge and Heddon were found to contain remains of shore crab and seafish, indicating the coastal foraging of the otters.

Otters were reported from some of the sites as a fairly recent occurrence (ie. within the last three years).

Torridge

21/23 full survey sites were positive compared to 17 in 1986 and 15 in 1979, indicating that this system continues to support a stable otter population that is probably close to carrying capacity for the system. Numerous recent reports of sightings have included adults with cubs on several occasions.

Road mortalities continue to be a problem for otters from this system, with at least five animals killed since 1986 (M. Lane, pers. comm.).

Taw

58/70 full survey sites were positive compared to 43 in 1986 and 22 in 1979, indicating that this system continues to support a large and stable otter population that has consolidated and slightly expanded its range.

Recent reports suggest that animals are fairly frequently seen and, like the Torridge system, otter road casualties have occurred in recent years.

Caen

2/3 full survey sites were positive on this small system that runs into the Taw estuary. All sites were negative in 1986, suggesting a range expansion by otters from the Taw itself.

Exe

43/54 full survey sites were positive, compared with 25 in 1986 and only five in 1979. The Exe system has shown a steady improvement for otters as they have consolidated their range.

Lower Exe

(0/3) The three full survey sites on the Clyst and tributaries were negative, as in the previous surveys.

Otter

4/9 full survey sites and one spot check were positive. This compares with a single positive site in 1984 and none in 1977. All evidence was confined to the main river at its middle and lower reaches, where some excellent habitat remains. Sites were surveyed following spate conditions which may have reduced the amount of field signs.

Sid

(0/5) The five full survey sites, investigated during a spell of wet weather, remained negative.

Axe

3/12 full survey sites were positive along the lower Axe and Umborne Brook, but no evidence was located on the Coly. This compares to a single positive site in 1984 and none in 1977.

Lim

(0/3) The three full survey sites on this small system remained negative.

Tamar (including Tavy, Inny, Lyd, Thrushel, Wolf, Kensey, Carey, Ottery and Claw)

49/58 full survey sites were positive on this system, which offers excellent habitat for otters and continues to support a population possibly near to carrying capacity. The 1985 and 1978 surveys revealed 44 and 28 positive sites respectively. New sites gained in the present survey were located on the Tavy and lower reaches of the Tamar itself.

Camel (including Amble, Allen and De Lank)

11/14 full survey sites were positive on this system, which offers some excellent habitat along its middle and lower reaches. There has been a slight improvement both in the number of sites showing signs and the amount of signs found. During the previous surveys of 1985 and 1978, eight and nine sites respectively were positive.

Fowey, Looe, Seaton and Lynher

29/45 full survey sites were positive on these systems which flow to the south coast. Good habitat can be found throughout, even along some of the smaller streams. The number of positive survey sites has doubled since 1985 (14 +ve). None of these sites revealed otter signs during the 1978 survey, suggesting a good recovery over the 14 years.

Polmassick

4/4 full survey sites were positive on this small system, which provides good riparian habitat. This compares to three sites in the previous survey. Otters seem to be using the sea and coast in this part of Cornwall, as spraint was located in a sea cave at Porthluney Cove.

At another site, 10km further along the coast, fresh otter tracks were located on the beach (Pendower).

Fal Estuary (including Fal, Tresillian, Truro, Allen, Carnon, Kennal, Mylor and Percuil)

10/25 full survey sites and one spot check were positive, compared to 6/24 in 1985 and 3/24 in 1977. New sites were located on the Fal and the Allen. The otter population appears well established along this upper catchment of the estuary, where the water quality appears good and excellent riparian habitat is offered.

No evidence was found on the Kennal, which showed a single positive site at Stithians Reservoir in 1985.

Helford Estuary

3/10 full survey sites were positive on this system, which consists of a network of creeks and small streams. The sites showing the presence of otters were well spaced and the animals were probably travelling individuals since the signs were sparse. All sites were found to be negative in the previous surveys.

Cober

3/6 full survey sites showed otter evidence on this system, which provides some excellent looking carr habitat around Loe Pool below Helford. The uppermost site was noted as suffering from acid mine drainage, but lower sites had better water quality since natural filtering is provided by water meadow/fen habitats along its course. It is thought that otters have only recently arrived on this system, as the previous surveys have all been negative.

Lizard streams

(0/5) The five full survey sites were negative on the Lizard peninsula, where small coastal streams and coves offer habitat that is locally suitable.

Marazion and Hayle

5/5 full survey sites and one spot check were positive.

This suggests a very encouraging recovery for these rivers, which were both completely negative for otters in 1985 and 1977. Up to seven sprainting places and 28 spraints were found at a single site on the Hayle, possibly indicating a stable population is now resident.

Red River

(0/4) As in both previous surveys the four full survey sites were negative on this system, which is badly polluted by acidic waste water drainage from disused mines.

North coastal streams – Strat, Millook, Crackington, Valency and Bossiney

5/6 full survey sites were positive compared to three in 1986 and none in the baseline survey. Otter evidence was found on each stream, apart from the Bossiney, which runs through a rocky valley offering potentially suitable habitat.

Grand Western Canal

(0/1) The single full survey site was negative, as in both previous surveys, although the site and the canal in general do offer some potential for otters.

Complementary information from additional surveys on rivers in the adjacent 50km squares:

SS s/w - 40/48 full survey sites were positive on the Torridge and its tributaries and the Tamar and coastal streams (Neet, Stratt, Marsland valley and Strawberry Water stream). Otter evidence was locally abundant, with up to 32 spraints counted over one 600m stretch of riverbank (M. Lane, pers. comm.).

SW n/e - No systematic surveys are available for this square. However, otters are known to occur on the streams of the Camel estuary as well as on many of the small streams of the North Cornwall coast (except those polluted by waste water from disused mines). The upper Fal system is also known to support a small population of otters (M. Lane, pers. comm.).

SX n/e - 41/88 full survey sites were positive from this square, which includes the Exe, Creedy, Clyst, Teign, Taw, Tavy, Dart, Plym, Erme, Avon and other south Dartmoor streams (M. Lane, pers. comm.).

ST s/w - The upper reaches of the Culm, Otter and Axe were not surveyed systematically. However, these rivers have been visited by experienced surveyors over the last five years and otter signs have been recorded in their upper reaches on a regular basis (H. Marshall, P. Howell, pers. comm.).

Summary

259/386 (67.10%) full survey sites were positive, representing a continued range expansion and consolidation throughout the Region, with some systems probably near to carrying capacity for otters. This is reflected in the frequency of spraints found at many sites, especially on the Tamar, Taw, Torridge and Exe. The expansion of the otter population has been not only eastward into Wessex Region but also westward to the toe of Cornwall with many coastal sites now occupied. The 1984-86 survey found 169/386 full survey sites positive, whereas the baseline survey of 1977-79 revealed only 91/386 sites positive.

Rainfall and subsequent spate conditions during the survey period may have affected the results on some of the river systems.

5.9 SOUTHERN REGION

Full surveys were carried out in the following 50km squares – SU s/e* (79 sites), SZ n/w (28), TQ s/e (134), TR n/w (34).

* SU s/e includes 11 sites non-visited and assumed negative.

Description of Region

The south-eastern corner of England encompasses the gently rolling hills of the downlands of Sussex and Kent. These run west to east with the highest point not exceeding 250m above sea-level. Rivers in the Region tend to drain off the downs northwards or southwards, with the area between them, the wooded Weald of Kent, giving rise to the Medway, Stour and Rother. The New Forest is another area of extensive woodland at the westernmost part of the Region. Good riparian habitat for otters remains on most river systems.

The rivers of the Isle of Wight are also included in the Region.

The land use is predominantly mixed farming (sheep, cattle and arable), with fertile calcareous soils giving rise to high yields.

This Region is fairly congested with the infrastructure that supports a moderately high human population density. The largest centres are coastal and include Southampton (196,700), Portsmouth (183,800), Brighton (149,200), Margate (129,700), Dover (106,300), Winchester (96,000), Folkestone (87,600), Hastings (83,100) and Eastbourne (80,300). Other centres occur along the rivers of the Thames Estuary or as London overspill towns, namely Rochester (146,600), Maidstone (136,000), Canterbury (131,900), Sittingbourne (115,300), Tonbridge (101,200), Tunbridge Wells (99,300) and Dorking (76,200). Scattered throughout the Region are many smaller market towns, hamlets and villages. The population seasonally swells among the coastal resorts and this increases recreational pressure on many of the river systems.

Pollution problems

Water quality is generally good, with the majority of rivers classed as 1A, 1B or 2 (see Appendix 3). The Pevensey Haven, Rother and lower Medway show poor quality and are classed 3. Sewage effluent, agricultural run-off and a few industrial pollution incidents are responsible for this poorer quality.

Fisheries on principal rivers

Chalk streams in the Region support salmonid fisheries which meet designated biomass targets. Cyprinid fisheries also meet biomass targets and support a good diverse fish fauna, with healthy populations of eels throughout. Enclosed fisheries show high stocking densities, and surplus fish are moved to rivers.

Otter conservation effort

Over the last 20 years this Region has been largely neglected in active effort to conserve its remaining otters. However, the situation is now being addressed by the RSNC's Otters and Rivers Project in partnership with the Hampshire, Sussex and Kent Wildlife Trusts and the NRA. Survey, monitoring and the creation of otter havens are presently being carried out by the project officer (G. Roberts, pers. comm.).

In addition, suitable receiving sites for otter releases are being identified. The first restocking project took place on the River Itchen at the end of 1993, using captive-bred otters from the Otter Trust (G. Roberts, pers. comm.).

Results for 1991-94

Total full survey sites	275
Positive full survey sites	12
Positive spot checks	1

Comparison of results for the same survey sites 1977-79 and 1984-86

	1977-79	1984-86	1991-94
Full survey	5/241 (2.1%)	7/241 (2.9%)	9/241 (3.7%)
Spot check	0/7	0/7	1/7

Summary of results for each river system (see Figure 17)

New Forest rivers

(0/15) The 15 full survey sites on the Avon Water, Lymington River and lower tributaries of the Beaulieu River were negative as in 1986 (1/15 in 1979). Heavy rainfall and subsequent spate conditions may have affected the results on these systems, which are reported to hold otters from time to time.

Spraint had been found on the Lymington River prior to the survey period and on two of its tributaries subsequently (G. Roberts, pers. comm.).

Isle of Wight

(0/13) As previously, the 13 full survey sites were negative on watercourses on the western half of the island (upper River Medina, Newtown River, River Yar and tributaries).

Sparse riparian habitat was found at the sites surveyed and there were no recent reports from this half of the island. However, two animals have recently been sighted at Wooton Creek on the north-east coast (Isle of Wight Natural History Society reported to G. Roberts).

Test

(0/2) The two full survey sites investigated on the uppermost reaches were negative, as in the previous surveys. Otters continue to be reported from the middle reaches of this river (G. Roberts, pers. comm.).

Itchen

6/8 full survey sites on the upper reaches of the river showed the presence of otters. This is a slight improvement from five in 1986 and four in 1979, but the density of signs per 600m remains low (up to five spraints encountered at any one site). As with the Test, excellent habitat for the otter is still to be found on these well managed game fisheries.

Hamble and Meon

(0/12) The 12 full survey sites were negative as in the previous surveys.

Some suitable habitat remains, but the rivers were suffering from very low water levels. G. Roberts reports that two dead otters were found on a tributary of the Hamble following a pollution incident in 1990.

Wallington, Pagham and Lavant

(0/20) The 20 full survey sites showed no signs of otters, as in the previous surveys.

Arun and Rother

(0/29) The 29 full survey sites were negative, as in both 1984 and 1977. Much of the system shows suitable riparian habitat.

Otter signs had been located at a single site on the Rother's middle reaches by G. Roberts (pers. comm.) just prior to the survey period, perhaps suggesting a transient animal.

Cuckmere and upper Ouse

(0/17) The 17 full survey sites were negative, as in both previous surveys. Some suitable riparian habitat for otters remains.

Figure 17. Southern Region

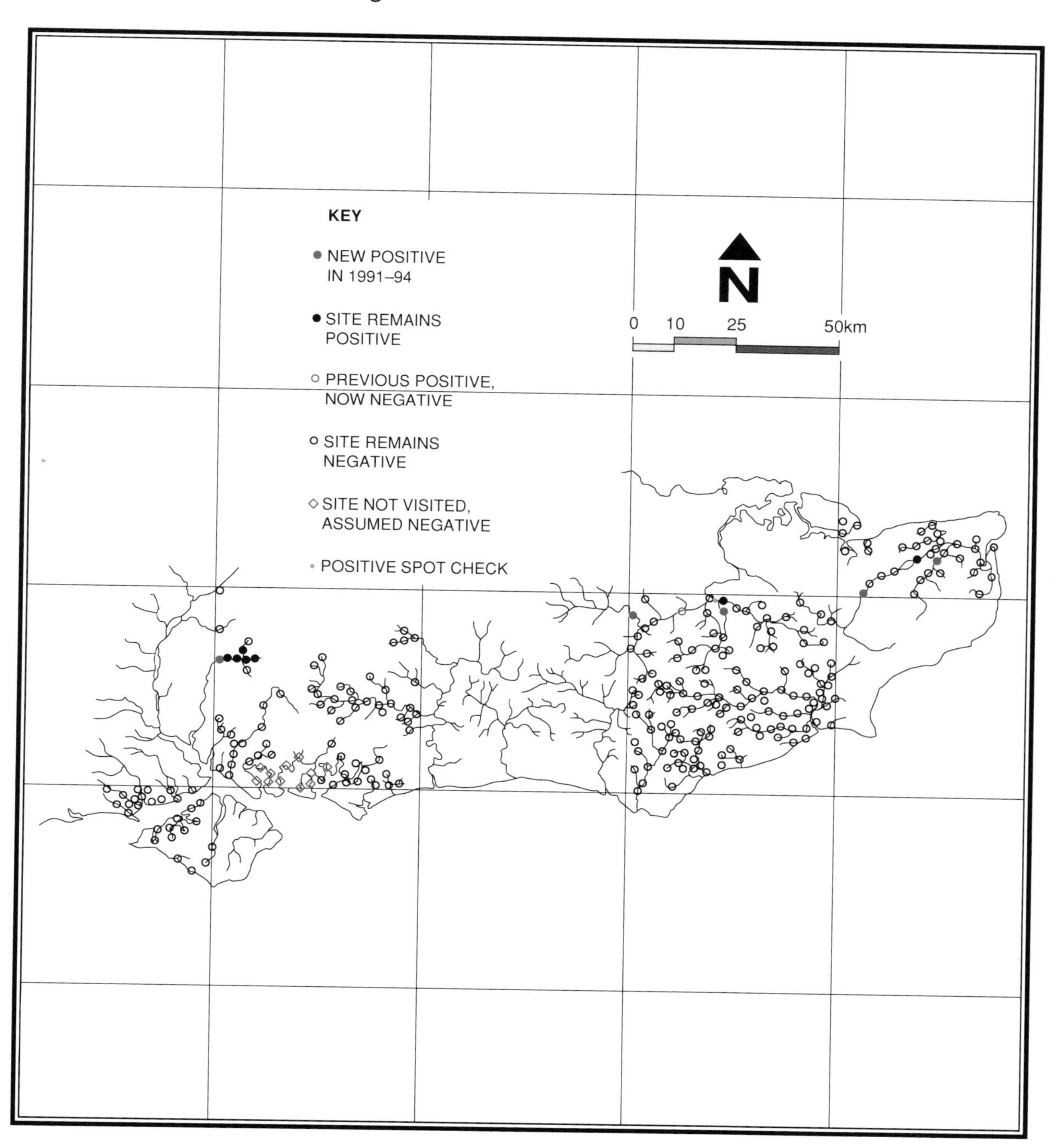

Pevensey Haven, Wallers Haven and Coombe Haven

(0/20) The 20 full survey sites, which offer only sparse cover, were negative as before.

Rother and Brede

(0/56) The 56 full survey sites remained negative for otters, despite some good local riparian habitat.

Medway (including Beult, Teise and Eden)

3/34 full survey sites were positive, as was a single spot check. This compares to two positive sites in 1986 and all sites negative in 1979.

The otter evidence was located on the middle reaches of the system in the same good habitat as recorded in 1986. Recent reports suggest a resident but small population remains on the system, but as yet no evidence of breeding has been confirmed (G. Roberts, pers. comm.).

Great Stour (including Little Stour, Sarre Penn and North Stream)

3/32 full survey sites were positive, compared to a single positive site in 1986. Only three sites on the Stour's uppermost reaches were examined in the baseline survey and all were negative. The catchment encompasses some excellent habitat for otters by way of flooded gravel pits, woodland carr and extensive areas of reedbed. Recent reports suggest that the otter population is small but resident on the middle reaches of the system. The presence of otters in the vicinity of Canterbury was confirmed during the spring of 1994 by G. Roberts (pers. comm.).

Thames Estuary/Isle of Sheppey

(0/6) The six full survey sites, which offer only sparse cover, were negative as in the previous surveys.

Complementary information from additional surveys on rivers in the adjacent 50km squares:

SU s/w - Surveys were carried out on the Test, Itchen and New Forest rivers (Beaulieu and Lymington) by G. Roberts. Otters were found to be using the middle reaches of the Itchen and parts of the Test. Otter evidence was also located on a number of the New Forest rivers in 1993 and 1994 (G. Roberts, pers. comm.).

TQ s/w - Systematic surveys were carried out in this square by G. Roberts during 1993, covering the rivers Arun, Adur, Ouse and Medway. The presence of otters was determined on the Adur at Chithurst lakes and Honeybridge stream as well as along the western part of the River Rother. Only single spraints were found, suggesting that these were from transient animals.

The upper Medway (including Eden) showed the presence of otters at Golden Green on the Bourne stream, where an otter (road casualty) was found later in that year (G. Roberts, pers. comm.).

TQ n/e - No systematic surveys were carried out in this square, although unconfirmed reports of otters were received from the lower Medway estuary, as in previous years, suggesting infrequent visitations by transient otters (Maidstone Museum, pers. comm.).

TR s/w - G. Roberts carried out a systematic survey of this square during 1993. Otter evidence was located only at a single site on the east Stour near Ashford (G. Roberts, pers. comm.).

Summary

12/275 (4.36%) full survey sites showed evidence of otters confined to the Itchen, Medway and Great Stour systems, as noted in 1986. The otter populations are thought to be very small and highly fragmented, casting doubt on their longer-term viability.

The Test and Itchen in Hampshire offer excellent riparian habitat and could act as a springboard for the natural recolonisation of this Region.

Weather conditions for surveying were generally favourable throughout the period during which this Region was examined.

5.10 WELSH REGION

Only the English parts of the Wye and Dee hydrometric areas were examined. Full surveys were carried out in the following 50km squares – SJ n/w (9 sites), SJ s/e (2), SO n/w (30), SO s/e (40).

Description of Region

Two rivers are described here, the River Wye and the River Dee, which rise in Wales and flow into England and then return to Wales. Parts of both rivers form the English/Welsh border.

The Wye enters England at Rhydspence and runs a natural meandering course, being joined just below Hereford by its two main tributaries, the Lugg and Frome. At Symonds Yat it forms the national boundary until it reaches the Severn Estuary below Chepstow. The underlying rocks of the Wye are predominantly sandstone, shale and limestone.

The Wye catchment is a rural one, with the human population centred on Hereford (49,000), Leominster (39,300), Chepstow (10,300), Ross-on-Wye (9,800) and Monmouth (8,600).

The fertile plain of the Wye provides good mixed farming with cattle, sheep and arable production. Along its lower reaches the river flows adjacent to the mixed woodland of the extensive Forest of Dean.

The Dee enters England at Shocklach and, from here down to Aldford and along its tributary the Wych Brook, it forms the national boundary. From Aldford to Chester the river is slow and wide and beyond Chester it is tidal, flowing back into Wales to its estuary along a canalised course. The geology of the Dee hydrometric area is complex, but in England the underlying rocks are predominantly sandstone.

Like the Wye, the Dee catchment is predominantly rural, with the human population of the lower Dee centred on Chester (117,100). The soils are fertile, providing mixed farming (MAFF Land Grade 2).

Pollution problems

The water quality of the Wye system is excellent, being Class 1B from Hereford downstream to its tidal limit (see Appendix 3). Few pollution incidents have resulted in fish mortality or serious problems, and the frequency of minor incidents resulting from farm run-off is around 12 per year on the Wye, Lugg and Arrow.

The headwaters of the Dee suffer mine-waste pollution (principally heavy metals), but a few incidents on the lower part have resulted in fish kills from industrial, domestic and agricultural sources in the vicinity of Wrexham and Chester. In addition, a major fish kill was recorded on the Wych Brook as a result of a slurry spill in 1989. Since that time resources have been invested into upgrading the water quality, and much of the catchment is now Class 1B/2.

Fisheries on principal rivers

The Wye is the most important salmon fishery south of the Scottish border, with extensive spawning beds which support a good fry to parr ratio. Allis shad also spawn in significant numbers, making the Wye nationally important for this species. Apart from the migratory salmonids, brown trout and a good mixed population of coarse fish occur, including eel, chub, dace, grayling, roach, perch and pike. Biomass targets for the river are easily met.

On the Lugg and Arrow good populations of coarse fish and eels are recorded downstream of Leominster, but recruitment of salmonids was low during the 1992 fisheries monitoring programme.

Salmon and brown trout occur throughout the Dee, with a few sea trout. The Dee in England also supports a good mixed coarse fishery with good numbers of eels.

Otter conservation effort

Little direct conservation effort has been applied to the English parts of these systems, although the Wye Otter Conservation Group regularly monitors the distribution and relative abundance of the otter on the lower Wye (through the Forest of Dean).

Results for 1991-94

Total full survey sites	81
Positive full survey sites	51
Positive spot checks	5

Comparison of results for the same survey sites 1977-79 and 1984-86

	1977-79	1984-86	1991-94
Full survey	12/81 (14.8%)	24/81 (29.6%)	51/81 (63.0%)
Spot check	1/6	5/6	5/6

Summary of results for each river system (see Figure 18)

Wye tributaries, upper Lugg and Arrow

23/30 full survey sites and five spot checks were positive. This compares with 17 positive sites and five positive spot checks in 1986 and 12 positive sites and one positive spot check in 1979. These rivers and their tributaries showed evidence of a healthy otter population, with a slight expansion in range in this habitat that provides good tree-lined banks running through pasture and wooded valleys with low disturbance.

Recent records and reports were consistent over the interim period between surveys.

Wye (including Lugg and Frome)

23/40 full survey sites were positive. Otters now range throughout the entire length of the catchment area surveyed. This is a considerable improvement since the previous two national surveys (6/39 full survey sites were positive in 1986 and all sites were negative in 1979). Otter evidence was sparse on the middle reaches of the Wye, lower Lugg and Frome, where bank cover was poor offering few good sprainting sites. Sympathetic riparian management in this area would greatly enhance the potential for resident otters.

Dee (including Aldford Brook)

4/9 full survey sites were positive compared to a single positive site in 1986 and none in 1979. Spraint and tracks were located at two sites on the main river at/near Aldford as well as on the Aldford Brook, although the frequency of signs was low. This lower part of the Dee catchment offers sparse suitable riparian cover and is difficult to survey adequately owing to frequent flooding with deep muddy water and steep banks. The Otter Survey of Wales 1991 found that the otter distribution on the lower Dee had remained the same as in 1985 but did find an improvement on the River Alyn, a lower tributary.

Recent reports suggest that otters are still using the lower Dee infrequently.

Dee – Wych Brook

1/2 full survey sites showed evidence of otters along the upper reaches of this Dee tributary, which provides excellent riparian habitat further downstream. Both sites were negative in previous national surveys.

Figure 18. Welsh Region

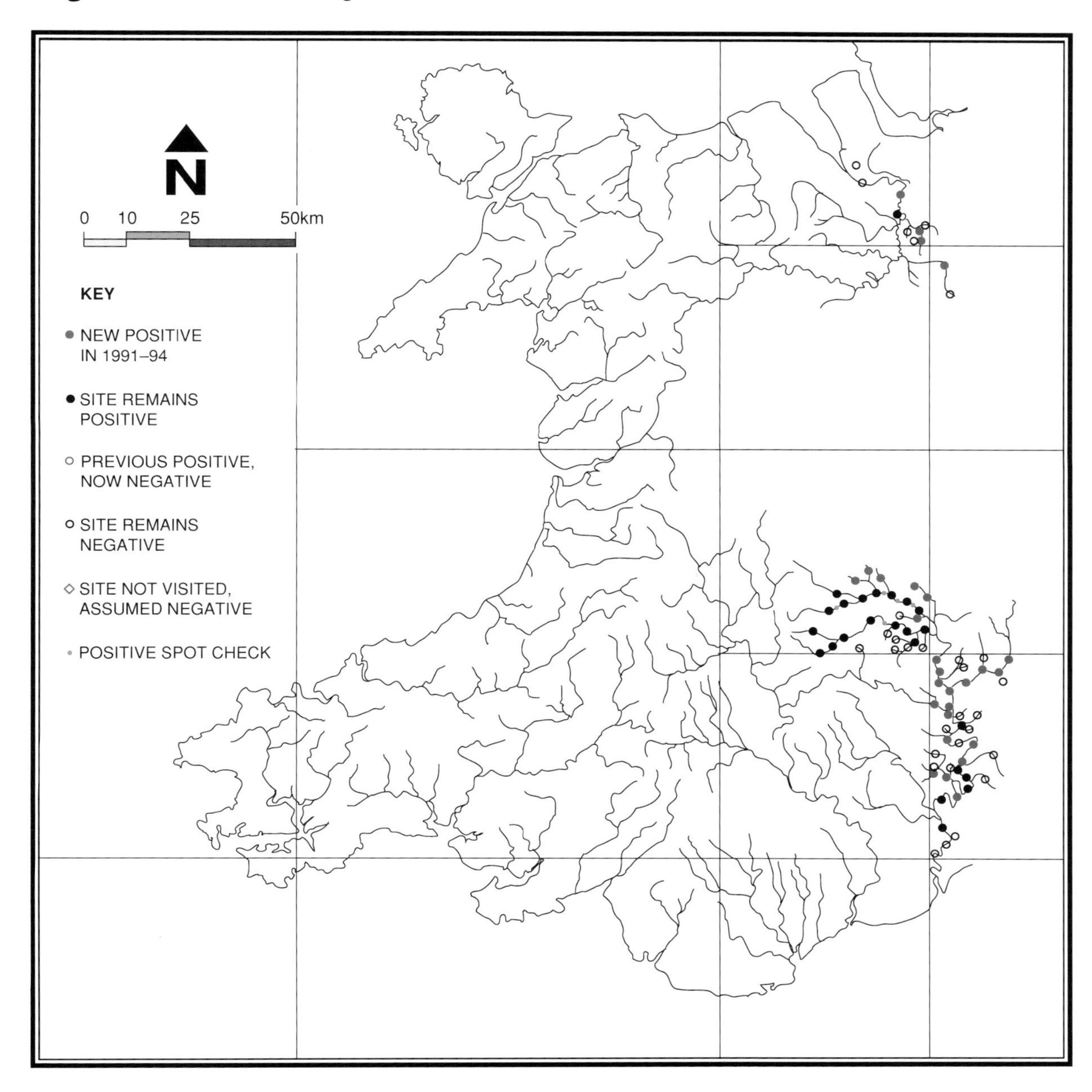

Complementary information from additional surveys on rivers in the adjacent 50km squares:

SO s/w - Concurrent surveys of the rivers in this square were carried out by The Vincent Wildlife Trust for the Otter Survey of Wales. The Wye and Monnow were found to support good populations of otters. (82% of the full survey sites were positive.)

SJ s/w - Concurrent surveys of the rivers in this square were carried out by The Vincent Wildlife Trust for the Otter Survey of Wales. The Dee was found to support good evidence of otters. (49% of the full survey sites were positive.)

Summary

51/81 (62.96%) full survey sites were positive, representing a range expansion and consolidation along both systems, as mirrored in the Otter Survey of Wales 1991 (Andrews *et al*, 1993).

This figure compares with 24/81 in 1984-86 and only 12/81 in the baseline survey of 1977-79.

Weather conditions were favourable for surveying in each period of investigation.

6. FURTHER ANALYSIS

6.1 Spraint density

6.1.1 The use of spraint density counts

The main results of this survey covered in the previous two Sections have been based on sites being found either positive or negative (ie. with or without signs of otters). Comparisons between survey years are then made on the basis of changes in the percentage of sites found positive in each Region and England overall. However, as remarked by Jefferies (1986), another important variant on the use of spraints in surveys is the counting of spraints and/or spraint sites found during the 600m walks at each positive survey site and then using these counts as a 'fine adjustment' on the broad finding of presence or absence. Such a variant (with slight differences) has been used by Green and Green (1980, 1987) in the first and second otter surveys of Scotland and by Strachan *et al* (1990) and ourselves in the second and third otter surveys of England. It was first considered that it could be a useful additional and supporting technique for showing that improvements had occurred in Scotland if there was but a small increase in percentage site occupation between two surveys (Green & Green, 1980; Jefferies, 1986).

However, doubts have been expressed over the value of this technique and the interpretation of any survey results based on spraint distribution and counts (Kruuk *et al*, 1986; Kruuk & Conroy, 1987). The main criticism arises because spraint numbers are known to show seasonal cycles in some areas (Erlinge, 1968a; Jenkins & Burrows, 1980; Mason & Macdonald, 1986; Kruuk *et al*, 1986; Macdonald & Mason, 1987; Conroy & French, 1991), though these do not have the same timing of their seasonal peaks in every area studied (Jefferies, 1986) and other areas (River Avon and River Taw in Devon, for example) do not show any seasonal cycles at all (Marshall, 1991). It may be thought that this would indeed invalidate the use of any technique dependent on spraint numbers. On the other hand, spraints have a 'life' of 3 to 15 weeks (Mason & Macdonald, 1986) and, provided the same river is examined in the same season in different years, the overall picture for a Region or a country over a two-year period should be valid for comparative purposes (Jefferies, 1986).

The second criticism from Kruuk *et al* (1986) and Kruuk and Conroy (1987) is that they could find no correlation between the numbers of otters and their activities and the numbers and disposition of their spraints in their studies of coastal otters on Shetland. Again, it must be agreed that absence of spraints does not necessarily mean absence of otters and many spraints can be produced for marking by one male otter. On the other hand, it is known that the presence of scats of the same species of mustelid will stimulate the considered territory owner to mark the area itself (studies on feral American mink; Dunstone, 1993). Also, there is an obvious progression from the large numbers of spraints and otters on Shetland to the very few spraints and few visible (or road casualty) otters in the past in East Anglia. It seems likely that intermediate numbers of spraints would indicate intermediate numbers of otters if sufficient sites were examined. Although individual variation may well be too great to use spraint counts as an indication of otter numbers in comparisons of two or three rivers, its use over the 1,189 miles covered during the two-year survey of 3,188 600m, largely riverine, sites in England would seem to be a different matter (Jefferies, 1986).

The value of being able to have an indication of animal density within occupied areas is considerable. It may be that this technique of determination of spraint density (by counting spraints per 600m) could provide one easily determinable way of looking at it. This may be the only way. We may have been misguided in discarding it as an important component of survey design, owing to the above early criticism. Consequently, we have taken a 'second look' at the data which can be obtained by counting spraint density now that further information is available from a third England survey.

First, we examine the likelihood that it is mainly male presence that we are demonstrating by spraint survey (Section 6.1.2). Then the data on spraint density from two

(1977-79 and 1984-85) 1,728 mile surveys of Scotland are examined (Section 6.1.3), followed by the data from two (1984-86 and 1991-94) 1,189 mile surveys of England (Section 6.1.4). Support for any conclusions made is sought in the comparative data from an 885 mile survey of the rivers of Ireland (Section 6.1.5) and from an examination of changes in spraint density with altitude in the three countries (Section 6.1.6). If the same relationships are present in the data from these three countries, or they show definite trends with time and altitude, then the technique may well be considered valid for use as a survey tool (see Conclusions, Section 6.1.7).

6.1.2 Different spraint marking by male and female otters

Wild-caught otters in Scotland have been injected with very small amounts of the radio-active isotope Zinc-65 before release (Mitchell-Jones *et al*, 1984). This causes the spraints they produce to be labelled with the isotope and so they are identifiable to an individual. The recovery rate for two females from Perthshire was 0.3 and 0.33 spraints/day (Green *et al*, 1984). However, that for three labelled males was 1.76 on Deeside (Jenkins, 1980) and 1.45 and 1.82 spraints/day on South Uist (J. Twelves & D.J. Jefferies; reported by Green *et al*, 1984). Thus, females produce far fewer findable spraints than the males (ratio, 1: 5.3). Presumably they defecate largely in the water.

In addition, studies of captive otters have shown that post-partum females deliberately reduce, or further reduce, their level of marking, either by ceasing to spraint on land or ceasing to spraint on visible sites (Östman *et al*, 1985). Consequently, when very young cubs are present in an area, otter signs become greatly diminished (Green *et al*, 1984; Jefferies *et al*, 1986). A sudden and temporary reduction in spraint density may be taken as indicative of possible local breeding (Jefferies, 1986).

Thus, most spraints found during otter surveys are from males and it is largely this sex that we are surveying. The disposition of these spraints on conspicuous features, such as rocks, logs and bases of bridges, suggests that they are most likely being used for territorial marking and as a means of communication.

6.1.3 Regional and temporal variation in sign density in Scotland

Green and Green (1980, 1987) recorded the number of signs (spraints and sets of footprints) first found at each positive site in Scotland in the surveys of 1977-79 and 1984-85. Twelve Regions were surveyed in 1977-79 and eight in 1984-85. Shetland, Western Isles, Orkney and Highland Regions were surveyed in 1977-79 only. The mean sign density for each Region has been extracted from these reports (see Appendix 4) and graphed in Figure 19. A linear relationship can be seen between Regional percentage occupation of survey sites and the mean initial number of signs per positive site in that Region. Calculation of the correlation coefficient and regression line for the eight Regions which were surveyed in both periods (15 points; see Appendix 4) shows that this relationship is statistically significant (equation: $y = 0.02690x + 1.4271$; $r = +0.8975$; 13 df; $p<0.001$). Thus, as the percentage of occupied sites increases from Region to Region, there is a related increase in the number of signs per site in those Regions.

Further examination of Appendix 4 shows that as the percentage occupation of sites increases in each Region with population recovery from 1977-79 to 1984-85, then there is an increase in the number of signs/site in that Region too. The latter can be quite large even though there may be little increase in percentage site occupation, eg. from 3.10 to 4.05 signs/site in Grampian with only a 2% change in site occupation (77 to 79%).

6.1.4 Regional and temporal variation in sign density in England

As noted in Section 6.1.5, the total number of signs was counted for the full 600m of the sites in the surveys of England. These were almost all spraints so we are determining spraint density.

The total number of signs and the mean number of signs/positive site for each of the ten English Regions for the surveys of 1984-86 and 1991-94 are listed in Appendix 5. They

Figure 19. The relationship between the mean initial number of signs (spraints and sets of prints) found at positive sites and the percentage occupation of sites in eight Scottish Regions surveyed in both 1977-79 (+) and 1984-85 (O) (data from Green & Green (1987); Appendix 4). The calculated regression line of equation $y = 0.02690x + 1.4271$ is also shown. (There were no positive sites in Lothian in 1977-79, so the calculation was made from 15 points.) The four extra points (*) relate to the four Scottish Regions only surveyed in 1977-79 (Orkney, Highland, Western Isles and Shetland). These were not used in the calculation of the regression line.

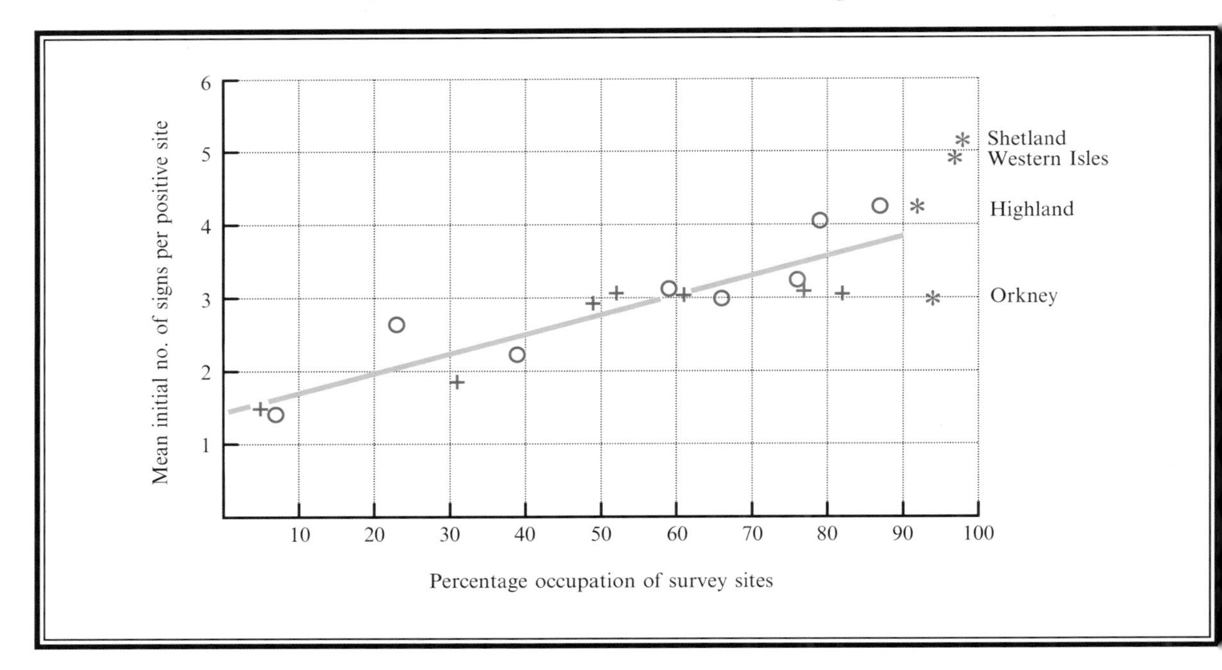

are shown plotted against the Regional percentage occupation of survey sites in Figure 20a. It will be seen that at site occupancies below 10% the sign density is very low and the resulting relationship is sharply curved. Transforming the x-axis to the $\log_{10}$ percentage site occupation +1 converts this curve to a linear relationship, allowing linear regression analysis (Figure 20b). As with Scotland (see Section 6.1.3), calculation of the correlation coefficient and regression line for the ten Regions surveyed twice (19 points; site occupancy was 0% in Thames Region in 1984-86) shows that this relationship is statistically significant (equation: $y = 2.8179x + 0.6442$; $r = +0.8037$; 17 df; $p<0.001$).

Again, as with Scotland, further examination of Appendix 5 shows that as the percentage occupation of sites increases in each Region with population recovery from 1984-86 to 1991-94, there is a shift upwards in the number of signs/site in that Region too.

6.1.5 Initial and total sign density at Irish sites

It will be noted that the mean number of signs/site is apparently lower at Scottish survey sites than those in England (compare Appendices 4 and 5). This is an anomaly due to the fact that the Scottish figures are the initial numbers of signs found, whereas the English figures are the total numbers of signs found at a site. In the Scottish surveys, the only way of completing the large number of widely spread sites within a two-year period was by counting each site positive at the initial finding of spraints or prints and then leaving for the next site. A large number of sites were then completed because of the very high rate of site occupancy. In England, on the other hand, searching and completing the full 600m of the site was the norm because of the very low rate of site occupancy. Consequently, it was decided that every site should be fully searched for 600m and the total number of signs counted, even if spraint was found within the first ten metres. This was in order to gain fully comparative information on this intriguing question of sign density.

Figure 20. (a) The relationship between the mean number of signs (largely spraints) found at positive sites and the percentage occupation of sites in ten English Regions surveyed in both 1984-86 (+) and 1991-94 (O).

(b) The data shown plotted with the x-axis transformed to the logarithm (log_{10}) of the percentage occupation of sites +1, in order to produce a linear relationship. The calculated regression line of equation $y = 2.8179x + 0.6442$ is also shown in both figures. (There were no positive sites in Thames Region in 1984-86, so the calculation was made from 19 points.)

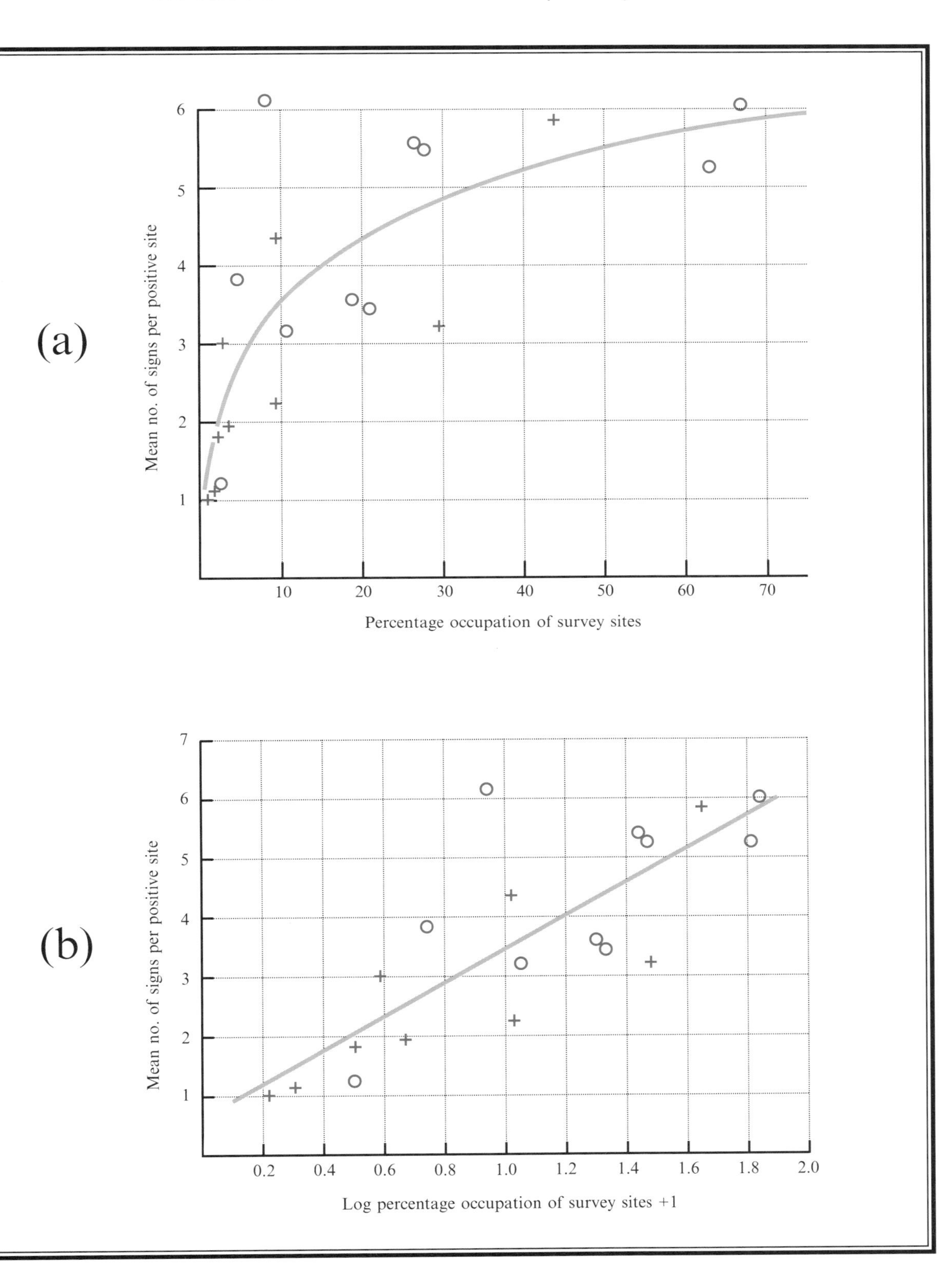

The 1980-81 survey of Ireland (Chapman & Chapman, 1982) was carried out in the same way as the Scottish surveys, again because of the large number of sites and the very high rate of site occupancy. However, in the case of Ireland, Chapman and Chapman examined the difference between initial and total number of signs/site at 36 occupied sites. These were 12 sites on the River Clare, Co. Galway, where the percentage site occupancy was 93.1% positive, and 24 sites in West Mayo, with 98.9 and 99.1% (mean 99.0%) of sites occupied. The relationship was found to be extremely variable, but taken overall one can estimate a correction factor. Thus, the ranges of initial and total signs found were 1–48 and 1–151 respectively, and the mean proportion (± standard error) of the total signs was 0.3538 ± 0.0555. The correction factor to obtain total from initial signs/site is therefore x 2.83. This would suggest that the overall total signs/site for Scotland in 1984-85 may be as high as 5.7 and 11.5 at 20 and 90% site occupancy (see Table 4), much higher than those in England (see Appendix 5 and Table 4). However, even these are far below the counted mean total signs/600m site in Ireland, which reached 19.37 in West Mayo (99.0% site occupation) and 79.08 in Co. Galway (93.1% site occupation).

6.1.6 Changes in density of signs and percentage occupation of sites with increasing altitude

Green and Green (1980) were the first surveyors to report a marked reduction in the numbers of otter signs (largely spraints but also sets of footprints) with increasing altitude in Scotland. Combining the results for the three Regions (Highland, Western Isles and Shetland) with a high-density, "naturally-constituted" otter population, they found that signs decreased significantly from 6.67 to 4.03 to 3.30 to 2.45 at the coast, 5–100, 100–300 and above 300m, respectively. These were not total signs per 600m survey site length but the numbers of the first signs found (see Section 6.1.5). The pattern of change of spraint density with altitude was much more flattened for the remaining Scottish Regions with their reduced percentage site occupation.

Chapman and Chapman (1982) looked at the apparent effect of altitude on otter presence in Ireland by examination of the alternative statistic of percentage occupation of survey sites. They found that this was 86.0, 94.1, 90.3, 72.6 and 73.7% at 0m (coastal), 0–122m, 122–244m, 244–366m and 366–732m respectively, ie. there was a general decrease in percentage occupation with increasing altitude. The slope is too small (see Figure 21a) to show significance when analysed by regression. However, grouping the data and using chi-squared analysis shows (a) a significant decrease in percentage site occupation (from 94.1% to 90.3%) with increasing altitude from 0–122m to 122–244m ($\chi^2 = 7.6325$; 1 df; $p<0.01$); and (b) a further significant decrease in percentage site occupation (from 90.3% to 72.8%) with increasing altitude from 122–244m to 244–732m ($\chi^2 = 20.6187$; 1 df; $p<0.001$). Thus, the data on percentage site occupation (Ireland) support those for the number of signs per site (Scotland). On the other hand, unlike the Scottish data for number of signs per site, there was a significant increase in percentage site occupation (from 86.0% to 94.1%) with increasing altitude from 0m (coastal) to 0–122m ($\chi^2 = 22.8362$; 1 df; $p<0.001$). Thus, in Ireland the percentage site occupation was lower at the shoreline than immediately inland.

The data from two of the Regions surveyed in England over 1991-94 were examined for the possible presence of an altitude gradient in the number of signs per 600m site (as found in Scotland) and in percentage site occupation (as found in Ireland). The Regions chosen were South West and Severn-Trent. The former had the highest percentage site occupation of any Region in England (67.10%), whilst that of the latter was of intermediate level, being the fifth highest (20.66%). Both had sufficient high ground to provide the necessary range of altitudes. As is usual in surveys of England, the full 600m stretches were examined for signs, with all spraints counted. The resulting data are provided in Appendix 6 divided into nine groups with steps of 50m of altitude. The data are shown graphically in Figure 21a and b. The changes found in percentage site occupation with altitude in Ireland (data derived from Chapman & Chapman, 1982) are also shown in Figure 21a for purposes of comparison.

Figure 21. The influence of altitude on (a) the percentage of sites found to be occupied and (b) the mean number of signs (largely spraints) found per positive site for South West Region (with 67.10% sites occupied overall) and Severn-Trent Region (with 20.66% sites occupied overall). The total data have been divided into nine altitude groups (see Appendix 6) in each case. The graph relating changes in the percentage of sites found occupied in Ireland (with 91.74% sites occupied overall) with increasing altitude is shown for purposes of comparison (data derived from Chapman and Chapman (1982); see Section 6.1.6). The lines indicate that there is a significant decrease in (i) the percentage site occupation and (ii) the number of signs per positive site with increasing altitude. This is repeated in the two Regions of England. The forms of the lines are seen to be very similar for each statistic. The three lines of the upper Figure (a) demonstrate the series of changes to be expected in an expanding and recovering population. Thus, upland rivers or zones are not colonised until all lowland rivers or lowland zones are already occupied (see Section 6.1.6).

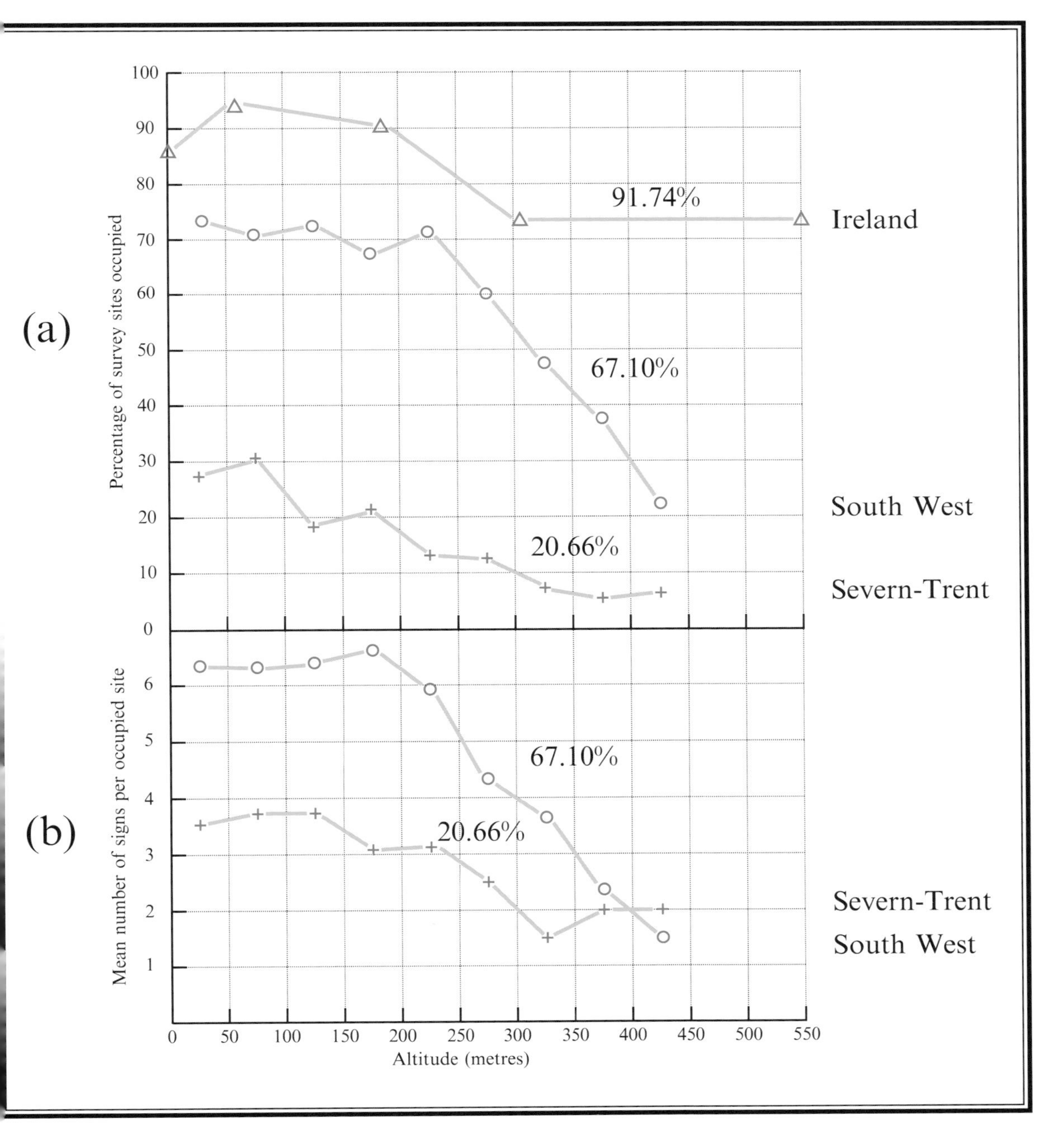

Regression analysis of the data in Figure 21 and Appendix 6 shows that there was a significant decrease in the percentage of sites occupied by otters with increasing altitude in Severn-Trent Region ($r = -0.9474$; 7 df; $p<0.001$; equation: $y = 29.962 - 0.0626x$, where x = altitude in metres and y = percentage site occupation) and in South West Region (taking the last five altitude groups because of the form of the line; $r = -0.9981$; 3 df; $p<0.001$; equation: $y = 126.394 - 0.242x$). Also, there was a significant decrease in the number of spraints per positive site with increasing altitude in Severn-Trent Region ($r = -0.8997$; 7 df; $p<0.001$; equation: $y = 4.042 - 0.00550x$, where x = altitude in metres and y = mean number of spraints per site) and in South West Region (again taking the last five altitude groups because of the form of the line; $r = -0.9922$; 3 df; $p<0.001$; equation: $y = 10.635 - 0.0218x$).

Thus, we have demonstrated that both percentage site occupation and spraint density decrease with altitude together in the same area. What is more, the forms or patterns of the declines with increasing altitude are remarkably similar for the two statistics for any one Region. (Compare lines for South West Region for percentage site occupation and spraint density in Figure 21a and b.)

What is the reason for these apparent decreases in otter presence at sites at the higher altitudes? Green and Green (1980) considered that the reduction in the number of signs they found was likely to be due to a reduced population density of otters because productivity in aquatic systems is reduced at the higher altitudes. In the River Tweed, for example, both the number of fish species and productivity decrease markedly with altitude from coast to upland river (Mills *et al*, 1978). Productivity is reduced by a factor of 14 times. Not only is productivity per unit surface area likely to be lower, but as the width of streams narrow with the higher altitudes, the number of fish per metre of bank is reduced as well, which may tend to cause otters to use greater lengths of waterway per individual. Also, studies of brown trout have shown that upland rivers tend to have large numbers of small fish, whereas there are small numbers of large fish in lowland rivers or lowland sections of the same rivers (Frost & Brown, 1967). Erlinge (1968b), working with captive otters, found that they showed a preference for the larger cyprinids, possibly because smaller fishes are more difficult to catch. Jefferies (1986) suggested that an additional factor, possibly helping to cause this reduction in upland spraint density, is that the higher altitudes may be populated largely by female otters. Radio-tracked females marked less frequently than males and have been found to live in headwaters (Green *et al*, 1984; also see Section 6.1.2). However, as both Green and Green (1980) and Chapman and Chapman (1982) based their findings on prints as well as spraints, this factor may be unimportant in the production of the altitude gradients which they found. Thus, it seems likely that upland rivers provide suboptimal otter habitat because of their reduced productivity and/or because they require greater energy expenditure to catch sufficient food.

If indeed otter density decreases with increasing altitude in England, Scotland and Ireland (and this would correlate with the likely reduced available food resource), then the corollary is that spraint or sign density is indicative of otter density.

The significant reduction in percentage site occupation at the coast in Ireland (see above and Chapman & Chapman, 1982) may indicate that the food resource there is slightly less than that available just inland, but there are few other data for use for comparison. Watt (reported by Kitchener, 1989) found that a quarter of the diet of young otters on the shores of Mull consisted of shore crabs. He suggested that this may be due to the otters there having access to fewer fish of fewer species than those on other Scottish coasts, such as Shetland. Thus, coasts may be suboptimal habitats in some areas.

A further interesting facet of the behaviour of an expanding and recovering otter population is demonstrated by the forms of the lines relating percentage site occupation and spraint density with altitude in Figure 21a and b. These form a series. Thus, when percentage site occupation is still comparatively low, as in Severn-Trent Region, the form of the line relating both site occupation and spraint density with altitude is linear and flattened. This form is similar to that described by Green and Green (1980) for the Regions of Scotland with reduced percentage site occupancy (see first paragraph of this Section 6.1.6). With a recovering and expanding otter population, as in the South West Region of England, ıe first river sites to fill up quickly towards their carrying capacity are the optimal lowland tes with the best food supply. The suboptimal upland sites are left 'trailing', with fewer tters living at lower densities. As the population expands towards the full carrying capacity f the area or country (as in Ireland), the high altitude 'tail' of the line lifts too, as probably ıe younger otters of the high-density population are forced to inhabit the suboptimal upland vers. This perhaps suggests that there are few aquatic sites that otters will not inhabit if ollution, disturbance and persecution are low, as in Ireland (see Section 7.3.5). Even those ith low and suboptimal cover and food supply will be occupied eventually.

.1.7 Conclusions: Spraint density relates to otter density

The data in Sections 6.1.3 and 6.1.4 show that there is a significant relationship etween the percentage site occupation in a Region and the mean number of signs or praints per occupied site in that Region, ie. as the former increases, so does the latter. This so for both England and Scotland, even though in Scotland this relationship was examined n an initial find basis only. In addition, in both England and Scotland, the number of signs er site shifts upwards for each Region as the percentage of sites occupied increases with ecovery. A secondary overall relationship exists within this system, and both percentage ite occupation and signs per site in England, Scotland and Ireland decrease with increasing ltitude. What is more, the forms or patterns of the relationships between percentage site ccupation and altitude and between sign density and altitude are very similar. Also, there is n ecological reason (ie. a reduced food resource) why decreases in both of these factors vith increasing altitude should be related to reduced otter density. The only conclusion vhich can be made from these consistent patterns is that density of signs (largely spraints) is ndeed indicative of the density of otters in an area and that the initial doubts (see Section .1.1) have not been substantiated. Thus, as more otters move into an area or Region (or ndeed the population increases by breeding), occupation of the available sites increases and o does the density of signs at each site. Lowland rivers or zones are occupied first, but even pland rivers or zones are occupied eventually (see Section 6.1.6).

Some additional support for this thesis is provided by the curve of the relationship etween percentage site occupation and sign density in England (see Figure 20a). Spraint ensity is relatively much lower in Regions with below 10% site occupation than it is above his level, when the relationship becomes linear as in Scotland. Lenton *et al* (1980) and efferies (1986) suggested that at very low densities otters may spraint less frequently in onspicuous places on land as they have less need to mark a territory. Thus, an alteration in he rate of decrease in sign density at very low occupation levels is to be expected.

Further examination of the data for individual Regional mean numbers of signs per site hows that there is no one level of sign density for each level of site occupancy (see Appendices 4 and 5). Thus, in England, Welsh Region had a similar level of site occupation 1984-86: 29.63%) to the North West (1991-94: 27.93%), but the mean sign density per ccupied site was 3.21 in the former and 5.52 in the latter. Similarly, Southern Region had a ow level of site occupation (1984-86: 2.91%) but a high number of signs per site (3.00), as lid Anglian Region (1991-94: 8.00% occupation; 6.17 signs per site). Some small differ-ences between Regions could be explained perhaps by cycles in spraint marking related to easonal differences in the timing of Regional surveys. Also, the high spraint density in Anglian Region at the few positive sites there may be a product of the past Otter Trust elease programme, ie. there are very high densities of otters in certain productive colonised ireas. However, the situation in Grampian Region, Scotland, may provide an insight. Here, he site occupancy only increased slightly from 77 to 79% between 1977-79 and 1984-85, out the mean number of signs per site increased considerably from 3.10 to 4.05 (Appendix 4; Green & Green, 1987). The time of the year was the same for both surveys. This suggests he possibility that the distribution of otters remained almost the same in the Region but heir density within the area increased, ie. there may have been more otters in Grampian at he time of the second survey.

The above suggestion could explain the disposition of the four extra points for Shetland, Western Isles, Orkney and Highland Region shown in Figure 19 (surveyed in

Table 4. The separate regression lines relating percentage site occupation and number of signs/site for the two Scottish (1977-79 and 1984-85) and two English (1984-86 and 1991-94) otter surveys. The x-axis was transformed to $\log_{10}$ percentage site occupation +1 before calculation of the two equations for England (see Figure 20(b)). Equations for Scotland derived from data by Green and Green (1980, 1987).

Number of Regions	Dates of Survey	Equation of regression line	Correlation coefficient	Percentage site occupation			
				$x = 10\%$	$x = 20\%$	$x = 70\%$	$x = 90\%$
SCOTLAND							
7	1977-79	$y = 0.02272x + 1.4969$	r = +0.8947 5 df. $p < 0.01$		$y = 1.95$ signs/site		$y = 3.54$ signs/site
8	1984-85	$y = 0.02938x + 1.4175$	r = +0.9255 6 df. $p < 0.001$		$y = 2.01$ signs/site		$y = 4.06$ signs/site
ENGLAND							
9	1984-86	$y = 2.6618x + 0.5115$	r = +0.8469 7 df. $p < 0.01$	$y = 3.28$ signs/site		$y = 5.44$ signs/site	
10	1991-94	$y = 2.4283x + 1.3643$	r = +0.6665 8 df. $p < 0.05$	$y = 3.89$ signs/site		$y = 5.86$ signs/site	

977-79; Green & Green, 1980, 1987). These were not used in the calculation of the regres-
ion line shown. Thus, the data for Highland Region on the mainland are in accordance with
s position at the upper end of the regression line (92% occupation; 4.23 initial signs per
ccupied site). However, although Orkney, Western Isles and Shetland have similar, very
igh percentage site occupation (94, 97 and 98% respectively), Orkney has a much lower
ign density than the mainland group, whereas in Western Isles and Shetland it is much
igher than might be expected. This situation is in accordance with the survey information
or these Regions provided by Green and Green (1980). Whereas otters are widely distrib-
ted and present at nearly all sites in all three Regions, they are only numerous, with
igh densities, in Shetland and the Western Isles. In Orkney they "appear to exist at lower
ensities than other marine-based otter populations in Scotland" (Green & Green, 1987).

This hypothesis can be developed further. The fact that there could be a 31% increase
n sign density in Grampian Region, Scotland, with little change in percentage occupation
see Appendix 4 and above), and that there is generally no one level of sign density for any
articular level of site occupancy (see above) suggests that, with a recovering otter popula-
ion, there could be a gradual increase in the level of sign density at each level of percentage
ite occupation. Thus, with time, the very high levels of signs per site, and so area otter
ensity, seen in Ireland may be achieved in Scotland (see Sections 6.1.3 and 6.1.5) and otter
ensity at English sites may be greatly increased too. In order to see if there was any firmer
vidence of such an increase in sign density with time in the available data, the overall
nformation for Scotland and England was reanalysed after division into separate survey
eriods. Thus, the data for the 15 Regional combinations (percentage site occupation/signs
er site) used to calculate the overall regression line for Scotland (see Figure 19) were
ivided into seven Regions (1977-79) and eight Regions (1984-85) and two regression lines
alculated. The results (see Table 4) show that there was indeed a small but consistent shift
pwards in the number of signs per site for each level of site occupation over the seven-year
ap between the two surveys. This is not due to a different number of Regions being used
or the two lines, because removing Lothian Region from the 1984-85 data set and then
ecalculating the regression line results in a similar answer (seven Regions; $y = 0.02609x$
$+ 1.6468$; $r = +0.8611$, 5 df, $p<0.02$; $x = 20\%$ occupation, $y = 2.17$ signs/site; $x = 90\%$
ccupation, $y = 3.99$ signs/site).

The same shift upwards in signs per site was found on division of the English data into
ine (1984-86) and ten Regions (1991-94), which provides further verification for the
ypothesis (see Table 4). Again this shift upwards is not due to a different number of
Regions being used for the two lines. A similar answer is obtained if Thames Region is
emoved from the 1991-94 data set and the line recalculated from nine points. However, the
orrelation is no longer significant without the lowest point (nine Regions; $y = 1.4169x$
$+ 2.8527$; $r = +0.4358$, 7 df, $p>0.10$; $x = 10\%$ occupation, $y = 4.33$ signs/site; $x = 70\%$
ccupation, $y = 5.48$ signs/site). These upward shifts cannot be due to an anomaly, such as
easonal spraint cycles, as each Region was resurveyed at the same time of year. They can
nly indicate an increase in sign density, and so otter density, with time and recovery at each
evel of percentage site occupation.

Our conclusion is that it is well worth carrying out a spraint density survey, and counting the spraints and other signs over the full 600m walk, at the same time as the original
standard survey is completed, based on percentage occupation of survey sites. Both provide
valuable, mutually supportive, indicators of the status of the recovering population. As
Jefferies (1986) concluded, the spraint density provides an important "fine adjustment" on
the data based on presence/absence at survey sites. Thus, over all England, the number of
positive sites (out of 3,188 surveyed) increased from 286 in 1984-86 to 706 in 1991-94
(a factor of x 2.47). However, the number of signs found increased from 1,293 to 3,630
(a factor of x 2.81). This was because the number of signs per occupied site increased by
13.7%, from 4.52 to 5.14 (see Appendix 5). Unfortunately, we cannot as yet calibrate this
spraint density in terms of actual numbers of otters per kilometre of waterway.

Further, as noted in Section 6.2 on recovery, the above analysis of information from
spraint density indicates that the time required before the British population of *L. lutra* has

the density of that of Ireland is even longer than that calculated from percentage site occupation alone.

The data analysed here and the hypothesis developed can perhaps be used to clarify the situation in Wales. There has been concern that percentage occupation of survey sites by otters in Wales has remained the same at two hydrometric areas between 1984-85 and 1991 (Severn) or between all three surveys (Glaslyn), whereas all other hydrometric areas have shown an increase (Andrews *et al*, 1993). However, the present Section shows that an increase in otter density could have occurred without there being an increase in percentage site occupation at survey. Also, whether suboptimal upland sites are occupied in these areas of Wales may depend on the availability of better sites nearby (perhaps in neighbouring hydrometric areas) (see Section 6.2.1).

6.2 Recovery

6.2.1 Calculation of the recovery curve for predictive purposes

The overall percentage site occupation in England increased from 5.78% in 1977-79 to 9.66% in 1984-86 and 23.37% in 1991-94 (Lenton *et al*, 1980; Strachan *et al*, 1990; this report). That of Wales increased from 20.54% in 1977-78 to 38.99% in 1984-85 and 52.48% in 1991 (Crawford *et al*, 1979; Andrews & Crawford, 1986; Andrews *et al*, 1993). Also, the 2,650 sites examined in both the first (1977-79) and second (1984-85) surveys of Scotland showed an overall percentage site occupation increasing from 57.02 to 64.79% (Green & Green, 1980, 1987).

It will be noticed that the third overall result for England (23.37% site occupation) is close to that of the starting (1977-78) percentage for Wales (20.54%) and that the third result for Wales (52.48%) is close to the starting (1977-79) percentage for Scotland south of the Highland Region (57.02%). If these data and their connecting lines showing increases in overall percentage site occupation with time in the three countries are plotted superimposed on the same graph (Figure 22), they indicate the form of the recovery curve for the species in Britain. The form of this unrefined, composite curve appears to be sigmoid, as might be expected, as the percentage increases on the baseline figure for each seven-year interval are not uniform, but are low at low site occupation levels, then increase before decreasing again at high occupation levels.

This composite curve can be refined and an equation calculated for predictive purposes by calculating the percentage increase in site occupation in each Region after each seven-year interval. This has been done for all Regions with such an increase in the English surveys of 1977-79 and 1984-86, the Welsh surveys of 1977-78 and 1984-85 and the Scottish survey of 1977-79. The resulting data are reported in Appendix 7. If the logarithm_{10} of the percentage increase (y) is plotted against the starting percentage site occupation (x) for each of these surveys, then the result is a linear relationship (Figure 23). Regression analysis shows that this linear relationship and the negative correlation between x and y are highly significant ($r = 0.8551$; 43 df; $p<0.001$). As the starting percentage site occupation increases, then the percentage increase in site occupation after seven years decreases. The equation of this line derived from all 45 Regional combinations from England, Wales and Scotland is $y = 2.4328 - 0.02196x$.

The distribution of the data from the three countries along the same straight line and the significant negative correlation between x and y show that the form of the recovery curve is the same for England, Wales and mainland Scotland. This is confirmed by taking the overall data for percentage site occupation for these three countries and calculating the percentage increase on this after seven years, as before (see Appendix 8). Plotting these five points on Figure 23 shows that they too have a linear relationship and fall close to the previously calculated line. Their position below that line may be partly due to the inclusion of non-increasing Regions in the overall figure. This is particularly so for the England survey of 1977-79, where only seven of the ten Regions showed an increase in site occupation after seven years and two showed decreases.

Figure 22. The results for overall percentage site occupation at the first, second and third surveys of England and Wales and the first and second surveys of Scotland, each carried out at intervals of seven years, superimposed to show the approximate shape of the otter population recovery curve for Britain.

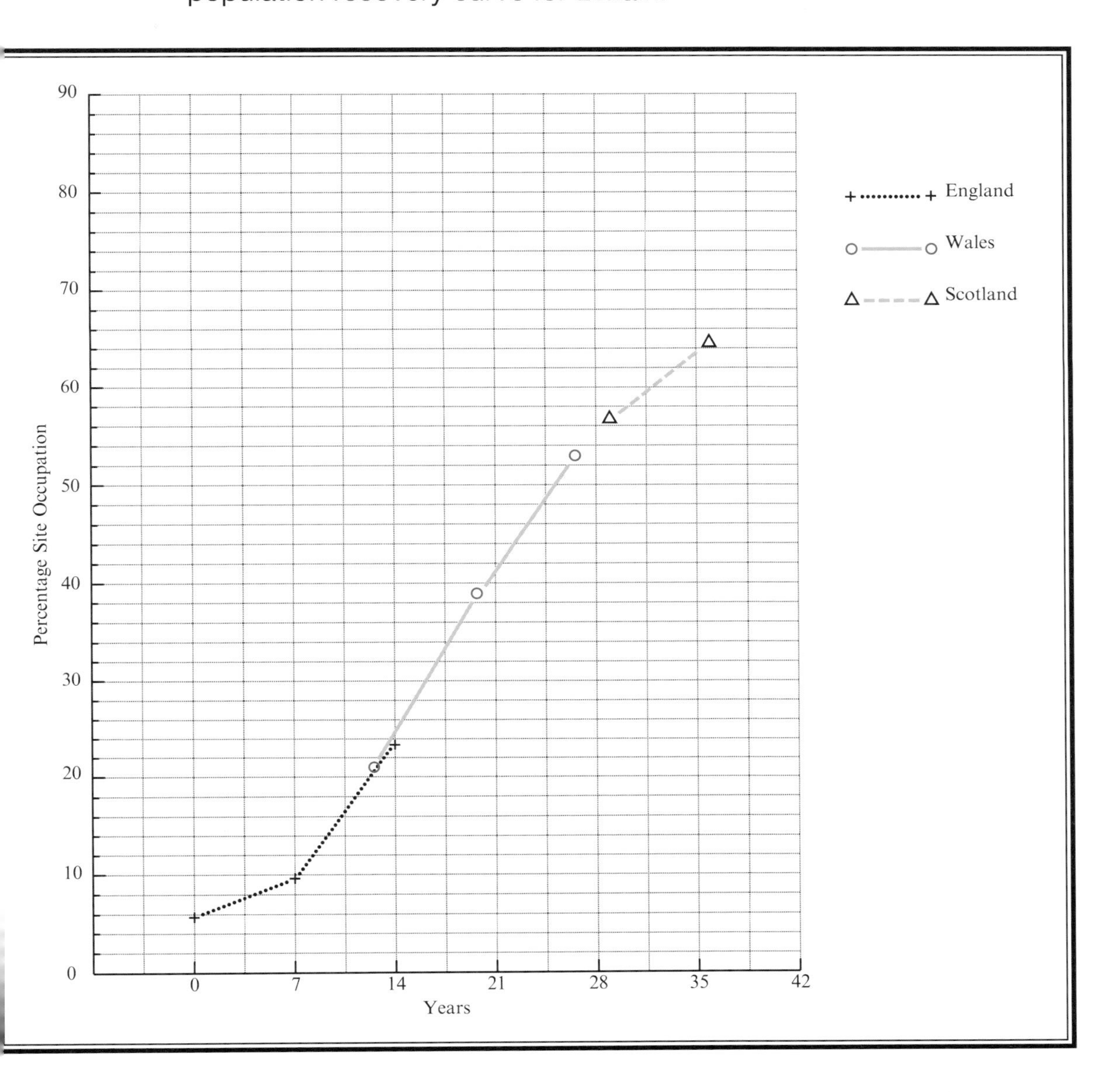

Using the above equation for this overall relationship between percentage site occupation at the start and its percentage increase after a seven-year period, a series of cumulative figures for percentage site occupation can be calculated from an initial figure of 1.0%. This cumulative series is tabulated in Appendix 9, together with the calculations from which it is derived. The recovery curve for the species under present conditions can then be drawn, using this series of percentages of sites occupied as datum points with seven-year intervals (see Figure 24). The resulting curve is sigmoid and extends over a surprisingly long period of 112 years from 1.0 to 99.9% of survey sites occupied.

The form of the recovery curve suggests an initial period of very slow increase in area occupation of a Region, perhaps owing to the presence of only a few itinerant animals, possibly young males, moving in and through. After about seven years the rate of area occupation increases, presumably with animals settling and breeding in the best sites and their young dispersing. Between 14 and 35 years the rate at which new sites are occupied is at its highest. This is possibly due to both an increase in locally bred dispersing young and a flood of dispersing young from the neighbouring Region to the west, where the best sites may be nearly all occupied. (The three surveys of England suggest a wave of recovery proceeding

Figure 23. The relationship between the Regional percentage site occupation at the first or second survey and the logarithm$_{10}$ of percentage increase after seven years. The data are from the first and second surveys of England (+), the first and second surveys of Wales (o) and the first survey of Scotland (Δ) (Appendix 7). Also shown is the calculated regression line (equation: $y = 2.4328 - 0.02196x$) derived from all 45 points. The positions of the five datum points (*) relating overall percentage site occupation for England, Wales and Scotland and overall percentage increase after seven years (data from Appendix 8) show that four of these too fall on the same line as that derived from the Regional data.

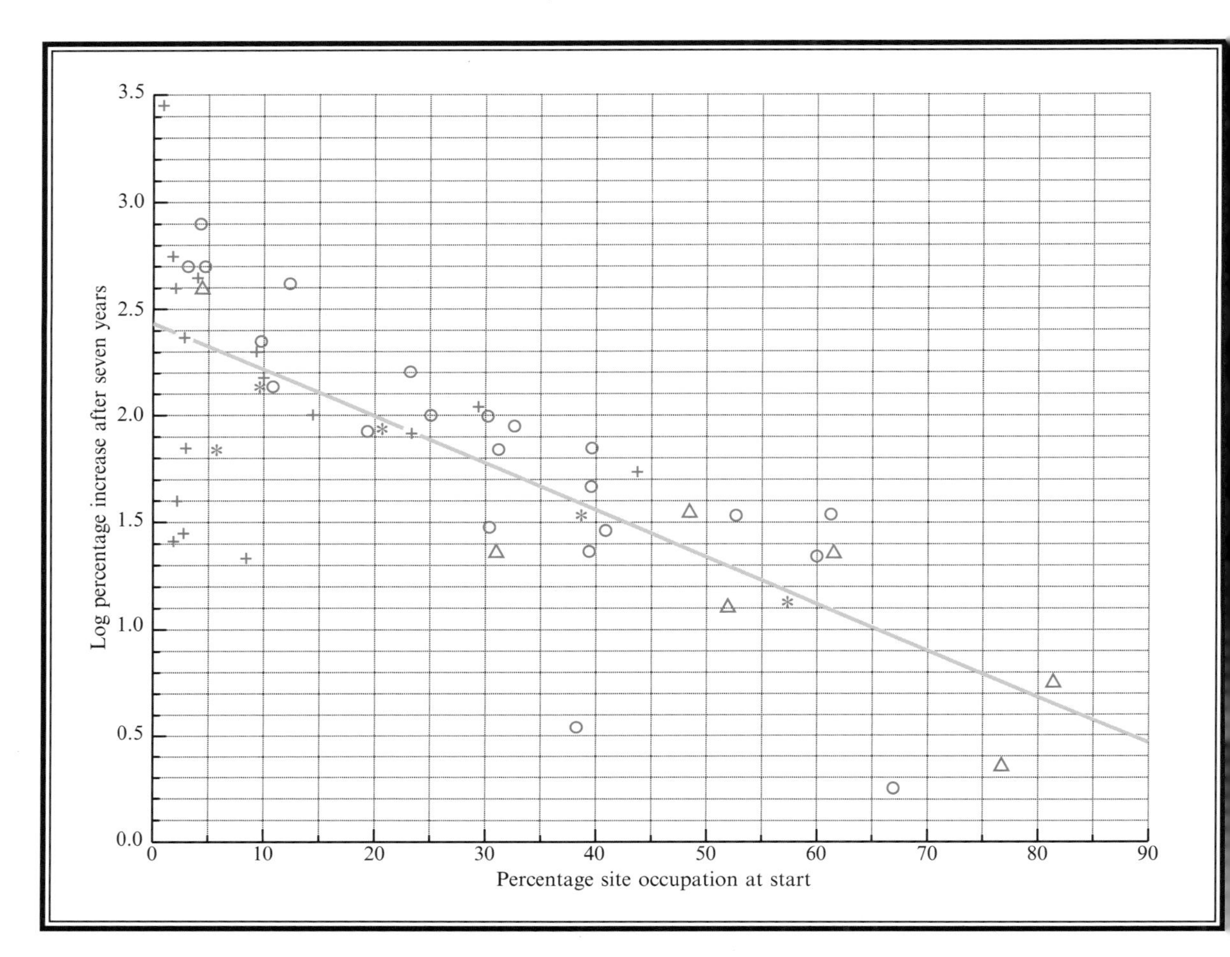

from the west to the east.) After 35 years, by which time site occupation has reached nearly 60%, the rate at which new sites are occupied decreases markedly and there is a long 'tail' to the curve, occupation of the last 40% of sites taking a further 77 years. This may be due to the remaining sites being poor and suboptimal. However, examination of changes in the effect of increasing altitude on percentage site occupation (see Section 6.1.6 and Figure 21a) showed that though the first river sites to fill up to capacity are the optimal lowland sites with the best food supply, with a rising population otters are eventually forced to inhabit the suboptimal upland sites as well. These have the lowest productivity and poorest food supply (see Section 6.1.6). Examination of data from Ireland (derived from Chapman & Chapman, 1982) shows that such upland sites are indeed inhabited where the population is strong and numerous, though they are as yet very poorly used in present day England (see Figure 21a).

Recovery of some western Regions, particularly in Wales with much suboptimal upland habitat, may be continuously depressed for several years because most of the annual cub production may move out eastwards into neighbouring Regions and into England, while many prime sites and habitat remain empty or under-used there.

Figure 24. The estimated recovery curve for the otter in Britain under present conditions. A series of cumulative figures for percentage site occupation at seven-year intervals was calculated from an initial figure of 1.0% at year 0 by use of the equation $y = 2.4328 - 0.02196x$ derived from Figure 23 (data tabulated in Appendix 9).

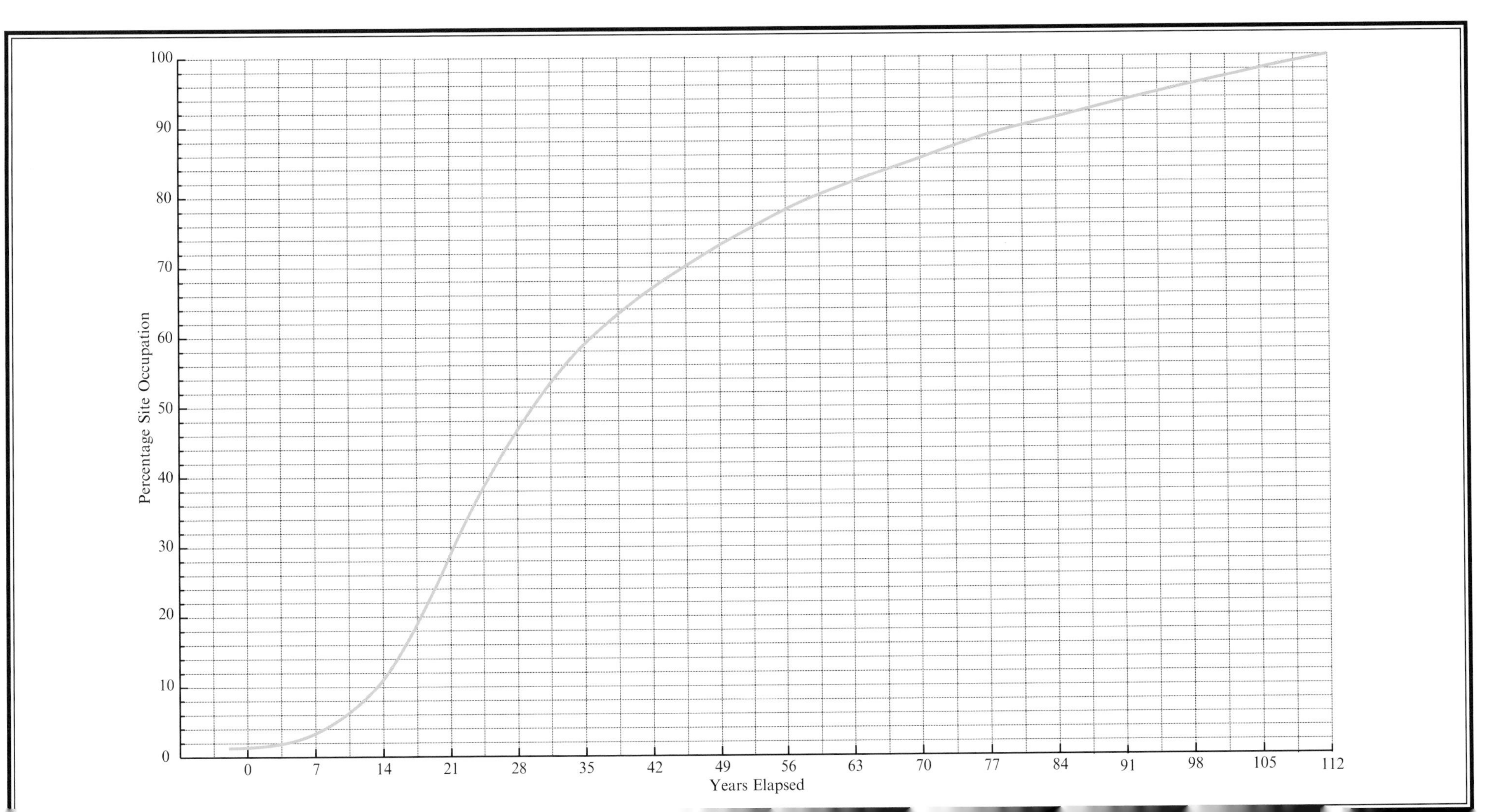

Table 5. The present percentage site occupation of otters in the ten English Regions and England overall as shown by this survey is provided in column 2. The final four columns list predicted percentage site occupation at seven-year intervals into the future, using the same method of cumulative computation and the equation $y = 2.4328 - 0.02196x$ as was used to estimate the recovery curve shown in Figure 24. This recovery potential is based on conditions and recovery during the last 14 years and the accuracy of predictions based on these data depends on conditions over the next few decades remaining the same or similar. For purposes of computation, the timing of year 0 for each Region was taken to be 1992, as much of the survey was completed over 1991-93. Individual Regions may vary a year in advance or behind in their actual dates of surveying.

	Present	**Future**			
Region	**1991 – 1994 (Taken as 1992)**	**1999**	**2006**	**2013**	**2020**
North West	28.70	46.92	58.77	66.93	73.07
Northumbrian	26.79	45.51	57.86	66.26	72.56
Yorkshire	11.06	28.19	46.55	58.53	66.75
Severn-Trent	21.16	40.82	54.86	64.14	70.92
Anglian	8.35	23.17	42.62	56.00	64.94
Thames	2.35	8.01	22.49	42.03	55.62
Wessex	18.83	38.52	53.40	63.12	70.15
South West	67.10	73.21	78.10	82.18	85.67
Southern	3.73	12.11	29.89	47.76	59.32
Welsh	62.96	70.03	75.53	80.02	83.81
England overall	**23.37**	**42.79**	**56.11**	**65.01**	**71.59**

6.2.2 Timing of recovery

Using the same method of cumulative computation and the equation y = 2.4328 – 0.02196x as for drawing up the form of the recovery curve (see Section 6.2.1), one can calculate the percentage site occupation in each Region of England after further seven-year periods into the future (see Table 5). The dates from which the seven-year periods run will be different for each Region, depending on when it was surveyed within the 1991-94 survey period. However, taking the mid-survey point of 1992 (much of the survey was completed over 1991-93), then it can be calculated and predicted that the Anglian Region, for example, could reach 25%, 50%, 75% and 92% site occupation by the years 2000, 2010, 2032 and 2066 respectively, even if no more otters are added by release. 92% is the known very high level of site occupation seen in Ireland in 1980-81 (Chapman & Chapman, 1982). In the South West Region 25% and 50% site occupation were reached by 1979 and 1987 respectively, and it can be predicted that 75% and 92% could be reached around 2002 and 2036. In the same way, these four site occupation levels may be reached over all England by 1993, 2003, 2025 and 2059.

Obviously, such a recovery curve and equation are based on countrywide conditions and recovery potential during the last 14 years, and predictions using them will be possible only if conditions over the next few decades remain the same or similar. The fact that the same curve covers situations in all Regions of England, Wales and Scotland suggests that it already encompasses a wide range of environmental and biological conditions. However, if freshwater pollution, for example with PCBs or mercury, or mortality from, say, road accidents, should increase markedly, then the actual percentage site occupation may fall short of the predictions. A similar effect would be seen if a new adverse factor should develop, such as a new biologically active pollutant, or a new serious cause of mortality such as a lethal disease should occur. Indeed, if actual site occupations over the next 14-21 years should fall markedly short of the above predictions (see Table 5), then this may be used as an indication of such a new adverse factor or factors already in operation.

On the other hand, should an adverse factor be ameliorated or should releases of captive-bred or rehabilitated otters (see Section 7.2) continue or be started in any Region, then this may considerably speed up the rate of site reoccupation. For instance, the East Anglian otter population was declining rapidly and could have disappeared altogether if releases by the Otter Trust had not started in 1983. After releases and between surveys two and three, the percentage increase in site occupation in Anglian Region was 552% from a starting figure of 1.28% occupation. This is much larger than the percentage increases for Southern (40), Severn-Trent (69) and North West (244) Regions, which started with higher site occupation (2.07, 2.29 and 2.78%, respectively) (see Appendix 7). The other Region which has received releases, Yorkshire, showed a percentage increase of 25% on a starting percentage occupation of 1.77% between surveys one and two. However, the percentage increase was much higher, at 400% on a starting percentage occupation of 2.21% between surveys two and three and after releases had started. Whereas the former increase is below average and below the regression line in Figure 23, the latter, following releases, is considerably above it.

It should be remembered that this Section on recovery and the predictions in it are based on site occupation. The analysis of changes in spraint density in Section 6.1 suggests that the population density of otters at occupied sites may increase slowly for some time after site and river occupation are first recorded. Thus, there would be a further delay after each of the predicted dates before the otter population of occupied sites in England reached the densities found at sites in Ireland or Scotland.

Whatever the accuracy of these individual calculated predictions proves to be, the form of the overall recovery curve shows that countrywide and complete recovery is still several decades away. It is a much longer period than may have been expected by conservationists, given the encouraging signs of otters in every Region, and shows that vigilance, habitat improvement and monitoring will have to be maintained for many years to come.

6.3 Otters and mink

6.3.1 The possibilities of interaction

Feral American mink were first noted as breeding in the wild in Britain in July 1956, when a female and kits were sighted on the River Teign in south Devon (Linn & Stevenson, 1980). The consequent spread of the mink and the decline of the otter led many people to suggest that the mink was driving the otter out. However, Chanin and Jefferies (1978) compared the timing of the otter's decline in Britain with that of the spread of the mink in large numbers and concluded that the former had preceded the latter. Thus, this theory has no basis in fact, though it still persists (as noted by Birks, 1990). While there was no doubt that otters were not driven out by mink, Chanin and Jefferies (1978) recorded another of the concerns of the time. This was that a well established mink population in much of Britain could impair the ability of the otter population to re-establish itself should other adverse factors be improved, ie. that otters may move on through an area with a high density of mustelid territorial signs rather than colonising it. The opposite point of view was also expressed for the first time by Chanin and Jefferies, that it was possible that the spread of the mink was facilitated by the low otter population in many areas in the 1960s and 1970s. In this case, a recovering otter population may be more likely to reduce the mink population and so, incidentally, aid the declining water vole population by removing one of its major predators (Strachan & Jefferies, 1993). Jefferies (1992b) extended this theory as another possible explanation for the sudden increase in the rate of spread of polecat distribution in Wales and western England in the late 1950s ie. the absence of the otter may have facilitated the spread of another mustelid as well as the mink.

None of the above theories could be substantiated or resolved in 1978 because of the lack of relevant information from any country. However, since then there has been occasional circumstantial evidence suggesting that, although there may be no adverse effect of mink on otters, this may not be true in reverse.

On the one hand, the spread of the otter in the South West Region between 1977-79 and 1984-86, reoccupying rivers in an area with the highest mink occupation density in England (76.12% and 83.58% of 10km squares occupied in 1977-79 and 1984-86, respectively; Lenton *et al*, 1980; Strachan *et al*, 1990), would suggest that (a) mink did not drive out otters at any stage in the latter's decline, and (b) the presence of the former does not appear to impair the ability of the latter to re-establish itself.

On the other hand, Birks (1990) collected independent circumstantial evidence from three widely separated areas of Britain (pers. comm. from P. Nicholson, A. Crawford and J. Green) regarding a possible reduction in mink numbers in recent years following some degree of recovery of the otter populations in those areas (Strachan *et al*, 1990; Andrews & Crawford, 1986; Green & Green, 1987). Thus, mink populations were regarded as "lower than in earlier years" in the South West Region of England. Also, they were considered to be "struggling" in mid-Wales and to have "become scarcer" in several parts of southern Scotland.

As noted by Jefferies (1992b), some support for these informed observations is provided by examination of data on the numbers of mink killed by gamekeepers (published by Tapper, 1992). After increasing steadily from 1961, the numbers killed per km^2 reached a peak for Britain overall in 1984 and then showed a plateau, being no higher in 1989 than they were in 1982. If the data for southern England are examined separately they appear to suggest a decline. Here the numbers killed increased from 1961 to 1982, when they reached a peak of 0.48/km^2, before they decreased to 0.40/km^2 by 1989.

Further support is provided by research carried out by Erlinge (1972) in Sweden, which suggested the possible exclusion of mink by otters. Thus, the Nobyan stream was densely populated by mink during the summer, whilst otters visited it only infrequently as they were hunting in the nearby lakes. However, when these lakes froze over in winter and otters moved to the stream, the mink numbers there decreased considerably. Erlinge considered that this could have been due to interference by the otters and that the mink had moved to suboptimal peripheral areas.

Finally, an examination of the distribution of feral mink in Ireland may provide some comparative information. Here, the otter population remained at its highest population density (91.7% of survey sites occupied) in western Europe, whilst that of Britain was declining (Chanin & Jefferies, 1978; Chapman & Chapman, 1982). Consequently, the feral mink population was attempting expansion along rivers already fully occupied by a larger aquatic mustelid. American mink were first ranched for fur in Ireland in 1950, and by 1969 feral animals had been sighted in 11 counties and they were known to be breeding in at least one of them, Tyrone (Fairley, 1984). They have since bred or have escaped and spread over a considerable part of Ireland. By 1983, they were present in 89 10km squares, largely in the north and east of the country (Arnold, 1984). Later, Dunstone (1993) wrote that they had been recorded in at least 16 counties by the 1990s, with the possible exception of Galway in the west. Their percentage occupation of survey sites, on the other hand, was found to be rather low in 1980-81 (6.15% overall; Chapman & Chapman, 1982). These figures can be compared to mink presence in 890 10km squares in Britain in 1983 (Arnold, 1984) and a percentage occupation rate of otter survey sites as high as 69.69% in the South West Region of England in 1984-86 (see Section 6.3.2). In addition, the distribution of the mink in Ireland is uneven and patchy. Thus, the north of Ireland, particularly County Donegal, the midlands and the eastern coastal area contain the largest mink populations, whereas its distribution remains restricted in western and southern counties (Dunstone, 1993). There are "reportedly substantial numbers in the district around Dublin" (Fairley, 1984). This pattern is remarkable in that it is the opposite to that recorded for otters in Ireland in 1980-81. Chapman and Chapman (1982) found that the highest percentage survey site occupation for otters occurred in the 50km squares of the west and south of Ireland (mean 96.76%).

However, in the north and east of Ireland the site occupation rate was appreciably lower (mean 81.77%). The 50km square with one of the lowest rates of site occupation by otters was that around Dublin (62.0%).

Taking all the above information together, it provides some circumstantial evidence that, although there appears to be no direct detrimental effect of the mink population upon the otter and its recovery, there may well be an interaction between the species. This may extend to prevention of the build-up of a high-density mink population where otters are many (as in Ireland) and reduction or limitation of an established population of mink when and where the otter population is recovering. Firmer evidence of such an interaction would be of considerable interest. Consequently, the data for mink and otter presence at a range of English sites surveyed in 1984-86 and 1991-94 were re-examined with a view to throwing light on this possibility (see Sections 6.3.2, 6.3.3 and 6.3.4).

6.3.2 Changes in site occupation by mink and otter in South West Region

6.3.2.1 Regional results and changes

The South West Region is the area where recovery of the otter population has been most marked in England (see Section 5.8). It now has the highest percentage site occupation by otters in this country (Table 2). However, in addition, it was the area where breeding of feral mink was first discovered and it has had a very high mink density for several decades (Lenton *et al*, 1980; Strachan *et al*, 1990; this report). Thus, it was considered that if there was an interaction between otter and mink, resulting in a change in the population of the latter, then this may show up first in this particular Region. Consequently, 386 individual site data sheets (see Section 3.2) for both the 1984-86 and 1991-94 surveys of the South West (772 in all) were re-examined with regard to past and present occupation and absence of the two species. It was found that, whilst Regional percentage site occupation by the otter increased significantly ($\chi^2 = 42.4717$; 1 df; $p<0.001$) from 43.78% (169/386) to 67.10% (259/386) (Regional occupied site increase = 90) in the seven years between surveys, that of the mink decreased even more markedly from 69.69% (269/386) to 34.97% (135/386) (Regional occupied site loss = 134) (see Appendix 10). This decrease in Regional mink site occupation rate is significant ($\chi^2 = 93.2390$; 1 df; $p<0.001$).

Further, if the 386 survey sites are divided into 16 groups (15 of 25 sites; 1 of 11) in the order in which they were completed, it can be seen (Appendix 10) that this phenomenon of otter increase and mink decrease occurred in every area of the south-west, ie. it was not a local effect. Grouping the data shows that the spread of mink presence was very even over the South West Region in 1984-86, with low variability (and variance) in occupation levels per 25 sites (see Appendix 10). This coverage was lower as well as slightly less even and with a slightly higher variance by 1991-94. Otter presence over the South West Region, on the other hand, was very variable, with a wide range in occupation levels per 25 sites and a very high variance in 1984-86. This variability was reduced by 1991-94, with a slightly more even coverage.

The above marked decrease in mink site occupation appears to have been closely coincidental in space and time with the rise in otter site occupation. The relationship between these two apparent population changes was examined by two separate approaches, with a view to trying to determine whether the former was brought about by the latter. These approaches are covered in the next two Sections (6.3.2.2 and 6.3.2.3).

6.3.2.2 Relationships between mink and otter presence

(a) First, the total database for otter and mink presence in each of the 16 25-site groups for the two surveys of 1984-86 and 1991-94 was analysed to test for positive or negative relationships (32 points; see Figure 25). Use of these data spanning seven years shows that there is indeed a significant negative correlation between otter and mink presence; as presence of otters increases, there is a concomitant decrease in the presence of mink ($r = -0.4929$; 30 df; $p<0.01$; equation: $y = 19.959 - 0.505x$, where x = otter presence within

Figure 25.

(a) The relationship (+) between the presence of otters *(x)* at survey sites in the South West Region in 1984-86 and the presence of mink *(y)* at these same sites at the same date.

(b) The relationship (O) between the presence of otters *(x)* at survey sites in the South West Region in 1991-94 and the presence of mink *(y)* at these same sites at the same date. Each point represents the presence of otters and mink within a group of 25 survey sites (n=16) at those dates. Two points (Δ , ●) are derived from the 16th group comprising only 11 survey sites rather than 25, so here the co-ordinates given are the calculated numbers of occupied sites per 25 surveyed.

The calculated regression lines are shown by two separate broken lines for the two survey dates (see text).

(c) The calculated regression line using all 32 points and relating changes in otter and mink presence over the seven-year period between the surveys is shown by the central unbroken line.

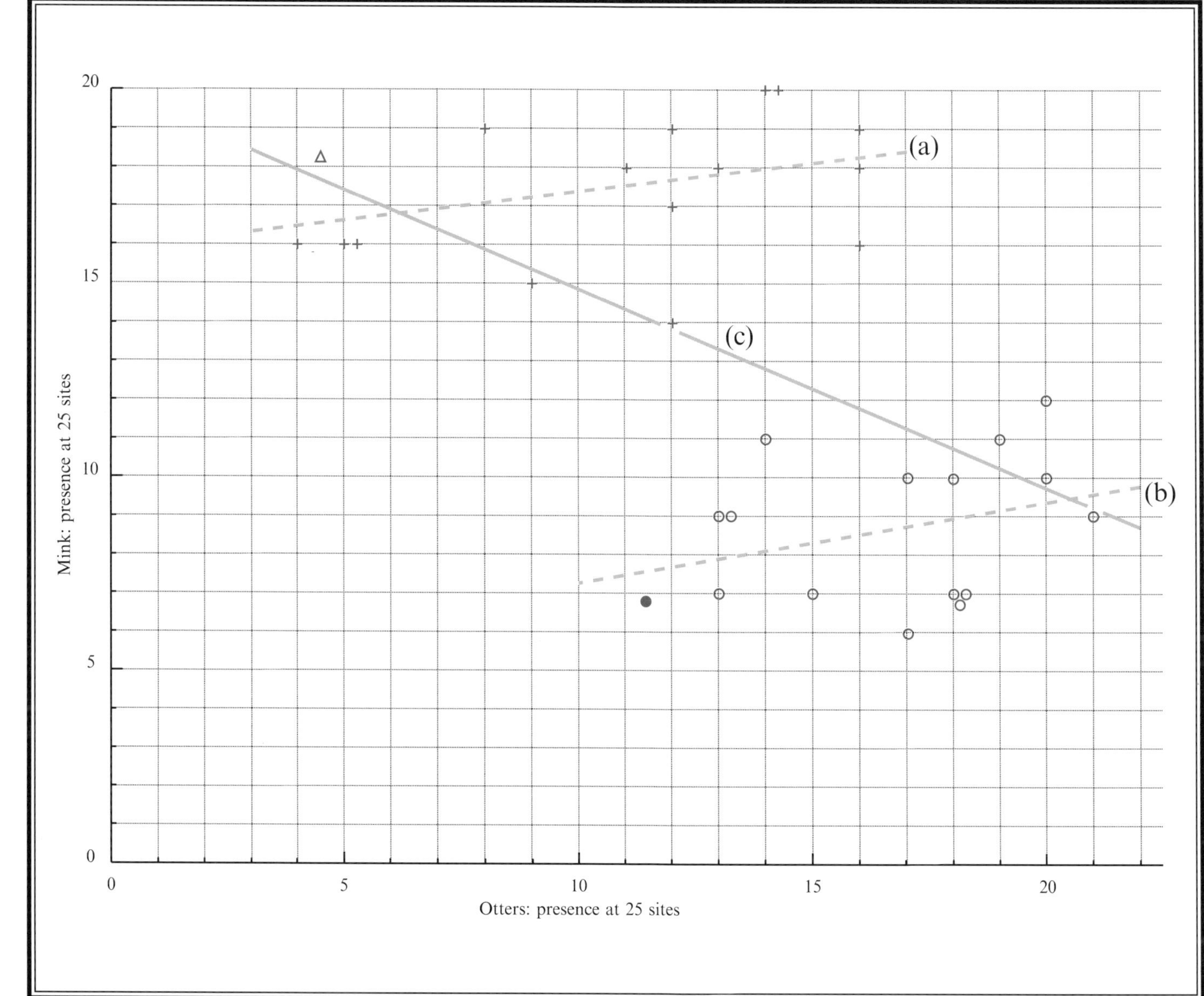

Figure 26. The lack of relationship between the percentage gain in otter-occupied sites in the South West Region between 1984-86 and 1991-94 *(x)* and the percentage loss in mink-occupied sites in the same seven-year period *(y)*. Each point represents percentage gain and loss within 16 groups of 25 survey sites.

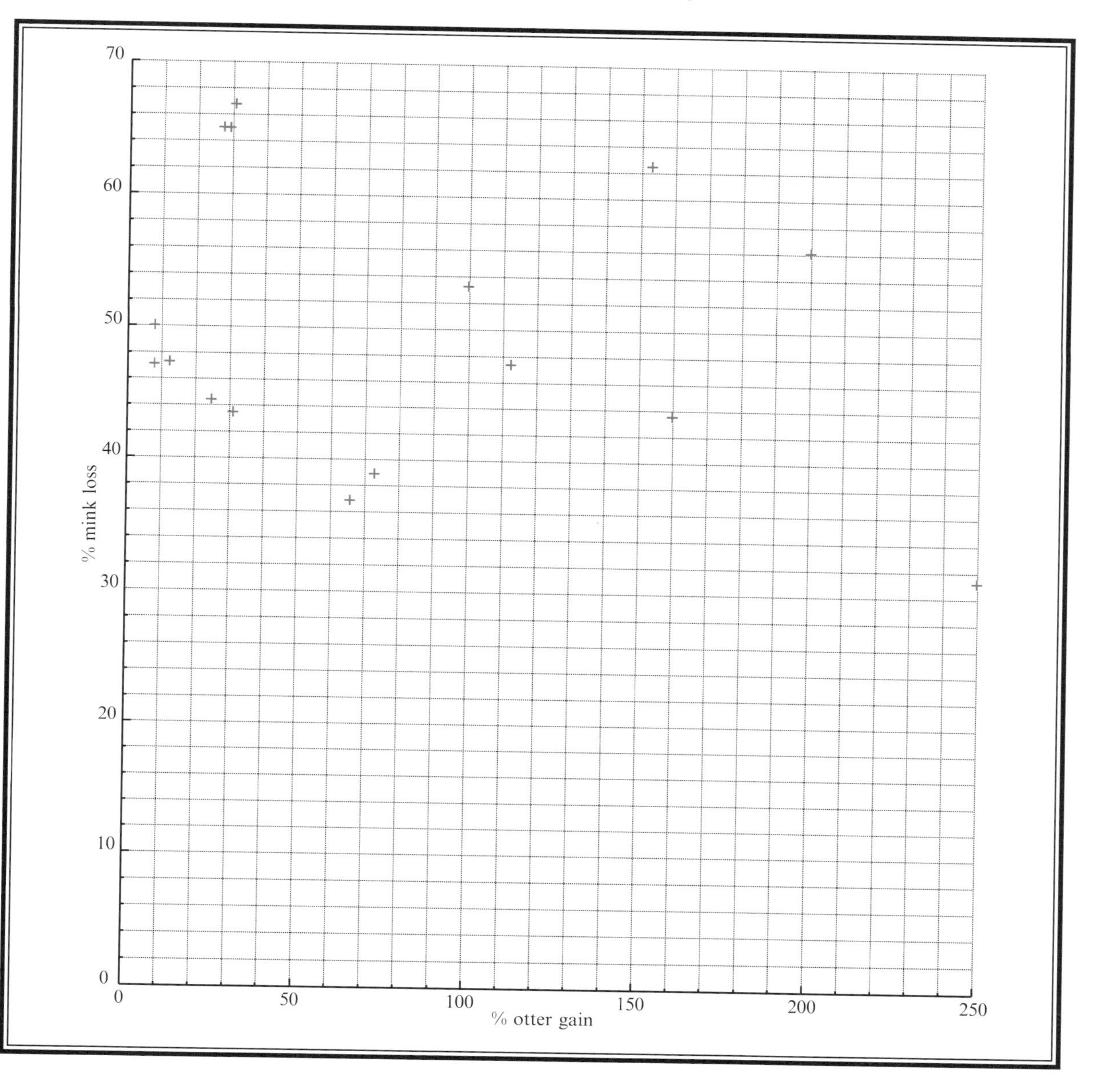

a group of 25 sites and y = mink presence within the same group of 25 sites; see regression line (c) in Figure 25).

However, as can be seen from the Figure, this significant relationship is brought about by the data separating into two groups for the two surveys (one with low otter and high mink presence and the other with high otter and low mink presence) rather than a progression with steadily increasing otter and decreasing mink presence. Thus, the result obtained may indicate that the two changes in occupation level were just a coincidence. We need to know whether the degree of mink disappearance was in proportion to the rise in presence of the otters (as with water voles and mink, shown by Strachan & Jefferies, 1993) to be more sure of a direct link between the two phenomena.

(b) Consequently, the second analysis compared percentage gain of otter-occupied sites and percentage loss of mink-occupied sites in the seven years between 1984-86 and 1991-94. The relationships between the individual gains and losses for each of the 16 groups of South West survey sites are shown plotted in Figure 26. The distribution of these data is at

Figure 27. The relationship between the presence of mink in 16 groups of 25 sites in the South West Region in 1984-86 *(x)* and the number in each group of 25 which had been lost by 1991-94 *(y)*.

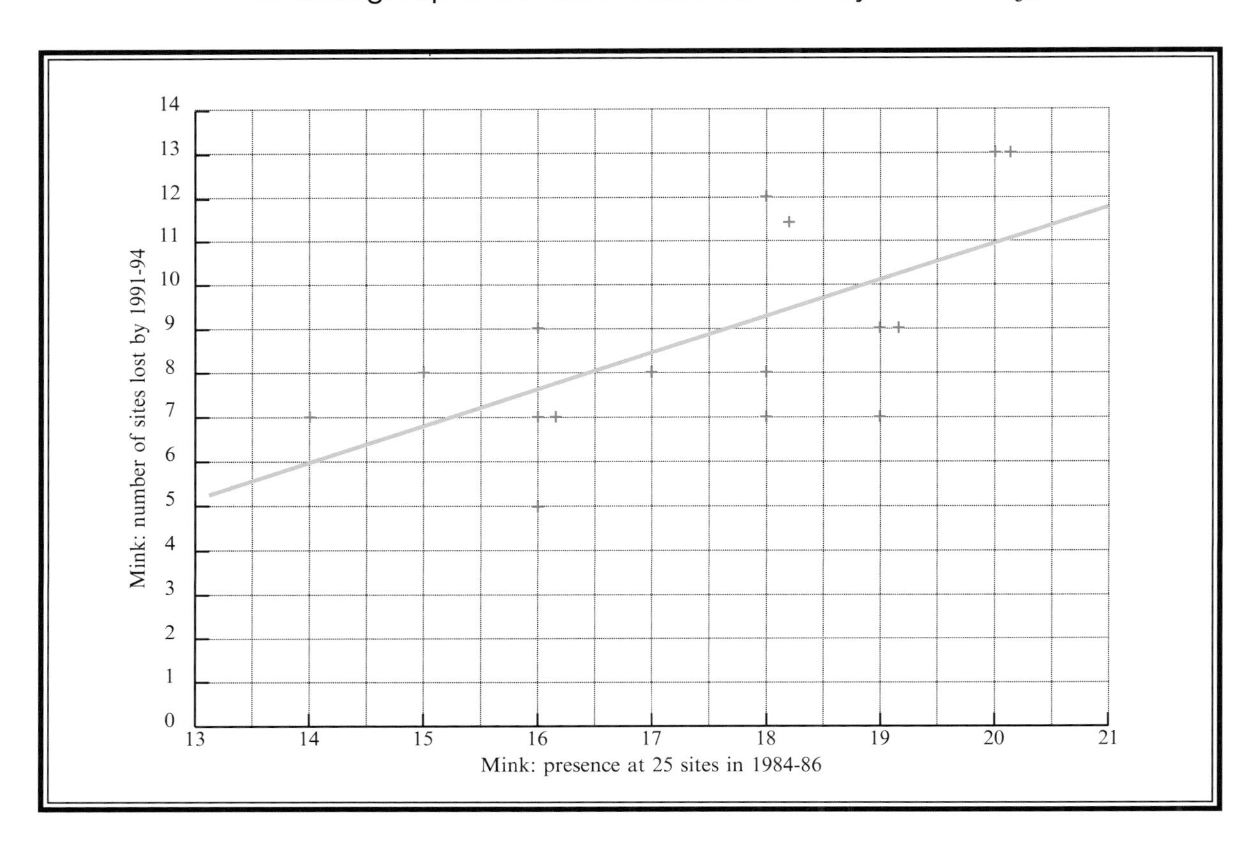

Figure 28. The relationship between the presence of otters in 16 groups of 25 sites in the South West Region in 1984-86 *(x)* and the number in each group of 25 which had been gained by 1991-94 *(y)*.

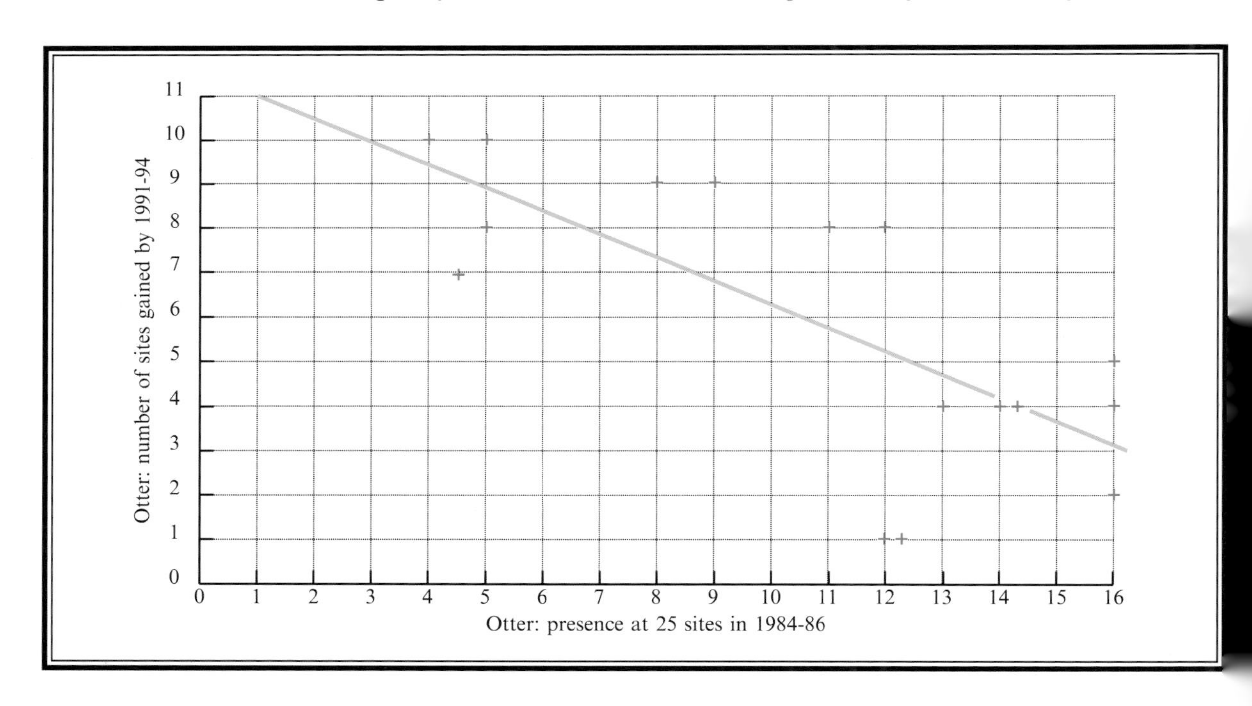

first surprising and shows the opposite tendency to that expected. Thus, the highest percentage mink loss (66.7%) was in a group with a comparatively small percentage otter gain (30.8%). Similarly, the highest percentage otter gain (250.0%) was related to a comparatively small percentage loss of mink-occupied sites (31.2%). However, the complete data set shows a wide scatter (see Figure 26) and there is no statistically significant correlation between percentage gain of otter-occupied sites and percentage loss of mink-occupied sites in the same groups (r = -0.2614; 14 df; p>0.10).

Dividing the data into two equal groups above and below a level of 50% gain in otter sites provides a similar picture. Group 1 has a mean otter gain of 21.66 ± 3.59% related to a mean mink loss of 53.66 + 3.55%, whilst Group 2 with the highest mean otter gain (139.40 + 22.57%) has a lower mean mink loss (46.27 + 3.75%). However, again there is no significant difference between the two groups in terms of percentage mink loss (t = 1.4302; 14 df; p>0.10).

This apparent anomaly is explained below (see Section 6.3.2.4).

(c) Examination of the data relating otter and mink presence in each 25-site group for each of the two surveys separately (Figure 25) suggests a positive relationship and that the numbers of the two species present increase together.

Although first examination by regression analysis of the relationship between otter and mink presence in the 1984-86 survey (group and line a in Figure 25) shows the correlation and slope to be not statistically significant (r = +0.3494; 14 df; p>0.10; equation: y = 15.882 + 0.146x), re-examination by grouping indicates that a significant correlation is indeed present. Thus, dividing the data into three equal-sized clusters of five 25-site groups (omitting the single 11-site group; see Section 6.3.2.1) by using the sequence along the above regression line we find that a progression of increasing mink and otter presence is obtained. The three groups of mean mink/otter numbers per 25-site group (± standard errors) so determined are: Group 1, 15.40 ± 0.40/7.00 ± 1.52; Group 2, 17.80 ± 0.58/11.80 ± 1.28 and Group 3, 19.00 ± 0.45/14.60 ± 0.60. The difference between Groups 1 and 3 is significant both for otters (t = 5.4563; 8 df; p<0.001) and mink (t = 5.9999; 8 df; p<0.001).

The above finding is strengthened by an exactly similar result being obtained for the 1991-94 survey. First, a regression analysis of the relationship between otter and mink presence in 1991-94 (group and line b in Figure 25) shows that the scatter of the results prevents the slope and correlation being found statistically significant (r = +0.3280; 14 df; p>0.10; equation: y = 5.251 + 0.206x). However, dividing the data into three clusters of five 25-site groups again produces a progressive and significant increase in mink and otter presence. The mean mink/otter numbers per 25-site group (± standard errors) in these three groups are: Group 1, 7.60 ± 0.60/14.20 ± 0.80; Group 2, 8.40 ± 0.87/17.00 ± 0.77 and Group 3, 10.40 ± 0.51/19.60 ± 0.51. As before, the increase between Groups 1 and 3 is significant both for otters (t = 5.6920; 8 df; p<0.001) and for mink (t = 3.5559; 8 df; p<0.01).

The finding of a positive relationship between mink and otter presence in the 1984-86 survey results shows that the area with the highest density of mink sites also has the highest density of otter sites. This suggests that the habitat features selected by one are also ideal for the other (see Section 6.3.5.2). The similar finding from the results of the 1991-94 survey, although mink are now lower in density whilst otters are higher, provides additional confirmation.

(d) An examination of the changes which follow when both species are at high density together may provide a clearer picture of events.

When the numbers of mink sites in each of the 16 25-site groups in 1984-86 are plotted against the numbers of mink sites lost in these groups by 1991-94 (see Figure 27), it can be seen that the two have a positive relationship. This relationship is statistically significant (r = +0.6329; 14 df; p<0.01; equation: y = 0.840x - 5.887, where x = mink presence in 25-site groups in 1984-86 and y = number of mink sites lost per group by 1991-94). Using the above equation, it can be calculated that when mink presence was 11 out of 25 sites in 1984-86, then the loss by 1991-94 was 3.35 sites (or 30.50%). However, when presence was twice as great, at 22/25 sites, then the loss was almost four times greater at 12.60 (or 57.26%). This means that most mink sites were lost (whether expressed as numbers or

Figure 29. South West Region site histories: the number and percentage of mink plus otter (++), empty (- -), mink-only (+ -) and otter-only (- +) sites out of the 386 sites surveyed in 1984-86 and the change in their constitution by 1991-94.

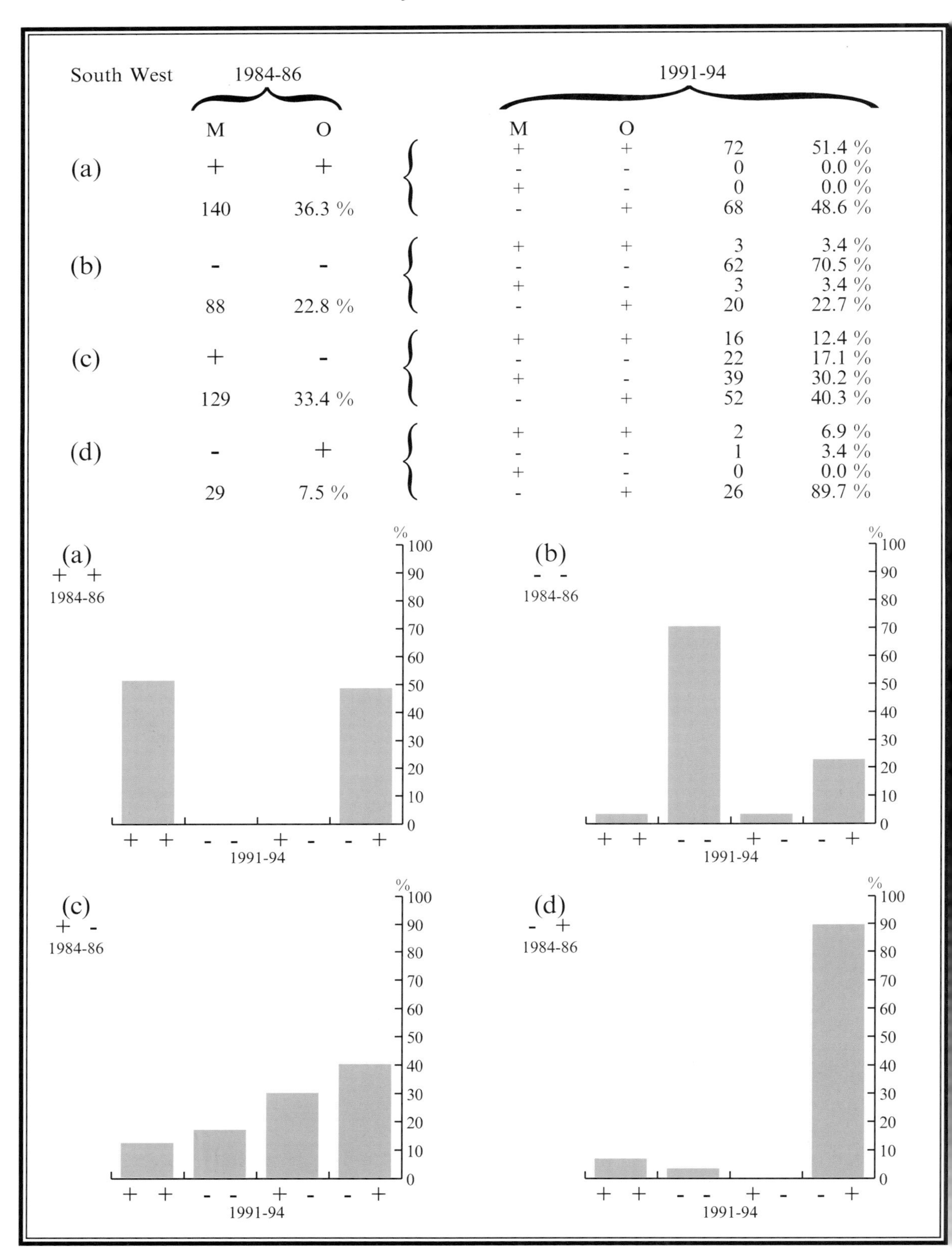

percentages) in areas which had the highest mink site density. The corollary is that the mink site loss rate must have been highest in those areas where occupied otter sites were at their highest density (as mink and otter site density increase together; see Section 6.3.2.2c).

When a similar form of analysis, plotting the number of otter sites in each of the 16

25-site groups in 1984-86 against the number of otter sites gained in these groups by 1991-94, it is found that the two have a negative relationship (see Figure 28). This relationship is again statistically significant ($r = 0.7220$; 14 df; $p<0.01$; equation: $y = 11.499 - 0.525x$, where x = otter presence in 25-site groups in 1984-86 and y = number of otter sites gained per group by 1991-94). Using the above equation, it can be calculated that when otter presence was high at 16 out of 25 sites in 1984-86, then the gain by 1991-94 was 3.10 sites (or 19.37%). However, when presence was only a quarter of this density, at 4/25 sites, then the gain was three times higher at 9.40 sites and the percentage gain very high indeed (235%). This means that most otter sites were gained (whether expressed as numbers or percentages) in areas which had the lowest otter site density before, ie. the otters appear to have been filling-in the almost empty areas rather than moving into those which had been more highly occupied or favoured previously.

6.3.2.3 South West: Site histories

If the changes occurring in site occupation are examined on the basis of individual site histories, then greater detail can be observed regarding the circumstances of the mink disappearance. Thus, of the 386 sites surveyed in 1984-86, (a) 140 were found to have both species (+ +), (b) 88 had neither (- -), (c) 129 had mink but not otter (+ -) and (d) 29 had otter but not mink (- +). The changes in the disposition of the species occupancy of sites included within these four groups can then be examined using the data sheets for the 1991-94 survey. All groups showed considerable changes (see Figure 29).

Group (a): In seven years, mink disappeared or moved out from nearly half of this group of sites which originally had both species. Otters then stayed on alone. The other half of these sites remained as they were, with both species present.

Group (b): The majority (70.5%) of these 'empty' sites remained 'empty', but otters moved into 22.7% of them. On the other hand, very few mink moved in (only 6.8% sites).

Group (c): 30.2% of these 'mink-only' sites remained as they were. In 12.4% otters moved in with the mink, whereas 40.3% became 'otter-only' sites (possibly after being two-species sites for a while). In a further 17.1% the mink disappeared for unknown reasons and the sites became 'empty'. However, otters may well have been moving through the latter sites although not permanently occupying them.

Group (d): Nearly all (89.7%) of these 'otter-only' sites remained as they were. Hardly any mink moved into these areas which otters had occupied alone in 1984-86 (only two sites; 6.9%).

The eventual distribution of the four types of site occupation by 1991-94 was both species (+ +) 93, 'empty' sites (- -) 85, 'mink-only' (+ -) 42, and 'otter-only' (- +) 166.

6.3.2.4 Regional conclusions

The above analyses provide the following picture of events in the South West Region. First, both species appear to select similar habitat features, so that the density of mink was highest where otter occupation was at its highest (see Section 6.3.2.2c). This was so both in 1984-86, when the otter population was still low, and in 1991-94, when it was much higher. When the otter population expanded between 1984-86 and 1991-94 there was a contemporaneous decrease in the number of sites occupied by mink.

Examination of the site histories for the South West (see Section 6.3.2.3) shows that the otters moved into 'new' sites (for them) whether mink were already there or not with equal facility. However, this was not the case in reverse. There were only two sites out of 386 surveyed where mink moved into sites already occupied by otters. There were also changes in site histories which suggest a progression from 'mink-only' sites through two species, as otters moved in, to finish as 'otter-only' sites.

Most mink sites were lost where the mink and also the otters were at their highest densities (see Section 6.3.2.2d). Most of the otter site gains, on the other hand, were where otters were originally at low densities in 1984-86, ie. they were 'filling in' and ending with a more even coverage in 1991-94 (see Section 6.3.2.2d). This is why, when looking at percentage site gains and losses for the two species (see Section 6.3.2.2b), we discovered the apparent anomaly that the lowest percentage gains for otters were linked with the highest percentage losses for mink. This was because both occurred in the areas of highest otter presence.

The disappearance of mink from 49.8% of their previously occupied sites in the South West Region suggests a loss of around half of the original Regional population. That is, unless they were present at double the density in the remaining occupied sites, which seems unlikely. There are no signs of mink moving out into previously 'empty' sites as, taking known movements into account, there is a net loss of 134 mink-occupied sites in the seven years' interval between surveys.

The pattern of the site histories, the loss of most mink sites where otters were at highest density, and the synchronous timing strongly suggest that the increase in otters in the South West was the cause of the decline in the mink of that Region. However, to be more certain we need to know that the British mink population is still capable of expansion elsewhere in the country where otters continue to be few and also that there is a progression through mink increase to mink decline with an increasing otter population and that the two are significantly correlated. Hence, we looked at mink and otter relationships in two more Regions to provide the necessary range of data. These were Severn-Trent and Anglian Regions.

6.3.3. Changes in site occupation by mink and otter in Severn-Trent Region

6.3.3.1 Regional results and changes

Changes in the numbers of occupied mink and otter sites and their relationships were investigated in the area of the Severn-Trent Region because here the status of the otter lay mid-way between that of the rapidly expanding and large population of the South West and the still low population of Anglian Region. Also, the maps published by Lenton *et al* (1980) and Strachan *et al* (1990) showed that there had been an increase in the mink population of the south-west of the Severn-Trent Region, in terms of area occupied, between the surveys of 1977-79 and 1984-86. Here it was thought that we might determine whether a 'medium-sized' otter population (fifth in terms of percentage site occupation by 1984-86) had any effect on a mink population known to have been expanding its area of occupation at that time.

Thus, the 610 individual site data sheets for both the 1984-86 and 1991-94 surveys of the Severn-Trent Region (1,220 in all) were re-examined with regard to past and present occupation and absence of the two species. As expected, Regional percentage site occupation by otters increased significantly ($\chi^2 = 83.1722$; 1 df; $p<0.001$) from 3.61% (22/610) to 20.66% (126/610) (Regional occupied site gain = 104 or 472.7%) in the seven years between surveys. Unlike the situation in the South West, with its declining mink site occupation, the percentage site occupation by mink in the Severn-Trent Region showed an increase too. However, this was relatively small, from 41.80% (255/610) to 49.34% (301/610) (Regional occupied site gain = 46 or 18.04%). This increase is significant ($\chi^2 = 6.9925$; 1 df; $p<0.01$).

Division of the 610 survey sites into 25 groups (24 of 25 sites; 1 of 10) in the order in which they were completed allows examination of the changes in distribution of the two species within the Region and the variability in the numbers of site gains and losses (Appendix 11). Grouping the data in this way shows that otters were unevenly distributed in 1984-86 and occurred only in six (or 24%) of the groups, largely in the west of the Region. By 1991-94 they had spread to 18 (or 72%) of the groups and thus were present over most of the area. Wherever they occurred, the number of occupied sites per group showed an increase. Mink, on the other hand, were already present in nearly all areas of the Region by 1984-86 (24 or 96% of groups) and in all of the Region by 1991-94. However, although there were gains in the number of occupied sites per group in 18 groups, there were still seven (or 28%) of groups which showed site losses. The end result was only a moderate population increase for the mink.

6.3.3.2 Severn-Trent: Site histories

Examination of site occupation in 1984-86 showed that of the 610 sites surveyed, (a) 18 were found to have both species (+ +), (b) 351 had neither (- -), (c) 237 were 'mink-only' (+ -) and (d) four were 'otter-only' (- +). Again, examination of the same site data sheets for 1991-94 showed considerable changes (see Figure 30).

Group (a): As in the South West Region, mink disappeared or moved out of half of this group of sites which originally had both species, whilst otters stayed on alone. The other half of these sites remained as they were, with both species present. Thus, otter presence remained the same at these long-occupied sites.

Group (b): The majority (63.0%) of these 'empty' sites remained 'empty', but otters moved into 27 (7.7%) of them. Mink moved into an even larger number, occupying 115 (or 32.8%) originally empty sites. This movement of mink into the empty sites represents most of the mink gain in sites in the area of the Severn-Trent Region (see Section 6.3.3.1), as elsewhere, in the other three occupation types (a, c & d), there was a net loss of 69 mink sites. The mink appeared to prefer completely 'empty' sites, as only 12 of the 115 new mink sites were occupied jointly with otters.

Group (c): The majority (51.5%) of these 'mink-only' sites remained as they were. In 21.9%, otters moved in with the mink, whereas 10.5% became 'otter-only' sites (possibly after being two-species sites for a short period). Thus this progression was the same as in the South West. In a further 16.0% of sites the mink disappeared for unknown reasons and the sites became 'empty'. It may be that as the mink population was expanding, albeit only slightly, in this Region, there were itinerant and dispersing mink about at both survey dates. These might have occupied sites for only a short time rather than being resident.

Group (d): There were very few (four) 'otter-only' sites in 1984-86 and, surprisingly, mink entered three of them. All four retained otters.

The eventual distribution of the four types of site occupation by 1991-94 was, both species (+ +) 76, 'empty' sites (- -) 259, 'mink-only' (+ -) 225, and 'otter-only' (- +) 50.

6.3.4 Changes in site occupation by mink and otter in Anglian Region

6.3.4.1 Regional results and changes

The Anglian Region was the third English Region for which the past and present data were examined in order to determine whether otter numbers have an effect on mink site occupation. Here, in this large Region, numbers of otter-occupied sites are still very low but are now expanding following releases (see Sections 5.5 and 7.2). Also, previous work (Lenton *et al*, 1980; Strachan *et al*, 1990; Strachan & Jefferies, 1993) has shown that the mink population colonised the eastern half of England much later than the west and with the large number of 'empty' sites, it was thought to be capable of expanding rapidly over the seven years since the last survey. If otters have indeed had an effect on mink numbers elsewhere in England (see Section 6.3.2), then their scarcity here could have facilitated such a considerable expansion.

Thus, the 725 individual site data sheets for both the 1984-86 and 1991-94 surveys of the Anglian Region (1,450 in all) were re-examined with regard to determining changes in site occupation by the two species.

As expected, Regional percentage site occupation by otters increased significantly ($\chi^2 = 38.1152$; 1 df; $p<0.001$), from 1.10% (8/725) to 8.00% (58/725) (Regional occupied site gain = 50 or 625.0%) in the seven years between surveys.

Examination of the data for mink occupation of these sites confirms the very large population expansion and colonisation of the east predicted above. Site occupation increased from 118/725 or 16.28% to 225/725 or 31.03% (Regional occupied site gain = 107 or 90.68%). This almost doubles the area occupied and is statistically significant ($\chi^2 = 43.7199$; 1 df; $p<0.001$).

Division of the 725 survey sites into 29 groups of 25 sites in the order in which they were completed, again allows examination of the changes in distribution of the two species

Figure 30. Severn-Trent Region site histories: the number and percentage of mink plus otter (+ +), empty (- -), mink-only (+ -) and otter-only (- +) sites out of the 610 sites surveyed in 1984-86 and the change in their constitution by 1991-94.

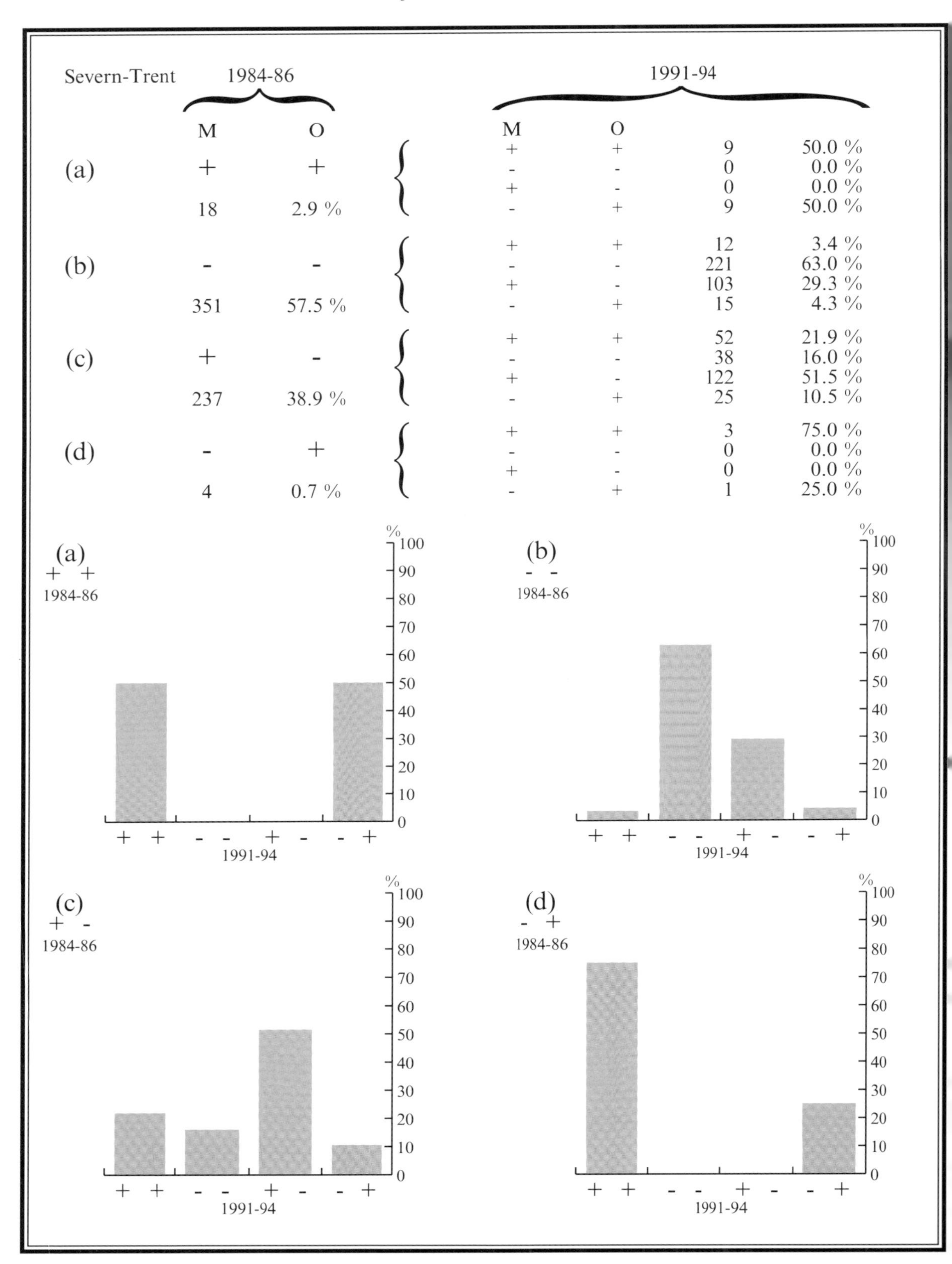

within the Region and the variability in the numbers of site gains and losses (Appendix 12). Grouping the data in this way shows that otters occurred in only five (or 17.24%) of the groups in 1984-86. However, by 1991-94 they had spread to many more groups (19 or 65.52%) and, wherever they occurred, the number of occupied sites per group showed an increase. Mink were already widespread by 1984-86, occurring in 75.86% of the groups but only at an occupied site density of 4.07 per group. By 1991-94, they occurred in all but one group (96.55%) and occupied site density had nearly doubled to 7.76 per group. Although there were gains in the number of occupied sites per group in 23 groups, there were still five groups (17.24%) which showed site losses. These were much lower in number than in Severn-Trent Region.

6.3.4.2 Anglian: Site histories

Examination of site occupation in 1984-86 showed that, of the 725 sites surveyed, (a) two were found to have both species (+ +), (b) 601 had neither (- -), (c) 116 were 'mink-only' (+ -) and (d) six were 'otter-only' (- +). As with the previous two Regions, examination of the same site data sheets for 1991-94 showed many changes (see Figure 31).

Group (a): There were only two sites with both species present in 1984-86 and by 1991-94 the otters had moved out of both of them, leaving them 'mink-only'. This might be expected in a Region with an expanding but still small otter population and many 'empty' sites, ie. otters are likely to be dispersing through the area rather than settled.

Group (b): The majority (71.0%) of these 'empty' sites remained 'empty', but otters moved into 41 (6.8%) of them. Mink moved into an even larger number, occupying 140 (or 23.3%) originally empty sites. As with Severn-Trent Region, this movement of mink into the empty sites makes up most of the mink gain in sites in the Anglian Region (see Section 6.3.4.1) as elsewhere in the other three occupation types (a, c & d), there was a net loss of 33 mink sites. The mink appeared to prefer to colonise 'empty' sites separate to those taken by otters, as only seven of the 140 new mink sites were occupied jointly with that species.

Group (c): The majority (64.7%) of these 'mink-only' sites remained as they were. In eight of them (6.9%) otters moved in with the mink, and three (2.6%) became 'otter-only' sites (possibly after being two-species sites for a while). In a relatively large number (30 or 25.9%) the mink disappeared for unknown reasons and the sites became 'empty'. This might be expected in this Region, with a rapidly expanding mink population and many 'empty' sites, as there would have been itinerant and dispersing mink about at both survey dates. These may have occupied sites for only a short time rather than being resident.

Group (d): There were only six 'otter-only' sites in 1984-86. These were unchanged in 1991-94 and there were no signs of mink attempting to occupy these long-established otter sites.

The eventual distribution of the four types of site occupation by 1991-94 was both species (+ +) 15, 'empty' sites (- -) 457, 'mink-only' (+ -) 210, and 'otter-only' (- +) 43.

6.3.5 Overall conclusions

6.3.5.1 The effect of otter density on mink density and rate of spread

The fact that the mink population is still capable of doubling its area of occupation in the almost adjacent Anglian Region, suggests that the decline in the South West is unlikely to be due to pollution or disease and that causes should be sought elsewhere.

The three Regions, (a) South West, (b) Severn-Trent and (c) Anglian, when taken together form a descending series for otters and an ascending series for mink. Thus, although all have expanding otter populations, the percentage of occupied sites in 1991-94 decreases from (a) 67.10% to (b) 20.66% to (c) 8.00%. Examination of the related mink populations at that date shows that their status ranges from (a) considerable decrease (49.81% loss of sites), with all group areas showing site losses, to (b) almost stationary (only 18.04% gain in sites), with 28.00% of groups showing losses, to (c) considerable increase (90.68% gain in sites), with only 17.24% of groups showing losses. In addition, site histories (Sections 6.3.2.3, 6.3.3.2 and 6.3.4.2) show that otters expand into all favourable sites whether originally occupied by mink or not. Mink, on the other hand, seldom move into otter-occupied areas (see also Section 6.3.5.3). These observations suggest a negative interaction between the two species and perhaps an avoidance of otters by mink.

Figure 31. Anglian Region site histories: the number and percentage of mink plus otter (+ +), empty (- -), mink-only (+ -) and otter-only (- +) sites out of the 725 sites surveyed in 1984-86 and the change in their constitution by 1991-94.

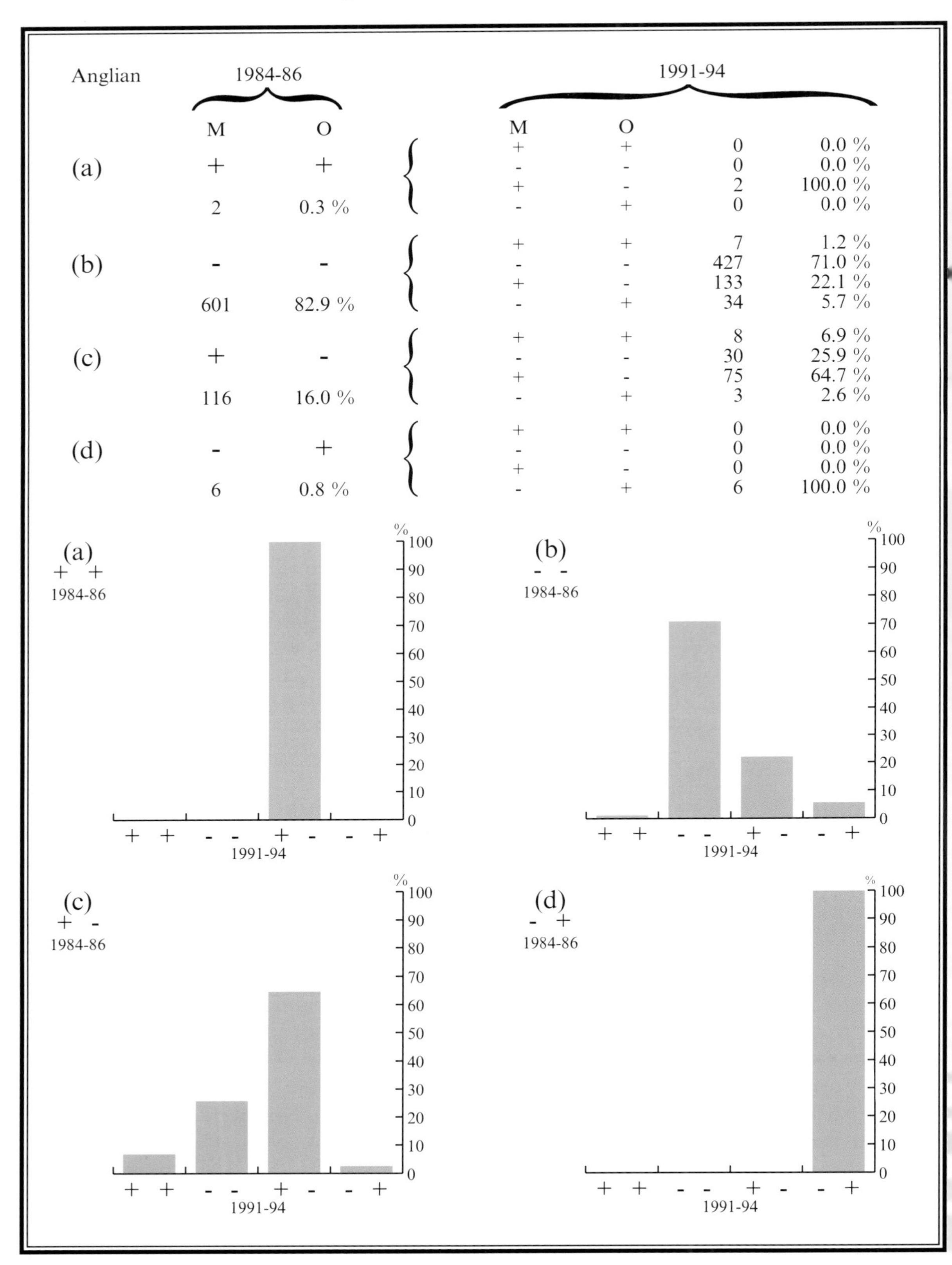

Finally, a significant negative correlation between mink population change and otter presence is obtained when data from the three Regions are combined (Figure 32). Thus, if the number of mink site losses or gains per 25-site group between 1984-86 and 1991-94 (y) (obtained from column 7 of the Tables in Appendices 10, 11 and 12) is plotted against otter

igure 32. The relationship between the presence of otters (occupied sites per 25-site group) in 1991-94 *(x)* and the difference between site occupation by mink (in the same 25-site groups) in 1984-86 and that in 1991-94 *(y)*. A total of 68 full groups was obtained from the data for three English Regions, Anglian (●), Severn-Trent (+) and South West (O). These data are listed in Appendices 10, 11 and 12. The calculated regression line is also shown. As the numbers of otters increased in the 68 areas, the numbers of mink-occupied sites decreased.

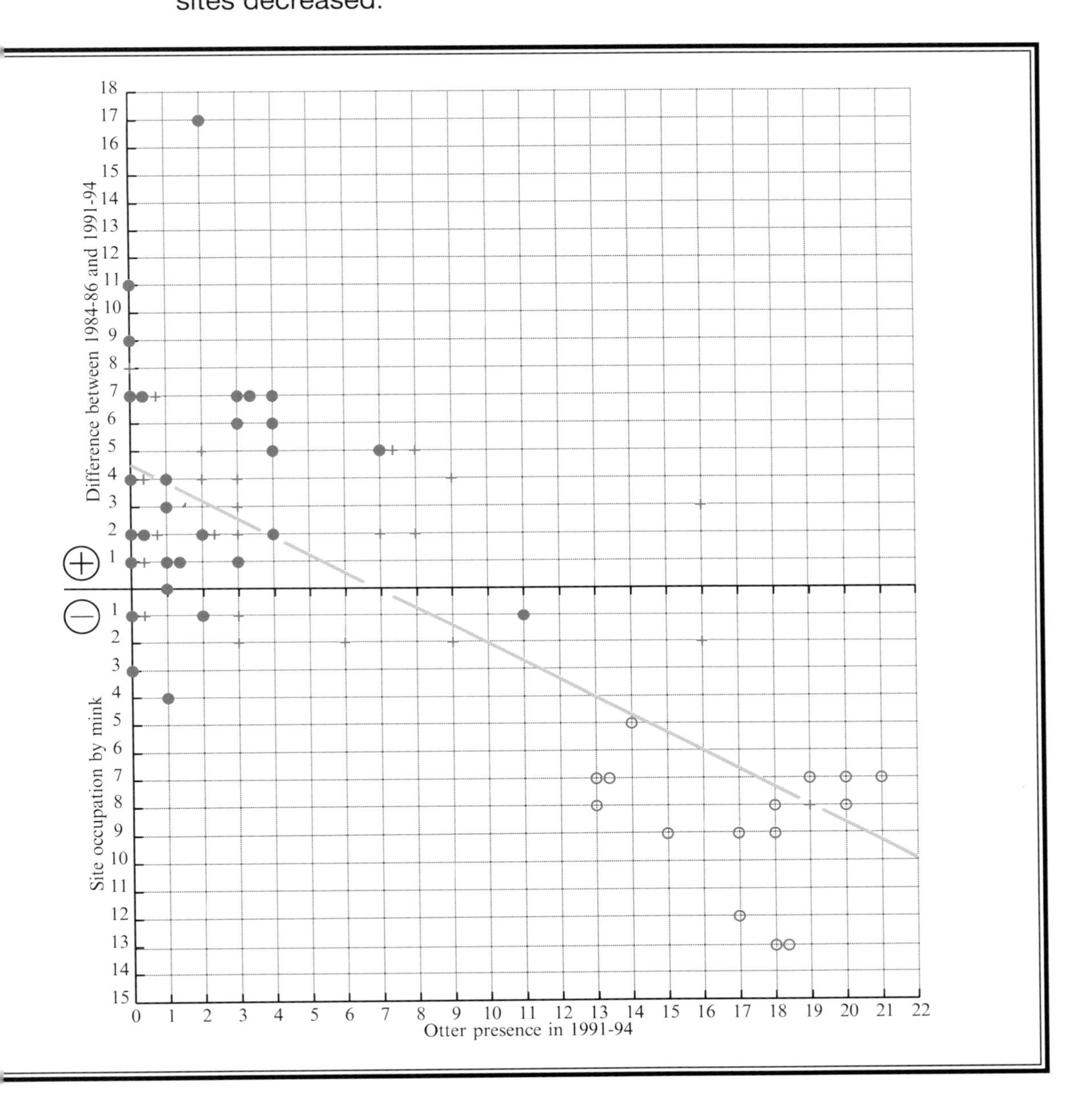

›resence in 1991-94 (x) (in terms of occupied sites in the same 25-site groups), then the data 'orm a decreasing linear relationship (as x increases, y decreases). This relationship is highly ›ignificant ($r = -0.7493$; 66 df; $p<0.001$; equation: $y = 4.587 - 0.660x$).

There is little doubt that the decline in the mink population of the South West, as shown ›y a reduction in occupied sites, is a direct effect of the increasing otter population. This :onsideration is supported by the information that the rate of loss of mink-occupied sites was ıt its highest in those areas where occupied otter sites were at their highest densities 'see Section 6.3.2.2d).

6.3.5.2 The possibility of resource competition

There were strong indications from the surveys of the South West Region that areas with the highest density of mink sites also had the highest density of otter sites. This indication was present separately in the data of both the 1984-86 and 1991-94 surveys of this Region (see Section 6.3.2.2c and Figure 25). However, these data could not be combined to provide sufficient information for a significant correlation. The possibility of correlation, which has important connotations, was further examined by linking the data from surveys of two Regions with long-established mink populations and expanding otter populations (South West and Severn-Trent) and analysing them using a different form of combination. Thus, if mink presence in 1984-86, before their decline, is plotted against otter presence in those same areas seven years later, when the increasing otter population has been able to spread into most suitable sites, it is found that they form a linear relationship (Figure 33). This positive correlation is highly significant ($r = +0.8644$; 37 df; $p<0.001$; equation: $y = 7.987 + 0.538x$, where y = number of occupied mink sites per 25-site group in 1984-86 and x = number of occupied otter sites in the same 25-site groups in 1991-94). There is little doubt that the otters are indeed occupying at their highest density those areas which formerly supported the highest density of mink-occupied sites. As noted in Section 6.3.2.2c, this suggests that the habitat features selected by one are also ideal for the other. The corollary is that competition between the two species for food and den sites could be greater than expected from previous work (see Section 6.3.5.4). Indeed, the results of some recent work carried out on the Outer Hebridean islands (Clode & Macdonald, 1995) would support this conclusion.

6.3.5.3 Avoidance of otters by mink

The site histories of the three Regions show that the expanding otter populations move into sites whether already occupied by mink or not (Sections 6.3.2.3, 6.3.3.2 and 6.3.4.2). There is also a correlation showing that otters select the same areas for highest density as did the mink before them (Section 6.3.5.2). However, when there are many 'empty' sites (ie. without mink or otter in 1984-86), as in Severn-Trent and Anglian Regions, then the expanding mink populations appear to select areas separate to those being occupied by the expanding otter populations. This can be seen by the very few jointly occupied sites within such originally 'empty' areas when re-examined in 1991-94 (see Figure 34). This again suggests the possibilities of either resource competition or antagonistic interaction.

6.3.5.4 Reduction or elimination of mink populations

Presumably, a lack of territorial signs is as indicative of absence or very low density of mink as it is of otters. If it is assumed, as it would appear, that the progressive reduction of mink signs, and so of mink, is directly related to the progressive increase in those of the otter, then how could such a relatively sudden and severe reduction in mink-occupied sites (49.81% in seven years in the south-west of England) be brought about?

There are two possibilities for potential resource competition between the two species; these are for food and for denning sites. Previous research on the diets of mink and otter has suggested that the level of competition for food may be low in Britain. Thus, Wise *et al* (1981) estimated a dietary overlap of only 40% between the two species on a eutrophic lake and a moorland river in south-west England. The otter is a fishing specialist, taking a diet largely of fish all year round. Mink take birds, mammals and fish, and fish predation is seasonal, occurring largely in autumn and winter when fish are easier to catch in cold water. In summer months, when fish are harder to catch, mink prey heavily upon rabbits. In Sweden, Erlinge (1972) estimated the dietary overlap to range from 50% in summer to 70% in winter. Ice cover in Sweden may last for three to five months and reduces the chances of fishing when fish are the main diet of both predators. The likelihood of competition occurring then is much stronger and mink may be limited by it.

igure 33. The relationship between the presence of mink in 1984-86 *(y)* and the presence of otters at the same sites in 1991-94 *(x)*. Each point represents the number of sites occupied by mink and otters within a group of 25 survey sites. A total of 39 full groups was obtained from the survey data for the two English Regions with long established mink and otter populations, the Severn-Trent (+) and South West (O). The calculated regression line is also shown. Both mink and otters are most frequent in the same areas.

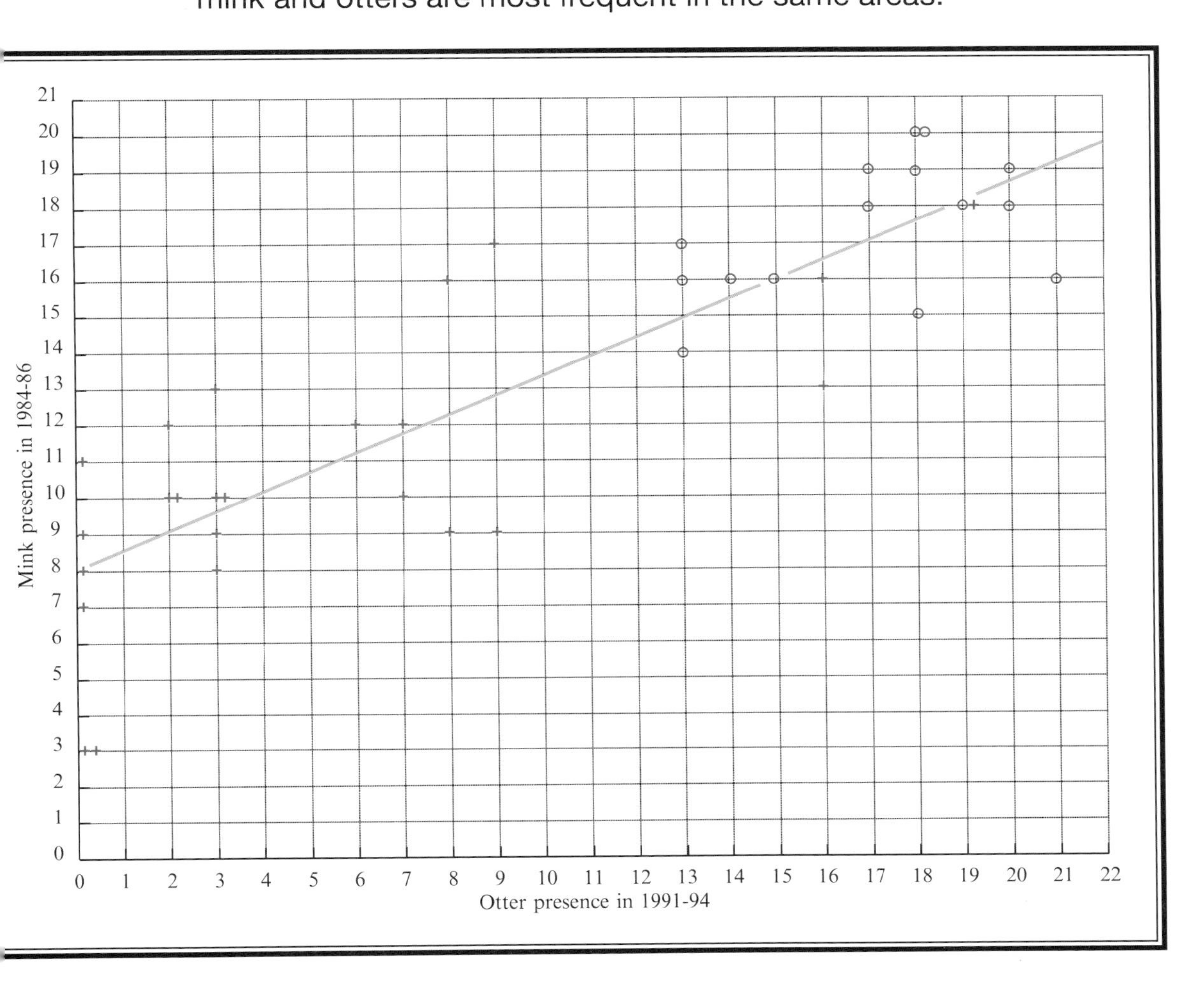

With regard to denning sites, the radio-tracking of otters (Green *et al*, 1984; Jefferies *t al*, 1986) in Britain has shown that they can find and use a large number (eg. 37 different olt sites by one male) of very varied sites, above and below ground, for resting during the ay. Also, the large distances over which otters range (eg. one male had a known home ange of 39.1km of waterway) should allow a large number of potential holt sites to be xamined, particularly as adjacent woodland and drainage ditches are visited as well as mmediately riparian sites. So it would seem unlikely that availability of holt or den sites vould be limiting and that this would cause competition between the two species, especially s they have been known to cohabit the same holt site (Green *et al*, 1986).

Consequently, with regard to competition for resources, Dunstone (1993) has noted hat "there is little reason to believe that feral populations of mink should not successfully o-exist with the otter" in Britain. Indeed, it seems unlikely that increasing competition for esources could have been the sole cause of a 49.81% loss of mink-occupied sites in only even years, taking into account the wide range of prey taken by the mink, the still small umber of otters and the fact that winters have not been more severe since 1984-86 than hey were in the previous 20 years when mink were increasing. However, the conclusion hat both species select and tend to live at highest densities in the same areas and habitats uggests that some resource competition may well occur and may be higher than earlier

Figure 34. The expansion of otters and mink into sites which were 'empty' of both species in 1984-86. This is shown for (a) Anglian and (b) Severn-Trent Regions, together with the combined results (c). There are few jointly occupied sites.

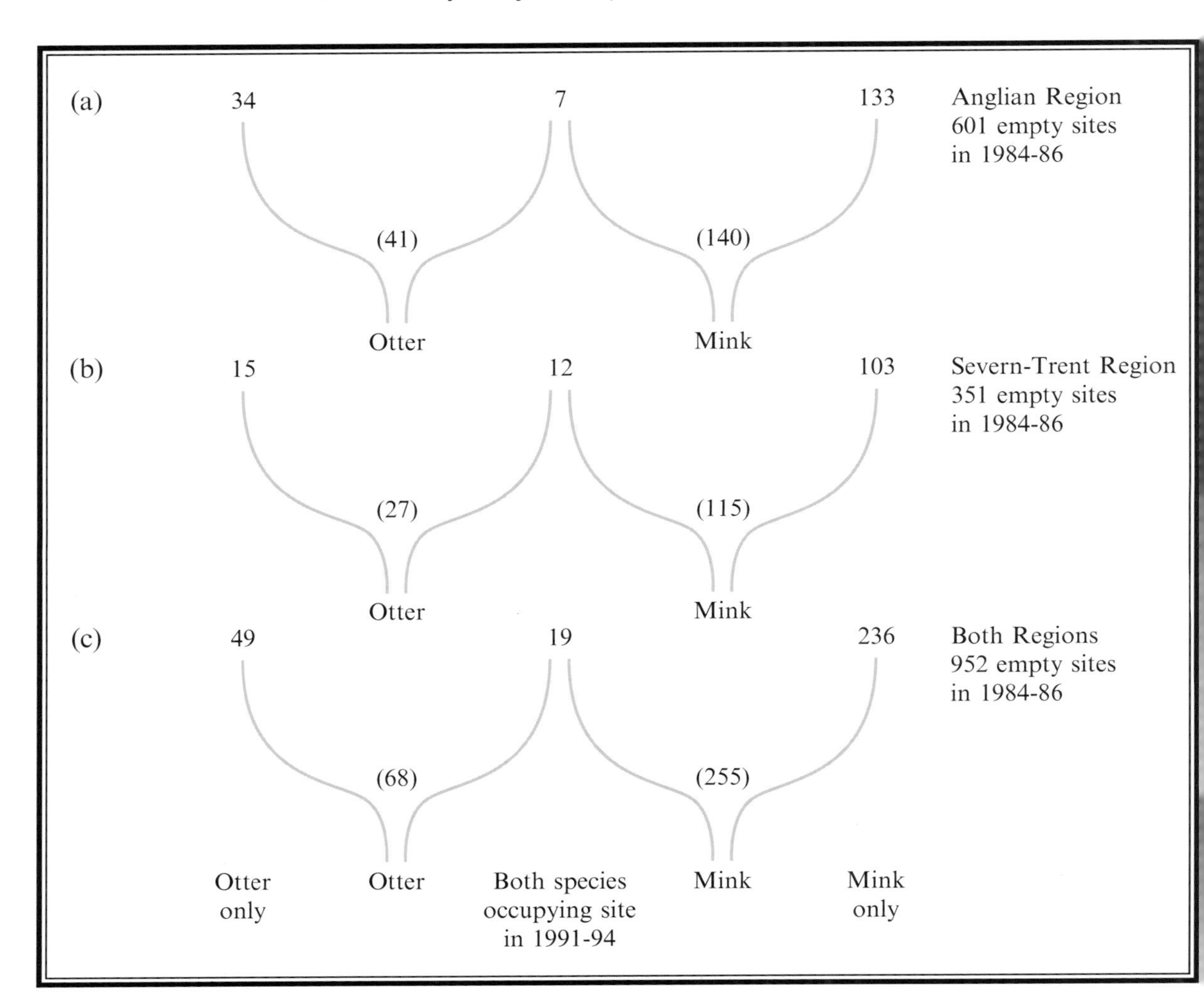

workers have supposed. It may be the basis for any interspecies antagonism as discussed in the next paragraph.

Thus, Jefferies (1992b) considered that a much more likely cause of any population declines in the mink would seem to be intraguild interference competition and predation. This is manifested in the marked antagonism seen between two carnivores of different sizes which live in the same habitats. It is seen most obviously between the domestic dog and cat, but has been noted between stoat and weasel (King, 1989), red fox and stoat (Mulder, 1990), red fox and pine marten (Hurrell, 1963, 1968; Strachan *et al*, in press), wolf and red fox (Macdonald, 1987), and even between carnivores as large as tiger and leopard (Thapar, 1986, 1989) and jaguar and puma (Hudson, 1892). This antagonism can take the form of the larger species directly interfering with the smaller species causing the latter to break off hunting and to hide (King, 1989), so becoming less efficient at feeding. Interference may also be indirect, taking the form of elimination of the smaller species' communication system. Thus, Wickens (1991) has found that otters interfere with mink scent-marking/ defecation sites by 'smearing' their spraint over the piles of mink scats. After this treatment the mink no longer uses those sites.

Again, however, interference to hunting and breeding alone appears unlikely to have caused such a severe crash in the mink population of the south-west of England in only seven years. Seven years is within the lifespan capability of individual mink (eight years: Dunstone, 1993) and would not encompass many generations. Also, if this milder form of ntraguild competition was the cause of the disappearance of mink from some areas, it might e expected that this disappearance would be due to displacement. That is, the mink might isperse to nearby suboptimal areas not yet inhabited by otters. That this is not the case can e seen in the south-west, where the otters colonise previously mink-only areas prior to the isappearance of the latter species, whilst adjacent presumably suboptimal areas without nink or otter remain empty (see Section 6.3.2.3). It is unlikely that mink would move all the vay over from the south-west to East Anglia (see Section 6.3.4).

This leaves the most severe form of intraguild interference as one of the most likely auses of the mink population crash in areas colonised by otters, ie. lethal fighting and redation. This is known among all of the pairs of large and small predators listed in the aragraph above, as well as with other carnivores.

Otters are known to kill American mink in Ireland (Smal, 1991). Also, Grigor'ev and gorov (1969) wrote that the otter is a "serious enemy" of the mink in Russia, where xamination of 880 otter spraints showed the remains of six mink. Novikov (1956) too eported that otters "vigorously hunt" mink in Russia. The lack of an increase in reported nink bodies in the south-west suggests the possibility that some of them may have been aten by the otters, though, of course, they could easily be missed. Otters have eaten smaller nustelids occasionally, as is shown by the mink hair in the Russian spraints and the weasel air in a spraint from Slapton Ley, Devon (Chanin, 1985). Also otter spraints consisting nainly of mink hair may be confused with mink scats.

It may seem at first sight unlikely that the presence of one predator should have such a narked effect on the population of a smaller predator. This is particularly so when it is onsidered that, despite the above known intraguild lethal fighting between native predators f disparate size, such as stoat and weasel, polecat and stoat, and fox and stoat, these species ave lived together for centuries in most parts of Europe without the elimination of the maller animal. Presumably they have achieved stable relationships. However, if the opulation of the larger species is greatly reduced by a factor not operating on the smaller, r indeed if the pressure is removed and the larger species returns after an absence, then the mbalance produced by the sudden change may have considerable effects on the smaller pecies. Thus, when the stoat population declined in Britain following myxomatosis in the abbit (Jefferies & Pendlebury, 1968; King, 1989), it was found that there was a sudden ncrease in the weasel population. Also, the stoats disappeared from an area of dunes in the Netherlands following the return of the fox several centuries after its removal by ersecution. Foxes were observed chasing and killing stoats and polecats in this area Mulder, 1990). In a similar way it can be seen that the feral American mink managed to get foothold in many areas of Ireland, despite the presence of an established otter population. However, the return of the otter to many parts of England probably had a much greater ffect on the mink population which had built up in its absence than if this otter population had been there all the time. This suggests the possibility that there may be some degree of ecovery of the mink population of England with the passage of time, as its relationship with the larger otter population stabilises.

The speed at which the mink population of the south-west was reduced shows the rate t which a new factor, when acting directly on adult mortality, can eliminate a low-density redator from large areas very rapidly. This may be indicative of the rate at which the otter could have been lost in Britain in the late 1950s, when the survival of breeding adults was suddenly reduced by dieldrin poisoning (Chanin & Jefferies, 1978; see Section 7.1.4).

7. DISCUSSION

7.1 The decline and recovery of the otter population of southern Britain

With the completion of the third England otter survey in 18 years, the additional data allow further, more informed, examination of the pattern of the decline and recovery of the species in southern Britain and so a further look at how this pattern correlates with the proposed causes (Chanin & Jefferies, 1978; Jefferies, 1989b).

7.1.1 Distribution and relative density immediately prior to the decline

It must not be thought that previous to the 'crash' of the otter population in the late 1950s (see Section 7.1.2) the species was evenly and densely distributed over the whole of England and Wales. There would have been areas where the population was below a river's holding capacity owing to persecution because of fisheries protection or organised hunting and other areas where it was low owing to reduced fish biomass through industrial or organic pollution. Yet other areas could have held high-density populations, eg. the Devon rivers, the Somerset Levels and parts of Norfolk and Suffolk, with their dense riparian cover, marshes and fens, a lack of industrial pollution and a high biomass of coarse fish, particularly eels (Jefferies *et al*, 1988). However, there were no structured surveys in those days to provide us with a picture of the otter distribution within these areas. The only comparative area data are the subjective reports of status collected by Stephens (1957) and the records of hunting success in the diaries of the packs of otter hounds.

Marie Stephens carried out the first field study of the biology of the otter in Britain over 1952-54 and also collected much useful information. She requested information on the status of the otter from the 34 River Boards then in existence in England and Wales. Replies were to be "educated estimates" by the area biologists based on an amalgam of numbers of sightings, trappings, casualties and hunt and bailiffs' reports. Of the 33 replies, 30% said that the otter was "very numerous", 36% said "numerous", 27% had "small populations" and only 6% (two) said they were "scarce". None said they were absent. The distribution of these four categories of status is shown mapped in Figure 35. It can be seen that the areas with the most numerous otters were said to be in central and west Wales, Devon and Cornwall, eastern Norfolk and Suffolk, Cumbria, Yorkshire and Scotland. The western half of northern England (Cumbria) had higher densities than the eastern half (Northumberland, Wear and Tees areas) north of Yorkshire. There were said to be only small populations in south-east England (Kent, Sussex) and south Wales.

The only numerical check of the veracity of this map is provided by an analysis of the hunting success of the 13 packs of otter hounds operating at this time (Chanin & Jefferies, 1978). The mean number of otters found per 100 days hunting is shown in Table 6 for the period 1950-55. Unfortunately, the hunt territories did not cover the whole surface of England and Wales (ie. much of the Midlands and Yorkshire were not actively hunted) and each is spread over several River Board areas (see Figure 36). However, it can be seen that the highest success rates, and so possibly the highest otter densities, were in southern Scotland, central and western Wales, Devon, Somerset and southern England. The lowest success was in Cornwall and south-east England (Kent and Sussex). In northern England success was relatively low, but slightly higher in the west than in the east. Thus, there is some degree of agreement between the two sources of information for the pre-decline period, Stephens (1957) and the hunt data. However, it would seem that the East Suffolk and Norfolk River Board may have overestimated its otter population in 1952-54, as, unexpectedly, the more objective hunting success record shows it to have been not particularly high (see this Section and Figure 36). One general pattern which emerges from both forms of survey is that even prior to the population crash otters appear to have been fewer in the south and east than in the north and west.

Figure 35. The results of the first otter survey carried out over 1952-54 by Stephens (1957). This was based on the subjective reports of status published by the 34 River Boards then in existence in England and Wales. Replies were based on "educated estimates" by the area biologists and were divided by Stephens into the four categories which are shown mapped below. Compare with the more objective data (see Figure 36) derived from the hunting success of packs of otter hounds over the same period and collected by Chanin and Jefferies (1978).

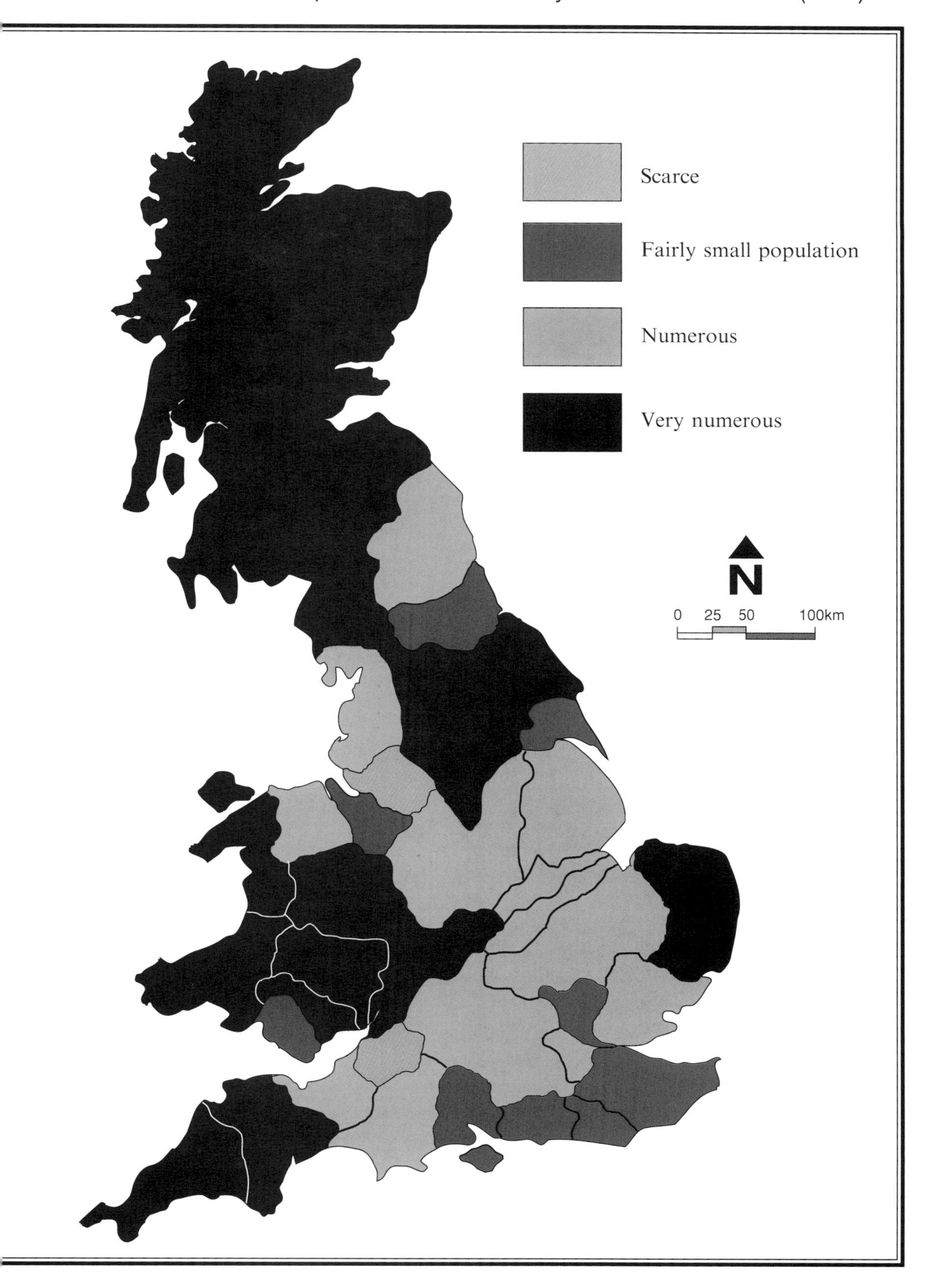

7.1.2 The decline of the otter population

Shortly after the Stephens' report was published, the first concerns were being expressed by the otter hunters. Lloyd (1962) reported that hunters were finding very few otters. One pack found only four in 20 days hunting. An initial analysis of hunting success (Anonymous, 1969) suggested that it had decreased between 1957 and 1967. It was decided by the Joint Otter Group, following a suggestion by Lord Cranbrook (1977), that a more detailed examination should be made. Consequently, Chanin and Jefferies (1978) examined the hunting records of the main 11 packs active before and throughout the decline period. Records were analysed in terms of 'finds' (otters seen and hunted) per 100 days hunting over the period 1950-76. The results (see Table 6) indicated that:

(i) All seven southern hunts showed marked and synchronised declines in hunting success.

(ii) Two of the three northern hunts showed less marked declines which again were synchronised to those of the south. The third, Border Counties O.H., changed its main hunting area in 1961 following poor results in 1959, so prolonging success and obscuring the original synchronised decline until that area too became depleted of otters (Jefferies, 1989b).

(iii) The sole Scottish pack, the Dumfriesshire O.H., showed a small but significant ($p<0.001$) decline like those in England and Wales.

(iv) The start of these marked declines (ie. when success first dropped below the mean rate for 1950-55 and did not rise above it again) was determined at 1956 (1), 1957 (2), 1958 (3) and 1960 (1) for the seven southern hunts (mean: 1957.7; Table 6).

(v) The mean percentage decrease in hunting success (in terms of finds/100 days hunting) from 1950-55 to 1966-71 was 51.32% in the southern and eastern hunts and 42.58% in the northern and western hunts (a 20.5% difference between north/west and south/east) (see Table 6). The difference is even greater (23.9%) if the Northern Counties O.H. (hunting on the eastern side of northern England) is placed within the northern and western group (south/east: 53.28% decrease; north/west: 42.99% decrease).

(vi) The combined results for all packs showed a rapid decline in hunting success from 1957 to 1963 followed by a period up to 1975 when success decreased more slowly (see Figure 39). This source of information ended in 1977 with legal protection in 1978.

Chanin and Jefferies (1978) concluded that there had been a severe decline (in fact a 'crash') in the otter population of England, Wales and southern Scotland, starting suddenly in 1957-58 and occurring simultaneously over the whole country. It appeared to have been most severe in the south and east. It was considered then that the main cause must have been a single new factor, suddenly introduced countrywide but to a higher degree in the south-east. A gradually increasing adverse factor would be unlikely to produce such a sudden and severe effect.

Between 1960 and 1974 most of the otters from the Midlands, south, south-east and northern England south of Lancaster were lost (see the two 10km square 'spot' maps for 1960-74 and 1975 onwards of Arnold, 1993; Jefferies, 1991b). These may have been the areas where the populations were at a lower density than elsewhere prior to the crash (see Figures 35 and 36), possibly because of earlier industrial and organic pollution as well as persecution (Jefferies, 1989b). From 1975, populations remained in any numbers only in Wales and the borders, the south-west and northernmost counties of England and in East Anglia (Lenton *et al*, 1980; Arnold, 1993). The populations remaining in most of the eastern counties were fragmented into isolated groups, many of which were too small to be viable

Figure 36. The hunting territories of the 13 active packs of otter hounds in the period just before the otter population crash. The mean hunting success in terms of finds per 100 days hunting over the period 1950-55 has been mapped using five categories (Chanin & Jefferies, 1978). These provide a more accurate indication of the relative status of the otter within the 13 areas hunted than the "educated estimate" of relative numbers provided by the River Boards (see Figure 35). Areas between the territories were not regularly hunted by organised packs.

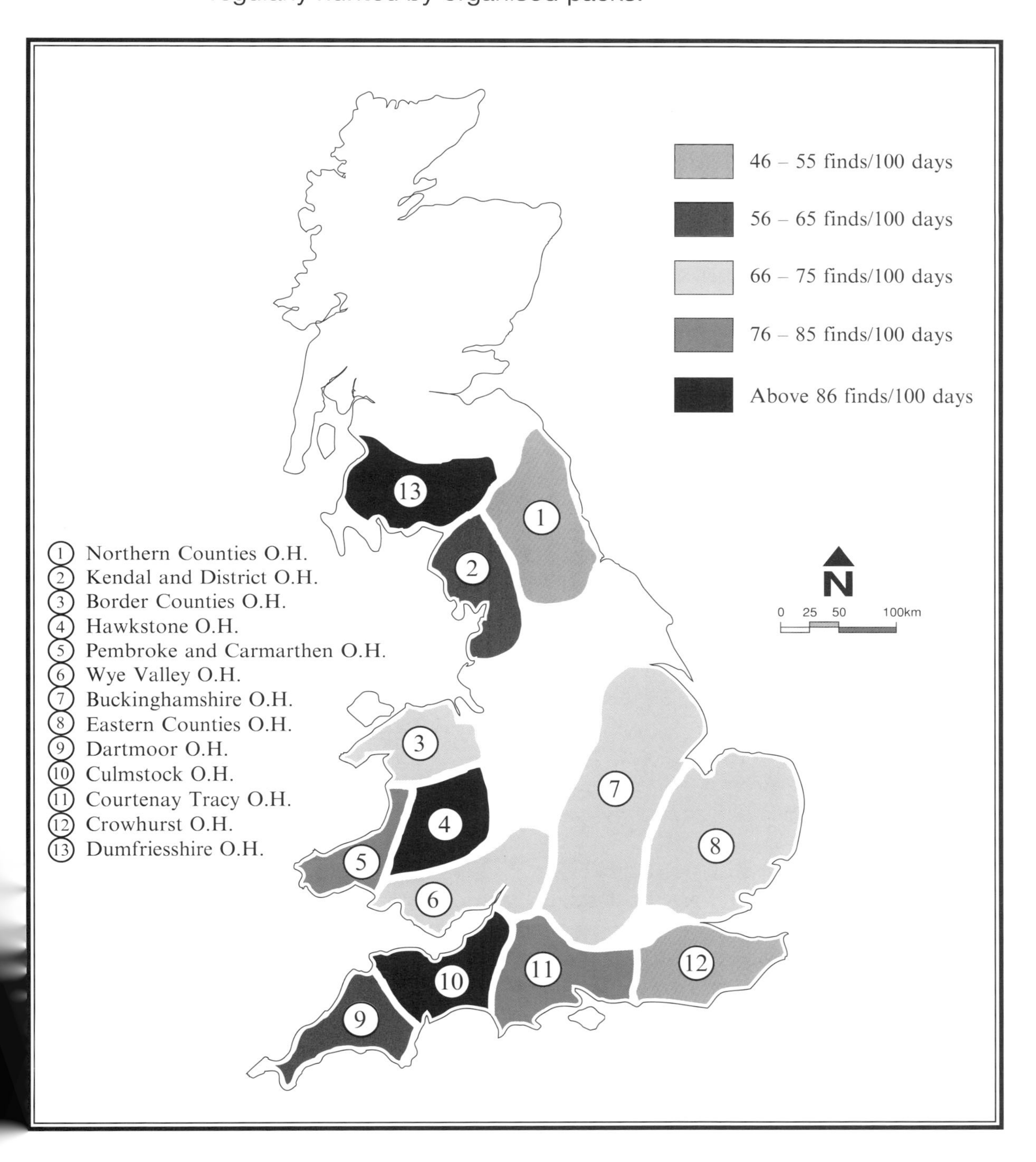

(Jefferies, 1989a). That of East Anglia continued to decline until that too was fragmented and by the early 1980s its survival seemed doubtful (see Section 7.1.3.2.7). The pattern and timing of the decline is important in determining its likely cause (Jefferies & Mitchell-Jones, 1993).

Table 6. The records of hunting success for the eastern and western packs of otter hounds for 1950-55 and 1966-71 showing the percentage decline, the date it started, the number of otters killed in 1950-55 and the number killed as a percentage of those found for each hunt territory (data and analysis from Chanin & Jefferies, 1978).

Several means are given for each of the two geographical groups, depending on the number of hunts used for their calculation. This is because there were insufficient data for the Wye Valley and the Crowhurst Otter Hounds. Also, the records of the Border Counties O.H. suggested at first that they had suffered no decline in hunting success (Chanin & Jefferies, 1978) but later analysis (Jefferies, 1989b) showed that this had decreased by 1959 but the small decrease was obscured by a change in kennels and hunting venues.

Note † : The dates of the start of the decline were derived by Chanin and Jefferies (1978) by taking the year when the success rate first dropped below the mean rate for 1950-55 and did not rise above it again. This is difficult to detect with a very gradual decline, but all hunts (except Border Counties) showed a decrease from the mean success of 1950-55 to that of 1970-71. Thus, the date of the start of the decline is likely to be the same in all cases (ie. soon after 1956).

Area in Figure 41 & Table 10	Otter Hunt	Otters found per 100 days hunting				No. of otters killed between 1950 and 1955	No. killed as a percentage of otters found	Date of start of decline
		1950-55	1966-71	Difference	% decrease on 1950-55			
EASTERN HUNTS								
Southern	Courtenay Tracy	76	36	40	52.63%	165	58%	1960
E. Anglia & E. Mid.	Eastern Counties	72	38	34	47.22	125	50	1957
E. Midlands	Buckingham	70	28	42	60.00	154	58	1956
North-East	Northern Counties	55	30	25	45.45	50	38	†
South-East	Crowhurst	47	–	–	–	–	–	–
Means for Eastern Hunts		**(a) 68.25 (4)** **(b) 64.00 (5)**	**(a) 33.00 (4)**	**(a) 35.25 (4)**	**(a) 51.32 (4)**	**(a) 123.50 (4)**	**(a) 51.00 (4)**	**1957.7**

Table 6 continued

Area in Figure 41 & Table 10	Otter Hunt	Otters found per 100 days hunting				No. of otters killed between 1950 and 1955	No. killed as a percentage of otters found	Date of start of decline
		1950-55	1966-71	Difference	% decrease on 1950-55			
WESTERN HUNTS								
Western England								
South-West	Culmstock	88	44	44	50.00%	84	35%	1957
South-West	Dartmoor	62	23	39	62.90	67	36	1958
North-West	Kendal & District	59	40	19	32.20	57	40	†
Wales & West Midlands								
Wales & West Mid.	Hawkstone	121	66	55	45.45%	256	62%	1958
Wales	Pembroke & Carm.	78	37	41	52.56	90	57	1958
Wales & West Mid.	Border Counties	71	71	see legend	small	85	46	†
Wales & West Mid.	Wye Valley	68	–	–	–	–	–	–
Southern Scotland								
Scotland	Dumfriesshire	89	78	11	12.36%	79	48%	†
Means for Western Hunts		**(c) 82.83 (6)** **(d) 81.14 (7)** **(e) 79.50 (8)**	**(c) 48.00 (6)** **(d) 51.29 (7)**	**(c) 34.83 (6)**	**(c) 42.58 (6)**	**(c) 105.50 (6)** **(d) 102.57 (7)**	**(c) 46.33 (6)** **(d) 46.29 (7)**	**1957.7**

7.1.3 The suggested main cause or causes of the decline and their east/west biases

Chanin and Jefferies (1978) examined all the possible causes and threats which could have initiated such a decline (see Section 7.3). The only new factor which fitted in time and area of greatest effect was that of the introduction of the persistent organochlorine insecticides, dieldrin and aldrin, as cereal seed dressings and sheep-dips in 1955 and 1956 (Moore, 1965). These new compounds were also lethal to vertebrates and could produce detrimental sublethal effects. They rapidly polluted waterways and became concentrated in fish (Prestt, 1970) and consequently in fish-feeding animals such as herons (Cooke *et al*, 1982) and otters (Jefferies *et al*, 1974). The populations of avian predators, such as the peregrine falcon and sparrowhawk, crashed, like that of the otter, soon after 1956. (This subject of pollution by organochlorines is covered in more detail in Section 7.3.1.1.) Chanin and Jefferies (1978) concluded that the immediate cause of the otter's decline was high mortality among breeding adults and reduced breeding success in the survivors owing to poisoning from dieldrin pollution of the waterways. As noted by Jefferies (1989b), "a species like the otter, living at low density in the environment which drained, channelled and concentrated the organochlorines from the land, with its main diet contaminated, is likely to have been seriously affected".

Since the 1978 paper, the above initial conclusions of Chanin and Jefferies have been amended. Chanin and Jefferies (1978) originally concluded that, as there was no evidence that the various forms of persecution had increased significantly during the mid-1950s, this factor was unlikely to have precipitated the decline. However, according to Jefferies (1989b), the numbers of otters killed through direct and indirect persecution through game preservation activities, as well as through organised hunting, could have been an important contributory factor to the eventual population crash although not by themselves precipitating it. He postulated that the effects of pollution (see above and Section 7.3.1.1) may not have been as severe if the otter population had been large and normally structured at that time. He suggested that, as it was, the population of south-eastern England was critically small and abnormally structured towards young otters because of the killing of large numbers of mainly adult, breeding-aged animals. This made the south-eastern population more sensitive to further stress. Thus, there could have been two main causes of the population decline, first persecution through game preservation and hunting and then severe pollution by agro-chemicals.

Both of the above factors would have had a very marked east/west bias and to a lesser extent a south/north bias. This can be seen from variation in the intensity of game preservation, which is highest in south-eastern counties, and from the polarisation of British agriculture into western livestock and eastern arable. The geographical biases of both of these industries are covered in more detail elsewhere (see Sections 7.1.3.1.5 and 7.1.3.2.1). Thus, in order to confirm the link between the sudden crash of the otter population and these two main causes, we need to examine the data closely for indications of any east/west biases in the pre-decline status of the otter, in the severity of the decline and in the timing of the recovery. The last should correlate with the various bans obtained regarding the cessation of usage of the organochlorine insecticides (Moore, 1965, 1987; Sheail, 1985). These too have different dates at which they became effective for their main eastern and western applications.

7.1.3.1 Persecution

7.1.3.1.1. Persecution of the otter prior to and during the population decline

Chanin and Jefferies (1978) concluded that, apart from those killed by organised hunting with hounds, it was impossible to assess the numbers of otters killed deliberately by man for fisheries and game preservation or for their pelts, or indeed how many were killed indirectly because of gamekeeping activities. However, as this factor had now assumed greater importance, it was decided to re-examine any available data on mortality caused by direct and indirect persecution by man. This was in order to see whether there were any area differences, any indications of the scale and level of importance of this mortality prior to and after 1957 and whether it could have seriously affected the population's size and structure at that time.

7.1.3.1.2 The contribution of the trade in furs

According to Havins (1981), native otter pelts were valued at around three shillings each in the fifteenth century, but by the end of the eighteenth century most trappers in Britain had been driven out of business by the furs flooding in from North America through the Hudson Bay Company. A revival in a native fur trade is reported to have occurred during the 1940s and 1950s, when a British otter pelt could fetch £2 (Stephens, 1957; Havins, 1981). 'The Yearbook of Otter Hunting' referred to by Chanin and Jefferies (1978) suggested that many otters were being killed specifically for the value of their pelts and that this was one cause of the decline. Havins (1981) reported that by the end of 1977 British otter pelts were fetching £25. However, when questioned on this matter, the British Fur Trade Association informed Lord Cranbrook that "the number (of pelts) is so small that the type has no real commercial value for our trade" (Cranbrook *et al*, 1976). One leading dealer had bought only five pelts in one year (1975) and at least two of these seemed to be road casualties (O'Connor *et al*, 1977). These pelts were probably derived from a small 'mixed bag' of otters shot or trapped for game preservation or picked up dead, rather than being killed specifically for their pelt value.

Thus, the taking of otters for their pelts, at least by the 1970s, appears to have been virtually non-existent. It would have been difficult for a person to make a living from trapping otters for their pelt value in Britain and it would seem more likely that, with the large numbers taken by game preservation interests (see Section 7.1.3.1.3), this incidental supply would have amply satisfied what market there was. On the other hand, its pelt value would have provided an extra incentive to take any otter found to be present on a 'keeper's beat. The increased pelt value and market in the 1950s may have been one reason why so few otter bodies were found by naturalists during the period of presumed high mortality after 1957. The trade in native otter pelts (even old ones without a licence) was made illegal in 1982 (under the Wildlife and Countryside Act 1981).

7.1.3.1.3 The contribution of direct and indirect persecution by game preservation interests

Otters would have been killed by gamekeepers and water bailiffs directly and intentionally in order to protect shooting and sport fisheries interests (Buxton, 1949) (and indeed to collect money for the pelts; see Section 7.1.3.1.2). This could have been by snaring, trapping or shooting. In addition, there may have been, and may still be, an important indirect effect of game preservation in that there would be a large amount of unspecific trapping and snaring of predators in areas where preservation is at high intensity. For example, otters are known to use rabbit burrows for day and night-time shelters (Green *et al*, 1984; Jefferies *et al*, 1986) and these may have traps set within them to catch rabbits, stoats or mink. Otters could then be affected incidentally.

The numbers of otters killed by the few packs of otter hounds are known in detail because of their diligent recording. These seem very large indeed (eg. the 307 killed by the Hawkstone O.H. between 1950 and 1956) and almost certainly had a detrimental effect on the local populations of some areas (see Section 7.1.3.1.4). However, the 'hidden' numbers of otters killed by the 'army' of gamekeepers and water bailiffs must have been very much larger, as they affected the numbers to be found by the otter hunters and recorders themselves (see this Section and Section 7.1.3.1.5). Compared to the 13 to 23 packs of otter hounds operating at any one time in southern Britain this century, there were 23,056 gamekeepers in England, Scotland and Wales when at their peak in 1911 (Tapper, 1992). Although the numbers of 'keepers were reduced after the 1914-18 World War because of the changed economy, there were still 14,500 in 1930, 9,500 in 1939 and 5,000 in 1950 (Tapper, 1992). In addition to their large numbers, they were trapping predators every day of the year and the whole area under their control was covered every day. The pressure must have been much greater than that occasioned by just hunting one or two rivers at a time for a small number of days in the summer (eg. the mean annual count of days hunted was 56.8 for the Eastern Counties O.H. in the period 1899-1977).

Looking back to the nineteenth century, both Howes (1976) and Jefferies (1989b) have attributed most of that century's otter population decline to the killing of otters for fisheries and game preservation. There are many descriptions in the Victoria County Histories of otter persecution by 'keepers occurring in most counties. The otter population had declined so severely from this cause that at least one Master of Otter Hounds in the south of England (Courtenay Tracy O.H.) wanted to resign in 1895 and 1896 owing to the scarcity of otters for hunting ('Polestar', 1983). The scale of 'vermin control' by 'keepers in Victorian and Edwardian England can be gauged by the 17,796 avian and mammalian predators killed in just one year (1903) on a Suffolk sporting estate (Elveden) with an area of 10,117ha (Turner, 1954). This year's total included 977 stoats and two otters. The otters were killed despite the estate being managed for the shooting of pheasants and, to a lesser extent, grey partridges. As many as 24,619 pheasants were shot in one year (1912/13).

The effect of removing the cause of this very large direct and indirect otter persecution was shown by Jefferies (1989b), who analysed hunting success before and after the 1914-18 World War. During the war many 'keepers went into the army and otter hunting with hounds ceased or was greatly reduced. One example shown was that of the numbers of otters found by the Border Counties O.H. for the five years before (1909-1913) and after (1919-1923) the war. These increased significantly ($p<0.001$) from 64.2 ± 4.8 to 86.7 ± 4.5 finds per 100 days hunting. This may indicate a 35% increase in otters in just four years of reduced persecution.

The level of 'keepering was reduced after the First World War because of economic reasons (see above). The effect of removal of this part of the total game preservation persecution can be seen in the period between the two World Wars. The numbers of otters killed by packs of hounds showed a consequential increase. 23 packs were able to kill 434 otters in England and Wales in one year in 1933 (Jefferies, 1989b). This is a mean of 18.87 otters killed per pack per year.

Finally, the part played by direct and indirect persecution by gamekeepers and water bailiffs in producing a high 'background' mortality among English otters was still considerable in the 1950s (see Section 7.1.3.1.5). Also, it did not cease with the population crash in 1957. Gamekeepers and water bailiffs confirmed to Woodroffe (1994) that otters were shot in Bilsdale in Yorkshire in the 1960s and as many as 13 were known to have been trapped in one locality in the north-west of the county over one 12-week period in 1964 (J. Paisley, in Woodroffe, 1994). This was whilst the population was still declining. The bodies of otters which have been killed by snares have been received even within the last ten years during The Vincent Wildlife Trust survey of causes of otter mortality (Jefferies & Hanson, 1987, 1988b).

7.1.3.1.4 The contribution of the otter hunts

The otter hunts themselves made a definite overall contribution to the weakened status of the otter population of southern Britain by the time of the early 1950s, just prior to the crash. Although this was not perceived at the time, critically large numbers of breeding-age adult animals were killed. Unfortunately, most of this pressure and mortality was concentrated on the eastern half of England. Here the otter population was already stressed by several other adverse factors, notably that of persecution by gamekeepers, which again was at its highest in eastern England. It was also to be the area most severely affected by the organochlorine insecticides. The effects of otter hunting before the decline and their area concentration can be appreciated from the following examples and analyses of available data from hunt records.

(i) As noted in Section 7.1.3.1.3, the reduction in 'keepering after the First World War probably enabled the sport of otter hunting with packs of hounds to develop to its highest

Table 7. The numbers of otters found and killed, the finds per 100 days hunting and the total numbers of days spent hunting for each of the four eastern hunts and seven western hunts with complete records available for the period 1950-55. This does not include the Crowhurst O.H. and the Wye Valley O.H. Note that the eastern hunts spent 30.09% more days hunting in this period than did the western hunts. The two different means for the 'kill as a percentage of otters found' in column (2) are derived from (i) the means of the individual hunt results given above, and (ii) the use of the totals for number killed and number found (columns (1) and (3)). The data in columns (1), (2) and (4) have been obtained from Chanin and Jefferies (1978) and those in columns (3), (5) and (6) have been calculated from them.

Hunt	(1) No. of otters killed 1950-55	(2) Kill as %age of otters found	(3) So, no. of otters found 1950-55	(4) Finds per 100 days hunting 1950-55	(5) So, no. of days hunting 1950-55	(6) Mean no. of days hunting per year
EASTERN HUNTS						
Courtenay Tracy	165	58	284	76	374	62.33
Buckingham	154	58	266	70	380	63.33
Eastern Counties	125	50	250	72	347	57.83
Northern Counties	50	38	132	55	240	40.00
Eastern: Total	**494**		**932**		**1,341**	
Mean per hunt	**123.50**	**(i) 51.00% (ii) 53.00%**	**233.00**	**68.25**	**335.25**	**55.87**
WESTERN HUNTS						
Hawkstone	256	62	413	121	341	56.83
Pembroke & Carmarthen	90	57	158	78	203	33.83
Border Counties	85	46	185	71	261	43.50
Culmstock	84	35	240	88	273	45.50
Dumfriesshire	79	48	165	89	185	30.83
Dartmoor	67	36	186	62	300	50.00
Kendal & District	57	40	142	59	241	40.17
Western: Total	**718**		**1,489**		**1,804**	
Mean per hunt	**102.57**	**(i) 46.29% (ii) 48.22%**	**212.71**	**81.14**	**257.71**	**42.95**
Overall: Total	**1,212**		**2,421**		**3,145**	
Mean per hunt	**110.18**	**(i) 48.00% (ii) 50.06%**	**220.09**	**76.45**	**285.91**	**47.65**

intensity. In 1933, 23 packs were hunting in England and Wales and their total kill was 434 otters (or 18.87 kills/hunt/year) (Jefferies, 1989b). A steady total cull of about 200 otters a year continued with a smaller number of packs up to the 1939-45 World War. After this war there were fewer packs, but even so killing rates were again very high and 11 active packs killed 1,212 otters between 1950 and 1955 (18.36 kills/hunt/year), just before the population crash (Chanin & Jefferies, 1978). Regrettably, because the actual situation was unknown at the time, the killing of otters by the hunts continued after the decline had started. Thus, a further 1,065 otters were killed by 11 active packs in the six years between 1958 and 1963 (16.14 kills/hunt/year) (Chanin & Jefferies, 1978). From 1964 a different policy was adopted of not killing the otters whenever possible. This was after the scale of the crash had been realised. Unfortunately again, the Eastern Counties O.H. killed a recorded 140 otters, largely in Norfolk and Suffolk, from 1957 to 1977 when hunting was stopped by the legal protection of the species (Downing, 1988). It seems that this level of killing applied to a population which was already very seriously hit by so many adverse factors (ie. 'keeper persecution, pollution, road mortality and fyke netting; see Sections 7.1.3.1.5, 7.1.3.2, 7.3.3.1 and 7.3.3.2.1) almost certainly made a definite contribution to the severity of the eastern crash. This is understandable if one considers the positive effect on the recovery of the Norfolk and Suffolk otter populations produced by the Otter Trust's insertion of only 37 captive-bred otters into these counties between 1983 and 1993 (see Section 7.2).

(ii) Another indication of the effect of the organised hunts killing the otters in their hunt territories may be provided by the records of the Hawkstone O.H. Whether owing to policy or type of habitat hunted, this pack killed the highest percentage of those otters found of any of the hunts in Britain (62%; Chanin & Jefferies, 1978). As their finding rate was particularly high to start with, this resulted in them killing 307 otters over the years from 1950 to 1956, before the crash, and a further 363 otters from a declining population between 1957 and 1975, when they ceased hunting (Greenwood, 1991). Ninety-one of these deaths occurred after the policy of not killing otters whenever possible had been introduced in 1964 (see above). The finding that the nadir of the decline curve (ie. before recovery started) for this particular hunt territory was not reached until after 1975 (ie. the last one of all the western hunts; see Section 7.1.3.2.6 (a) (iv) and Table 9) adds support for the hypothesis that the killing of otters by the hunts did have a contributory effect to the severity of the crash (Jefferies, 1989b).

(iii) A further indication of the contribution of the otter hunts to the severity of the 1957 crash of the otter population is provided by the following data. Jefferies (1989b) first recorded a significant difference ($p<0.001$) between the northern and southern hunts in the percentage of the otters 'found' which were subsequently killed during the period 1950-55. This was 43.0 ± 2.4% in the north and 50.9 ± 4.2% in the south (data from Chanin & Jefferies, 1978). If we examine this same statistic for eastern and western hunts (as divided in Tables 6 and 8(c)), we find a difference again.

Thus, the eastern hunts (four with data) killed a total of 494 otters (123.5 ± 25.9 otters per hunt) over the period 1950-55. This was a mean of 51.00 ± 4.73% of those 'found'. The western hunts, on the other hand, killed both fewer otters and a lower percentage of those 'found'. Seven western hunts killed a total of 718 otters (102.6 ± 25.9 otters per hunt) over the period 1950-55, which was a mean of only 46.29 ± 3.90% of those 'found'. The difference between the number of otters killed in the east and west is significant ($\chi^2 = 10.12$; 1 df;$p<0.01$). If the extraordinarily large kill of the Hawkstone O.H. (256 otters; 62% of those 'found') is removed from the totals for the western hunts, the difference is even greater (six hunts; 77.0 ± 5.1 otters killed per hunt; 43.67 ± 3.41% of those 'found'), as is the level of significance ($\chi^2 = 54.28$; 1 df; $p<0.001$).

The above east/west difference in the percentage of otters found which were subsequently killed could be brought about either by different regional hunt policies and practices on the matter or because the hounds found otters easier to catch and kill in the slower, deeper waters of eastern rivers compared to the faster, shallower waters of the more upland rivers of the west. The significant north/south difference (with a greater percentage kill in the south) noted already by Jefferies (1989b) would support the latter possibility.

Jefferies (1989b) suggested that the north/south difference in the number of otters killed by the hunts, for whatever reason, may well have contributed to the north/south difference in the severity of the population crash when the organochlorine insecticides came into use after 1955. The similar finding between the percentage kills of otters found in the east and west would provide further strong support for this suggestion. The eventual reduction of the British otter population was much greater in the south and east than in the north and west when the population crash started in 1957-58 (see Section 7.1.2).

(iv) Finally, although 'hidden' by the forms of analysis originally adopted by Anonymous (1969) and Chanin and Jefferies (1978), ie. recording the number of finds per 100 days hunting (or catch per unit effort), the actual hunting intensity in the western and eastern halves of Britain was very different in the early 1950s. Analysis of the total number of days spent hunting in 1950-55 shows that this was a mean (± s.e.) of 335.25 ± 32.55 days per hunt over six years in the east, but only 257.71 ± 20.43 days per hunt over the same six years in the west (see Table 7). This difference in the total number of days spent hunting (1,341 and 1,804 days by four eastern and seven western hunts respectively) is statistically significant ($\chi^2 = 53.52$; 1 df; $p<0.001$). Thus, the eastern hunts were putting 30.09% more effort into obtaining the results they recorded than were the western hunts. Presumably, this was because otters were fewer in the east owing to the 'hidden' attentions of the gamekeepers (see Section 7.1.3.1.5), though their own efforts were helping to produce the western bias in pre-decline otter density. The greater number of days spent hunting in the east does not alter their finds/100 days hunting, of course, or indeed the high percentage kill of those found (which, as noted in Section (iii) above, was higher in the east). However, it would increase the overall total of otters found and so of those killed over what would have been the results if they had hunted the smaller population only to the same extent and pressure as did their colleagues in the western hunts at that time (see Table 7). It means also that the smaller and more persecuted otter population of the eastern counties was disturbed and harried by packs of hounds much more frequently than elsewhere.

7.1.3.1.5 Use of hunt records to show the effects of the eastern bias in gamekeeping intensity on pre-decline otter numbers

Chanin and Jefferies (1978) used the records of hunting success of the packs of otter hounds, measured in terms of finds/100 days hunting, as indicators of local otter density. These changed with time and decreased as the otter population declined with pollution in the 1950s and 1960s. It is an interesting concept that these same detailed records of hunting success may also reflect the 'hidden' catches of the gamekeepers. Thus, before the days when toxic chemicals were providing the 'hidden' mortality which was being measured, the changes or area variation in hunting success indicated the levels of persecution caused by gamekeepers. When there were few otters found and killed by the hunts in any one area it could have been because the 'hidden' kills of the gamekeepers had been high in that area and vice versa. This has been mentioned briefly already in Section 7.1.3.1.3, ie. (i) when there were very few otters to be caught by the Courtenay Tracy Otter Hounds in 1895 and 1896, (ii) when the hunting success improved greatly after the reduced persecution of the 1914-18 World War years and (iii) when there were high numbers of otters available for hunting between the wars when gamekeeper numbers decreased with the changed economy. Consequently, we can use hunting success to examine indirectly the degree of persecution pressure exerted by gamekeepers in any one area.

Thus, if the territories of the 13 hunts active in 1955 are divided into eastern and western groups (as in Tables 6 and 8(c)), examination of their hunting records shows that the western hunts had a 24.2% higher hunting success rate in 1950-55 than that shown by their counterparts in the east. The two highest success rates, 121 and 89 finds/100 days (Hawkstone O.H. and Dumfriesshire O.H.), were in the west, and the two lowest, 55 and 47 finds/100 days (Northern Counties O.H. and Crowhurst O.H.), were in the east. The mean success rates were 79.50 ± 7.09 and 64.00 ± 5.54 finds/100 days hunting in the west and east respectively (see Tables 6 and 8(c)).

The eastern hunts, as we have noted in the previous Section 7.1.3.1.4, put in many more days hunting to find and kill the numbers of otters they recorded than did the western hunts (Table 7). A total of 1,341 days was hunted by four hunts over the six years 1950-55 in order to find 932 otters in the east, whereas a total of 1,804 days was hunted by seven

Figure 37. The intensity of game preservation in Britain in 1955 (redrawn from Moore, 1957). The data were originally derived from figures for membership of the Gamekeepers' Association and are not based on total numbers, ie. they provide a picture of relative density of gamekeepers in various parts of the country at that date. They show that by far the highest intensity of game preservation occurred in the south-eastern part of Britain.

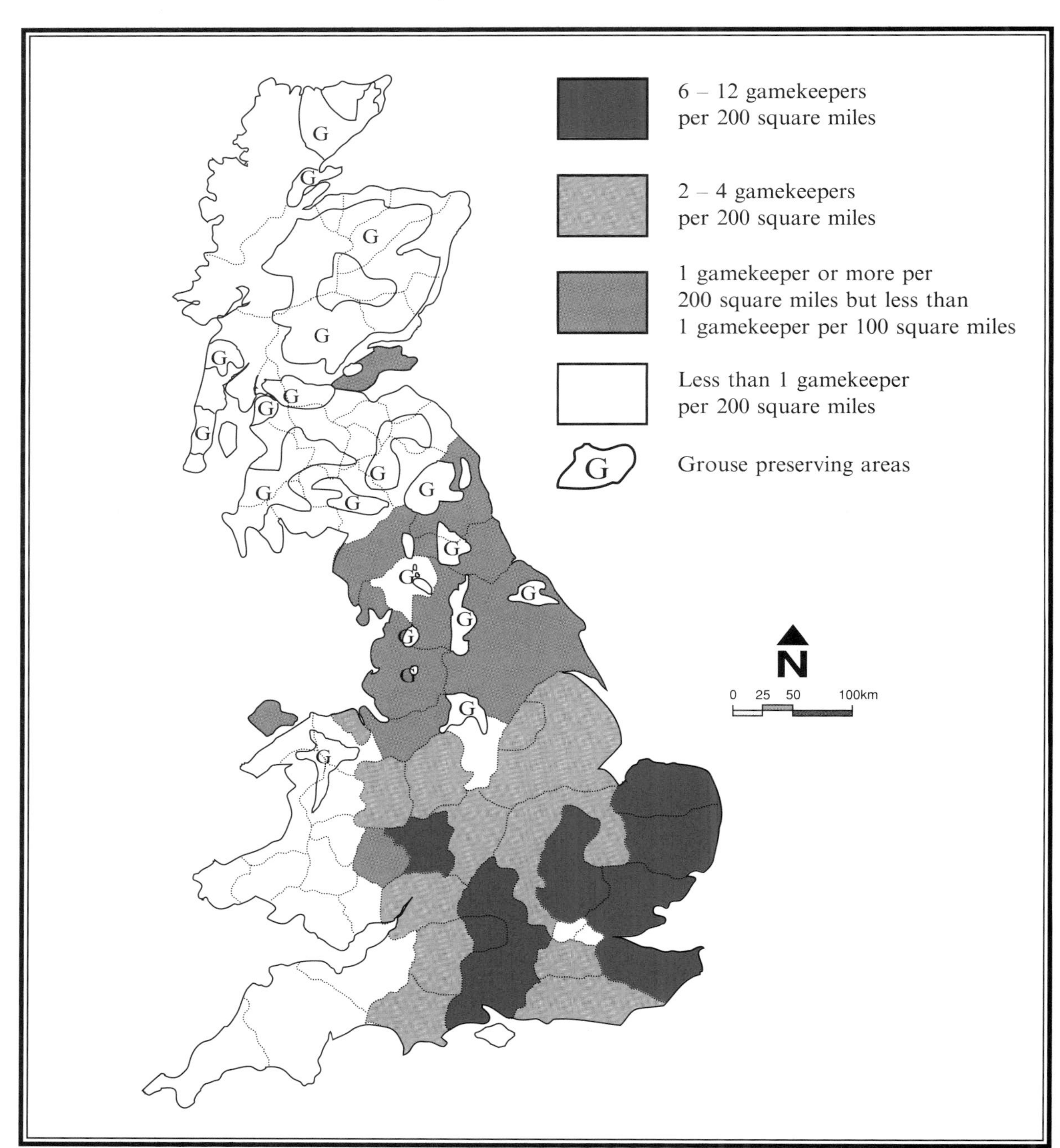

hunts over the same period in the west in order to find 1,489 otters.

The difference between these two rates of hunting success is significant ($\chi^2 = 16.99$; 1 df; $p<0.001$). This suggests a real difference in otter density in the two halves of southern Britain in 1950-55, only three to eight years before the population crash began, with its highest severity being recorded in the east.

It is unlikely that this marked pre-decline western bias in otter density is a 'natural' phenomenon caused by habitat differences, as the bias is in the opposite direction to that which would be expected. If upland waterways are suboptimal otter habitat compared to lowland rivers (as strongly suggested by the analysis in Section 6.1), then food supply, for

Table 8. The relationship between the hunting success of the 13 packs of otter hounds (in terms of finds per 100 days hunting) in 1950-55 and the intensity of game preservation within their hunting territories in 1955. (See text for method of scoring gamekeeper density.) A high score relates to a high density of gamekeepers. (a) The hunting success of the 13 hunts ranked in descending order of finds/100 days hunting. The corresponding scores for intensity of game preservation tend to show the opposite or inverse ranking; ie. as game preservation increases, hunting success decreases. (b) This is seen more readily if the 13 hunts are divided into three groups with ascending levels of gamekeeper density. (c) The 13 hunts divided into two groups, eastern and western, according to their geographical situation. The eastern hunts have the lowest success rates and are situated where there is the highest intensity of game preservation (see Figure 37).

(a) Otter Hunt	Finds/100 days 1950-55	Game Preservation Intensity Score
Hawkstone	121	13.7
Dumfriesshire	89	14.2
Culmstock	88	6.2
Pembroke & Carmarthen	78	5.2
Courtenay Tracy	76	51.6
Eastern Counties	72	84.6
Border Counties	71	14.4
Buckingham	70	47.4
Wye Valley	68	32.2
Dartmoor	62	5.0
Kendal & District	59	11.7
Northern Counties	55	16.2
Crowhurst	47	67.2

(b) Otter Hunt	Finds/100 days 1950-55	Game Preservation Intensity Score
Eastern Counties	72	84.6
Crowhurst	47	67.2
Courtenay Tracy	76	51.6
Buckingham	70	47.4
Mean	**66.25 ± 6.54**	**62.70 ± 8.45**
Wye Valley	68	32.2
Northern Counties	55	16.2
Border Counties	71	14.4
Dumfriesshire	89	14.2
Mean	**70.75 ± 7.00**	**19.25 ± 4.34**
Hawkstone	121	13.7
Kendal & District	59	11.7
Culmstock	88	6.2
Pembroke & Carmarthen	78	5.2
Dartmoor	62	5.0
Mean	**81.60 ± 11.18**	**8.36 ± 1.81**

(c)

Eastern Hunts: Otter Hunt	Finds/100 days 1950-55	Game Preservation Intensity Score
Northern Counties	55	16.2
Eastern Counties	72	84.6
Buckingham	70	47.4
Crowhurst	47	67.2
Courtenay Tracy	76	51.6
Mean	**64.00 ± 5.54**	**53.40 ± 11.37**

Western Hunts: Otter Hunt	Finds/100 days 1950-55	Game Preservation Intensity Score
Dumfriesshire	89	14.2
Kendal & District	59	11.7
Border Counties	71	14.4
Hawkstone	121	13.7
Pembroke & Carmarthen	78	5.2
Wye Valley	68	32.2
Culmstock	88	6.2
Dartmoor	62	5.0
Mean	**79.50 ± 7.09**	**12.82 ± 3.12**

example, is likely to be greater in the latter waters. It is much more likely to have been due to an east/west gradient in the level of 'hidden' persecution (as well as that provided 'openly' by the eastern hunts themselves; see Section 7.1.3.1.4). Indeed, close examination of the geographical distribution of various levels of intensity of game preservation uncovers just such a marked east/west gradient. This can be seen from the map of the numbers and distribution of members of the Gamekeepers' Association in 1955 (Figure 37). It was produced by Moore (1957) when studying the effects of gamekeeping activities on the presence of the common buzzard. Game preservation intensity was greatest in the south and east, with much less activity in the west and north.

In order to examine the relationship between hunting success and gamekeeping activities more closely, a scoring system for the relative intensity of game preservation in various areas of Britain was devised from the map in Figure 37. First, the outlines of the hunt territories were traced over the map of gamekeeper distribution and the proportion of each hunt territory falling into each band of gamekeeper density was estimated by use of a transparent square grid. Then, scores of 90, 30, 12.5 and 5 were attached to each of the four levels of gamekeeper activity and 25 to the small areas of grouse moors (see legend to Figure 37). By multiplying each of these density scores by the proportion of the territory including them, an overall score for each hunt could be obtained (eg. Wye Valley O.H.: $0.22 \times 90 + 0.33 \times 30 + 0.04 \times 12.5 + 0.41 \times 5 = 32.25$). Using this method, the maximum score for very high gamekeeper density and intensity of game preservation would be 90 and the minimum score for very little game preservation would be 5.

The calculated scores for the relative intensity of game preservation in each of the hunt territories are listed in Table 8(a). If we return to the two groups of hunts divided according to their geographical situation (eastern and western) in the second paragraph of this Section, we can now show that the game preservation scores for these two groups of hunt territories are significantly different ($t = 4.218$; 11 df; $p<0.01$). That of the east is very much higher (53.40 ± 11.37) than that of the west (12.82 ± 3.12) (see Table 8(c)). Thus, there is a significant difference in hunting success between two areas, east and west, which have significantly different levels of intensity of game preservation. This difference in hunting success was apparent several years before the population 'crash' started.

In Table 8(a) the hunting success rates for the period 1950-55 are listed alongside the corresponding calculated scores for game preservation intensity in each of the hunt territories. Examination of the individual pairs of data shows a tendency for an inverse ranking in the two columns. However, this can be seen more readily if the 13 hunts are divided into three groups with ascending levels of gamekeeper density (see Table 8(b)). There is a corresponding descending gradient in hunting success, ie. as game preservation increased area by area across southern Britain, then hunting success decreased. A higher level of correlation than this group ranking is unlikely between individual hunt territories and area game preservation scores because the otter hunt territories were very large (mean = 9,315km^2) and certain areas and rivers were hunted more frequently than others (Chanin & Jefferies, 1978). Thus, hunting success may not relate to the whole area covered by the game preservation intensity score. For example, the western and eastern regions of the Hawkstone O.H. territory would have very different scores (5.3 and 26.8, respectively). In addition, there will be an underlying variation in the otter density of hunt territories owing to factors such as past industrial pollution and/or present food supply. For example, the success of the Eastern Counties O.H. may be high in the rankings despite high persecution and because of the large supply of eels in the eastern rivers. If the problem of correlation is approached from a different viewpoint than that of variability in hunting success and the individual scores for game preservation intensity are compared directly with the dates at which population recovery started in each hunt territory, then a significant correlation between the two can be obtained (ie. high scores correlate with late recovery) (see Section 7.1.4).

It would appear then that, despite the great reduction in the number of 'keepers and water bailiffs after the First World War (see above), the remainder were still exerting an effect on the numbers of otters available for hunting purposes at least up to the time of the

Figure 38. The amount of arable land has changed over the last century. The polarisation of east and west into arable and stock rearing, respectively, is shown simplified in the map below for the situation as it was in 1988 (redrawn from Tapper, 1992). The two areas within the red lines are those known to suffer most acutely from attack by wheat bulb fly and will have been treated most intensively with organochlorine soil insecticides in the 1950s to 1970s (delineated by Bell (1975) and based on studies by Gough (1957)).

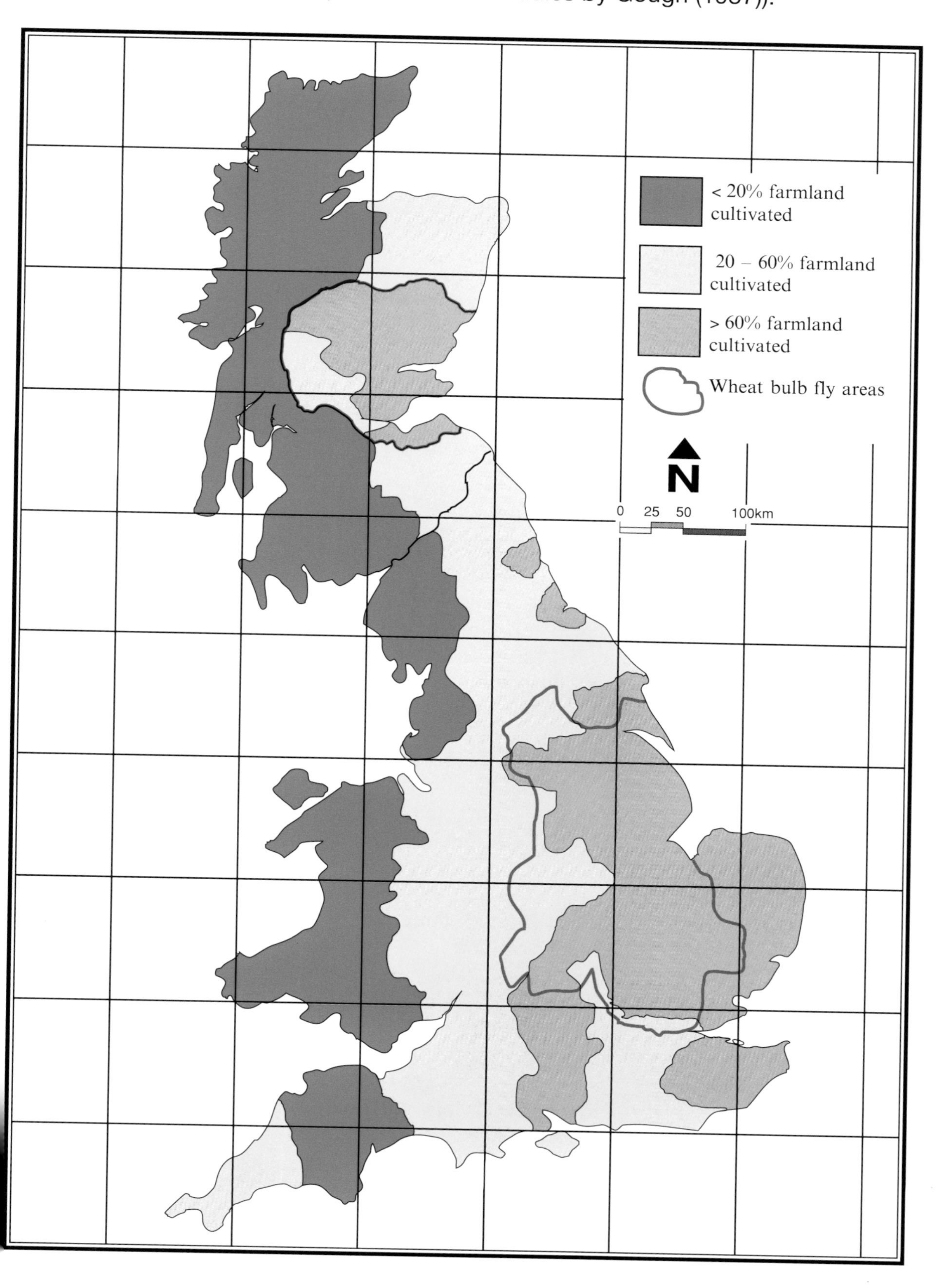

'crash'. Also, their greatest effect was in the east. There are parallels with certain other species persecuted by gamekeepers, eg. the common buzzard (Moore, 1957) and the badger (Reason *et al*, 1993), both of which have been almost eliminated from East Anglia. The proven effect on the otter shown in this Section would again lend support to the contention of Jefferies (1989b), that the otter population, particularly of eastern Britain, was reduced and stressed prior to the onset of organochlorine pollution in 1956 and that the former exacerbated the severity of the latter.

7.1.3.1.6 Further factors increasing the east/west differential in the level of man-made mortality before and after the start of the decline

There are two other major mortality factors which can be expected to have a differentially larger effect in the south and east of England than in the north and west.

These are:

(i) Road mortality: The density of people and road vehicles is highest in the south-east of Britain. Thus, road traffic accidents involving otters may be expected to be higher relative to the population size in that region.

(ii) Drownings in eel fyke nets: The major riverine eel fishing areas in Britain are situated along the east coast counties from Northumberland to Kent (Jefferies *et al*, 1988). Norfolk and Suffolk are particularly important in this respect. Thus, again, trapping in eel fyke nets is likely to kill a great proportion of the small, already stressed, otter population of eastern areas than would be the case in the west. The reports by Jefferies *et al* (1984), Jefferies (1985, 1990b) and Jefferies *et al* (1993) all provide case histories of drownings of otters from eastern counties of England, despite their very small populations. The six otters drowned in nets set in the River Bure system in one season in the late 1970s, for example, are considered to have virtually ended the population of that catchment (see Sections 7.1.3.2.7 and 7.3.3.2.1 and Table 12). Even some of the otters released by the Otter Trust and The Vincent Wildlife Trust have been lost in this way (Jessop & Cheyne, 1993; Woodroffe, 1993) (see Section 7.2.5.3).

7.1.3.2 Organochlorine insecticides

7.1.3.2.1 The east/west bias in the form of agriculture in Britain

One major east/west polarisation which could have affected the status and degree of decline of the otter population of the two sides of Britain is that of the form and intensity of local agriculture. Whereas in the late 1800s most farms were mixed, with incomes from both arable and livestock, this changed from the 1920s and 1930s onwards. Many western farms in hilly country completely abandoned their small areas of unprofitable arable crops and went into intensive dairy or livestock production. In flatter eastern England, on the other hand, the introduction of artificial fertilisers released farmers from the necessity of the traditional crop rotation and manuring to maintain soil fertility. Consequently, they were able to concentrate on profitable crops and to dispense with their livestock completely. This in turn led to larger and larger field sizes (and fewer hedgerows) to accommodate the machinery necessary for cultivation and harvesting. The gradual polarisation of Britain's agriculture into western livestock and eastern arable over the past century has been discussed and illustrated by Tapper (1992). The situation by the 1980s is shown in Figure 38.

7.1.3.2.2 The use of organochlorines on the western livestock and eastern arable farms

One consequence of this more intensive and dedicated form of agriculture in the eastern and western sides of Britain was the development and introduction of man-made chemicals into the originally 'organic' system. The development of herbicide sprays reduced the necessity for a large labour force on the arable farm. Also, in order to maintain the large areas of monoculture and control pests, such as wheat bulb fly, the persistent organochlorine insecticides were developed. Some of the most toxic were the cyclodienes aldrin and dieldrin.

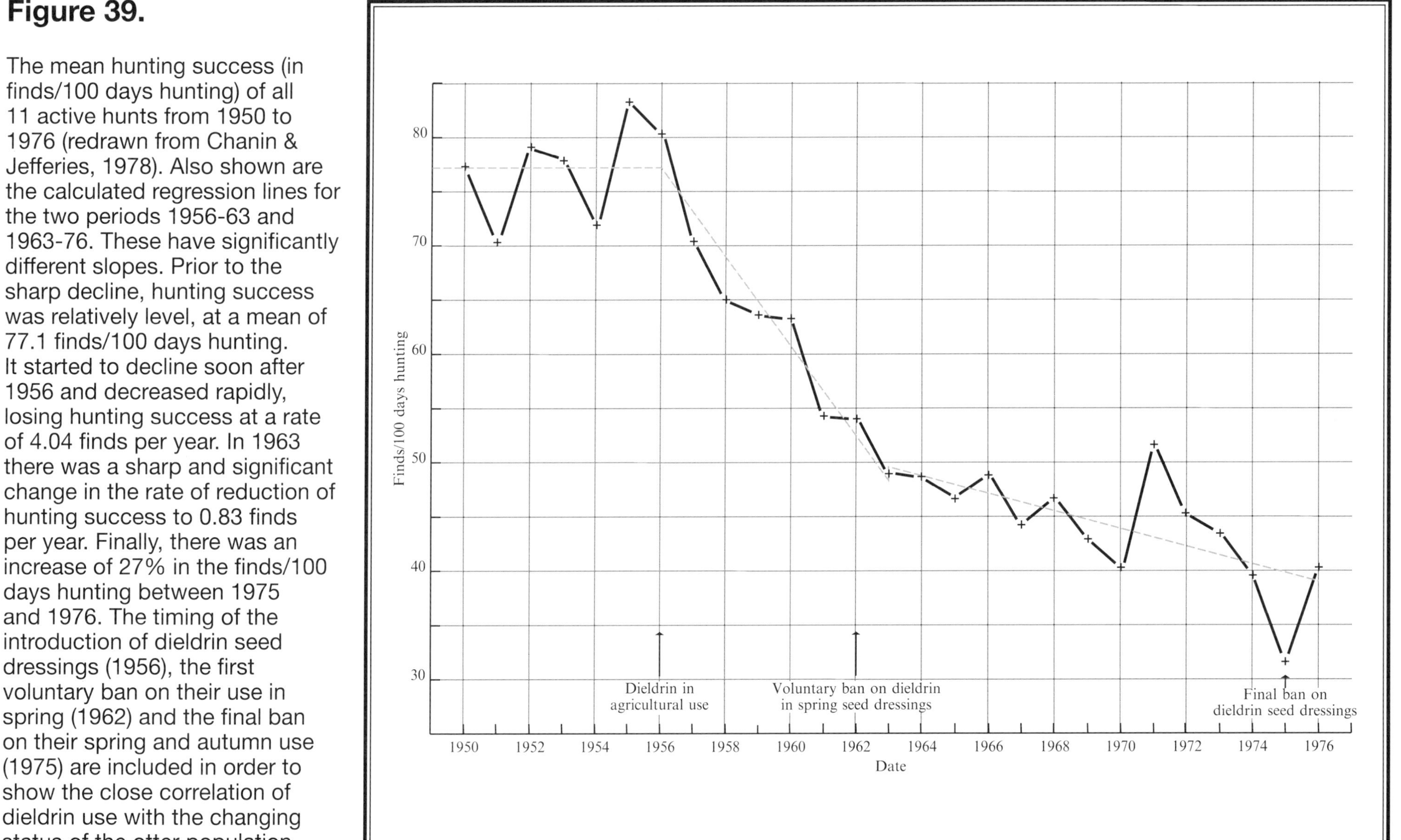

Figure 39.

The mean hunting success (in finds/100 days hunting) of all 11 active hunts from 1950 to 1976 (redrawn from Chanin & Jefferies, 1978). Also shown are the calculated regression lines for the two periods 1956-63 and 1963-76. These have significantly different slopes. Prior to the sharp decline, hunting success was relatively level, at a mean of 77.1 finds/100 days hunting. It started to decline soon after 1956 and decreased rapidly, losing hunting success at a rate of 4.04 finds per year. In 1963 there was a sharp and significant change in the rate of reduction of hunting success to 0.83 finds per year. Finally, there was an increase of 27% in the finds/100 days hunting between 1975 and 1976. The timing of the introduction of dieldrin seed dressings (1956), the first voluntary ban on their use in spring (1962) and the final ban on their spring and autumn use (1975) are included in order to show the close correlation of dieldrin use with the changing status of the otter population.

These man-made molecules were applied to seed as a dressing, together with an organo-mercury compound in order to protect it from both insect and fungus attack whilst in the soil. Many thousands of acres of eastern England were drilled with dressed cereal and sugar beet seed (see Table 14).

Sheep formed by far the majority of the livestock reared on the western farms. There were 7.7 million pigs, 13.8 million cattle and 28.1 million sheep in Britain in 1977 (Mellanby, 1981). These sheep were required to be dipped to control fly strike and the new insecticide dieldrin was so persistent that its use only required the sheep to be dipped once rather than twice a year. Consequently, much dieldrin was used in the sheep-dips which eventually polluted the western streams. Unfortunately, these new insecticides were in fact biocides and caused one of the largest wildlife mortalities and sudden population crashes of a wide range of predators that Britain had ever experienced (Cooke *et al*, 1982; Sheail, 1985; Moore, 1987; see Sections 7.3.1.1.3 and 7.3.1.1.4).

7.1.3.2.3 The date of the start of the decline of the otter population and the link with use of organochlorine insecticides

There are several factors which undoubtedly point towards the start of the use of the persistent organochlorine insecticides as the main cause of the initiation of the marked decline in the southern British otter population. These factors are listed in Section 7.1.4 and discussed in Section 7.1.3. The one factor which stands out is that of the timing, ie. the decline was first measurable all over the country in late 1957 and dieldrin, aldrin and heptachlor first came into use in 1955, with major usage by 1956. It is remarkable that their effect should have been so sudden and severe, with a noticeable decrease in population size within 18 months. This rapid effect occurred in other species too, eg. the sparrowhawk. The link between the two initial dates might be said to be circumstantial. If, on the other hand, there were changes in the rate of decline or in its cessation which correlated with the dates at which various uses of the organochlorine insecticides were banned, then one could say that the link was much firmer. This would be even more so if the bans had their greatest effects on the otter populations of certain areas and these correlated with the areas of greatest reduction in organochlorine insecticide usage. These links between the bans obtained on the use of organochlorine insecticides in seed dressings and sheep-dips and the changes in the fortunes of the western and eastern otter populations are examined in the next five sections.

7.1.3.2.4 The 1962 ban on dieldrin use on spring-sown cereals and its effect on the rate of decline of the otter population

The first, voluntary, ban on the use of aldrin/dieldrin dressings on spring-sown cereals took effect in 1962 and appears to have had a greater effect on the rate of decline of Britain's otter population than has been realised previously. Thus, Chanin and Jefferies (1978) noted, after summing the records of all active hunts, that the rate of decline appeared to be reduced after 1963. They suggested then that it could be due to a bias produced by hunters spending a disproportionate amount of time hunting in favourable areas when otters became scarce. However, there is no reason why this should have happened so suddenly and with such a marked change in rate in one year (1963).

Examination of the greater quantity and longer series of data now available shows an equally rapid reaction of the western and eastern otter populations following within two years of further major bans on dieldrin usage in sheep-dip (1966) and all seed dressings (1975) (see Figure 40; Sections 7.1.3.2.6 and 7.1.3.2.7). This led to a reappraisal of the earlier section of the decline curve (see Figure 39) to see if there had been any reaction to the first ban. The change in slope of decreasing hunting success in 1963 (one year after the ban) was indeed marked. The slope of the regression line fitted to the total hunt data from 1956 to 1963 ($r = -0.9610$; 6 df; $p<0.001$; equation: $y = 303.22 - 4.04x$, where y = finds/100 days hunting and x = date, ie. 60 for 1960) was significantly ($t = 5.9328$; 18 df; $p<0.001$) different to that for the period 1963 to 1976 ($r = -0.6722$; 12 df; $p<0.01$; equation: $y = 101.91 - 0.83x$). During the years 1956 to 1963 hunting success was decreasing at the rate of

Figure 40.

Diagrammatic representation showing the timing of the start of the decline and its 'bottoming out' or nadir, first in the west and then in the east. Almost certainly, the population of East Anglia would have continued to decline to extinction without the release of captive-bred otters starting in 1983 (see lowest line). The horizontal lines indicate the span of dates at which the nadir occurred in the different areas. It will be noted that the difference in the severity of the decline in the east and the west was not very great at the period 1966-71 (used in the analysis of Chanin & Jefferies, 1978) but the gap widened as the population of the west recovered and that of the east declined further. The x-axis provides the dates of introduction and important bans on the use of dieldrin in sheep-dip (in the west) and in seed dressings (in the east). There is a noticeable correlation between the status of the western and eastern otter populations and the use of organochlorines in agriculture in these two halves of southern Britain.

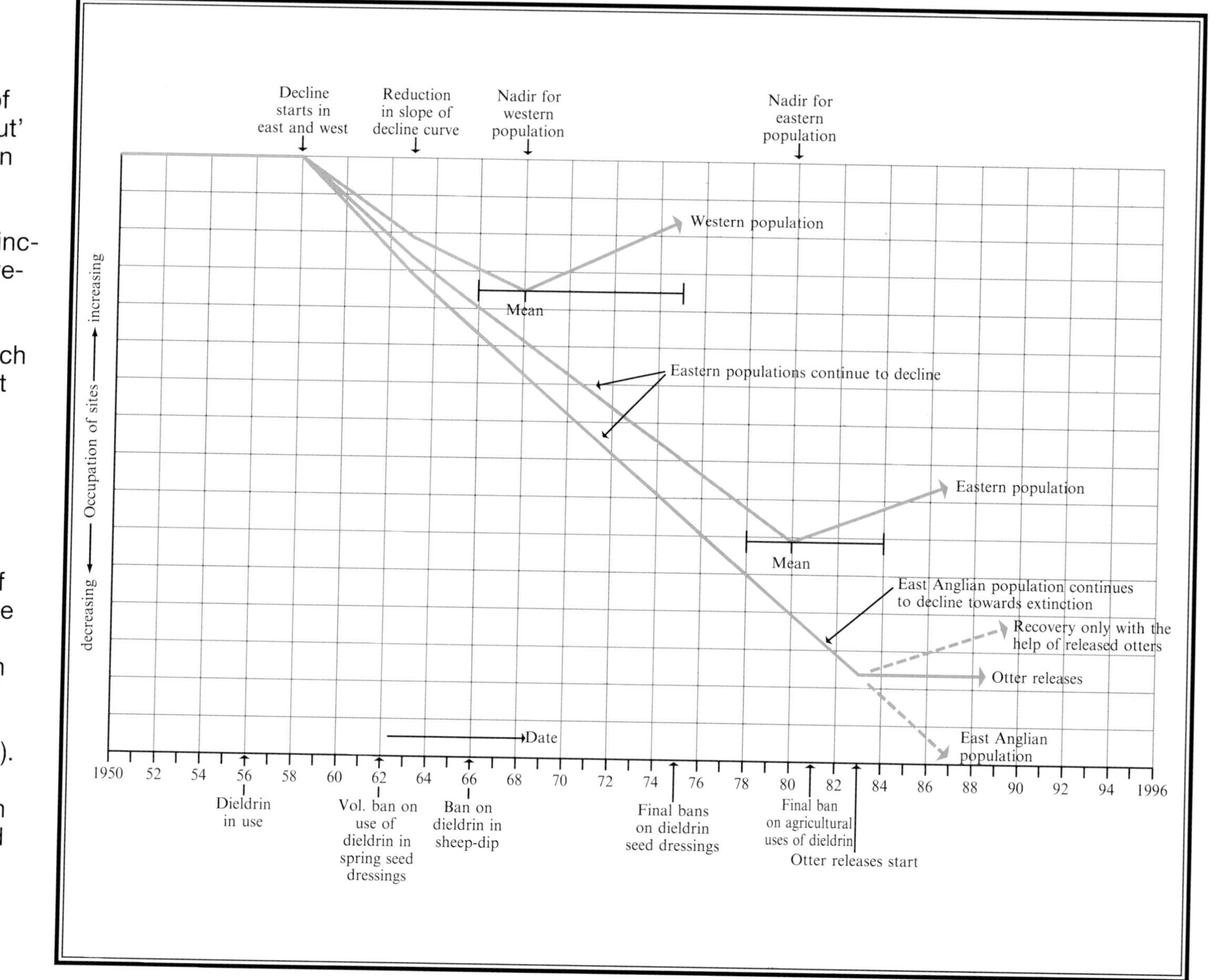

4.04 finds per year, but this suddenly changed to a decrease of 0.83 finds per year after 1963 (ie. only 20.5% of the former rate of loss). The timing correlates very closely with that of the first ban taking effect. One can only conclude that the two are connected. As noted in Section 7.3.1.1.5, this one ban considerably reduced dieldrin usage in Britain and the large kills of seed-eating birds observed in previous years.

A small reduction in the level of persecution also took place in 1964. This could only have helped the otter population, though it seems unlikely that it could have produced the remarkable change observed on its own. Thus, in 1964 the hunts changed their policy to one of not killing the otters found and hunted, whenever this was possible. However, by that time there were very few otters being 'found'. For example, the Eastern Counties O.H. killed a total of 48 otters in the four years 1960-63 and this was reduced to 29 in the years 1964-67. This should be compared to a total of 101 otters killed by the same hunt over the four years 1953-56 (Downing, 1988).

7.1.3.2.5 Variation in the progression of the decline and the timing of its nadir in various parts of England and Wales

The records of the various packs of otter hounds show conclusively that the otter decline started suddenly and at around the same date in all parts of the country hunted (Chanin & Jefferies, 1978). The mean date for the start of the decline is exactly the same on both western and eastern sides of southern Britain (1957.7; Table 9). Now that the surveys are showing recovery in the otter populations of England, Wales and Scotland, it is important to see whether it is now possible to determine the timing of the nadir or 'bottoming-out' of the decline. Was this again the same date in all parts of the country or did it vary, perhaps showing a progression from west to east? With a variable date for the nadir of the population curve, the eventual severity of the decline in each Region would depend partly on the population status and density before the start of the decline in the mid-1950s (this would have varied considerably with past persecution and the area food supply; see Section 7.1.3.1), partly on the steepness of that decline and partly on the duration of time the decline continued before recovery started (see diagrammatic representation in Figure 40). The dates at which the decline stopped and recovery started in each area of Britain is invaluable for providing a further pointer to the cause of the decline and so the confirmation of the hypotheses already developed (Chanin & Jefferies, 1978; Jefferies, 1989b). By comparison with the known dates and areas in which various environmental changes occurred, it should be possible to correlate cause and effect.

Here, the recent history of the sparrowhawk may provide an indication of the form and pattern of recovery to be expected from a species increasing in numbers after a population decline caused by the organochlorine insecticides. The population of this species, previously one of our most common birds of prey, 'crashed' between 1957 and 1960 (Newton, 1986). The decline was particularly marked in eastern districts, from some of which the sparrowhawk disappeared altogether (Prestt, 1965). Some 14% of 473 carcases examined and analysed at Monks Wood between 1963 and 1982 had dieldrin levels above 3 mg/kg^{-1} and the bodies with the highest residues showed similar lesions to birds dying of dieldrin poisoning in the laboratory (Newton, 1986). Thus, this species, like the peregrine falcon (Ratcliffe, 1993), shows the classic pattern of a population crash caused by organochlorine insecticides following rapidly after the date of the start of dieldrin use in 1956. In this matter, both avian species provide very similar patterns (almost mirror images) for the history of the otter population described here. One difference is that the otter population suffered more severely in the west because of pollution engendered by dieldrin sheep-dips than did either of these predatory birds. These received most of their dieldrin contamination from cereal seed dressings entering their seed-eating avian prey. Aldrin/dieldrin seed dressings were used at their highest area rates in the eastern counties of Britain, where the resulting pollution caused the highest losses among both predatory birds and otters (Chanin & Jefferies, 1978; Newton, 1986; Jefferies, 1989b; Ratcliffe, 1993).

Newton and Haas (1984) have followed the recovery of the sparrowhawk from this population 'crash' and, as with most ornithological incidents, there are many more data than are available for the much less studied and more secretive mammals like the otter. These authors divided the country into four zones for purposes of analysis. Zone 1 included south-west England, Wales, Dumfries & Galloway and most of central and highland Scotland. Zone 2 included southern England to West Sussex, west Midlands, most of northern England and the rest of Scotland. Zone 3 included Yorkshire, the Midlands, Buckinghamshire and East Sussex, and Zone 4 Lincolnshire, East Anglia and Kent. In Zone 1 sparrowhawks survived in greatest numbers throughout the 1950s and 1960s, with a population decline of less than 50% and recovery effectively complete before 1970. In Zone 2 the population decline was more marked than in Zone 1 but the species had recovered to more than 50% by 1970. In Zone 3 the population decline was more marked than in Zone 2 but the species had recovered to more than 50% by 1980. In Zone 4 the population was almost extinct around 1960 and little or no recovery was evident by 1980. However, by 1990 the recovery was said to be complete (Stroud & Glue, 1991) and the sparrowhawk is now (1996) breeding in the east Midlands in considerable numbers again (D.J. Jefferies). In general, recovery for this species seems to have occurred in a wave-like pattern spreading eastwards from Wales and the West Country.

In the following Sections we will examine the recovery pattern shown by the otter to see if there are indications of a similar pathway. This in turn will help confirm the original thesis (Chanin & Jefferies, 1978) that the sharp decline of the otter population was precipitated by the start of dieldrin use and provides another classic case of the near extinction of a wildlife species by the new intensive chemical farming of the 1950s. As with all predators, past and recent persecution comes into the equation, but this contributory cause has been covered in Section 7.1.3.1.

The basic data available for use in determining the dates of the western and eastern nadirs are (i) the records of the various packs of otter hounds from 1950 to 1977, (ii) the county surveys of Norfolk and Suffolk from 1969 to 1989, and (iii) the three national otter surveys of 1977-79, 1984-86 and 1991-94.

7.1.3.2.6 The timing of the nadirs of the declining western populations

An indication of a possible date for the 'overall' nadir of the decline in southern Britain may be given by the overall hunting success of the 13 packs of otter hounds from 1950 to 1976. Chanin and Jefferies (1978) found that this declined from 80.8 finds/100 days hunting in 1956 to 30.8 finds when it reached its lowest point in 1975. By 1976 it had shown a 31.2% increase to 40.4 finds/100 days hunting.

Looking at the west and the east separately shows a marked difference in the timing of their nadirs. The results of the three national surveys showed that all regional areas on the western side of Britain had already 'bottomed out' and were recovering by the time of the first survey in 1977, whereas all those on the eastern side did not do so until after 1977-79 (see Table 10).

Thus, the sole data from which to establish the dates of the nadirs for the western side of the country derive from the records of the otter hunts, as the national otter spraint surveys did not start until 1977. The nadirs for the east, on the other hand, occurred long after hunting had ceased, so we have to depend on the results of the surveys here. Fortunately, some area surveys started in the east (eg. for Norfolk and Suffolk) some years before the national surveys started, so the timing of the nadirs can be made with more certainty and accuracy here.

Chanin and Jefferies (1978) found that there were certain problems in using the records of Otter Hunts for determining dates for declines. First, Masters of Otter Hounds would naturally have placed their hounds where they believed they were most likely to find otters to hunt, ie. they were not acting in the manner of surveyors looking at each river in turn. Otherwise there would have been a rapid loss of interest by the hunt followers. Consequently, they spent a disproportionate amount of time hunting in favourable areas and rivers, so unintentionally masking the real extent of the decline and its nadir for the area as a whole. This was shown to have happened with the Eastern Counties O.H., who spent 65% of their hunting effort on 30% of the rivers hunted and only hunted 62% of the rivers in

Table 9. The year of the start of the otter population decline and the year of the nadir of that decline before recovery began in each of the territories of the hunts active in the early 1950s. The scores for the intensity of cultivation (high: arable; low: stock and grassland) and for the intensity of game preservation (highest density of gamekeepers with highest score) in each of these hunt territories is given for purposes of comparison and correlation (see Figure 43). The dates of the nadirs for the territories of the Eastern Side of England were obtained from the regional areas including them (see Table 11) as hunting had ceased by the time that the declines of the eastern populations had 'bottomed out'. Note that the areas of the hunts of the Eastern Side and those of the hunts of the Western Side of southern Britain used in this analysis are closely similar in total. Also, the mean dates of the start of the decline are the same for the Eastern and Western Sides of southern Britain.
*Notes: † The dates of the nadirs for these hunt territories were taken as the years in which those areas were surveyed within the national survey dates of 1977-79 and 1984-86. * There was no nadir in the decline of the otter population of the East Midlands and East Anglia, but a continuing progression towards extinction. This was only relieved by the release of captive-bred stock. ** This mean does not include the date 1986.*

Hunt	Area of territory (km²)	Year of		Score for		Notes
		Start of decline	Nadir before recovery	Intensity of Cultivation	Intensity of Game Preservation	
EASTERN SIDE						
Northern Counties	9,450	?	1978†	41.7	16.2	Date of nadir from nearest regional area
Eastern Counties	16,850	1957	post 1986* No nadir	80.0	84.6	
Buckingham	18,900	1956	1978†	68.5	47.4	
Crowhurst	8,100	–	1979†	56.8	67.2	
Courtenay Tracy	9,550	1960	1984†	53.6	51.6	
East (overall)	**62,850**	**Mean 1957.7**	**Mean ** 1979.7**	**63.78**	**54.97**	
WESTERN SIDE						
Dumfriesshire	7,050	?	1970	12.1	14.2	
Kendal & District	8,200	?	1969	12.9	11.7	
Border Counties	5,700	?	1966	13.9	14.4	
Hawkstone	9,050	1958	post 1975	22.8	13.7	
Pembroke & Carmarthen	5,800	1958	1966	10.0	5.2	
Wye Valley	6,650	–	?	25.1	32.2	
Culmstock	8,250	1957	1966	30.9	6.2	
Dartmoor	7,550	1958	1967	23.5	5.0	
West (overall)	**58,250**	**Mean 1957.7**	**Mean 1968.4**	**19.48**	**13.35**	

1977 which they had hunted in 1968 (Chanin & Jefferies, 1978). Second, the Border Counties O.H. was the only hunt not to show a decline in the late 1950s. However, Jefferies (1989b) showed that, after low hunting success in 1959 (ie. the start of their decline), this hunt changed its main hunting area in 1961, so prolonging success and obscuring the timing of the original synchronised decline found over the rest of the country. Nevertheless, hunting records form the only available data, and examination of them does indeed show likely early dates for the nadirs in the west and supports the conclusion that these dates were very much later in the east.

Data for the years 1950 to 1971 were available for all active hunts and to 1977 for a very few (Chanin & Jefferies, 1978; 'Polestar', 1983; Sagar, 1984; Downing, 1988; Greenwood, 1991). The decline started in all hunt territories in or around 1957 and a change in the trend, or nadir, when discernible, was usually noted within the final six years 1966 to 1971. The last dates (in brackets) before an upturn in hunting success were determined as follows.

(a) Western Hunts:

(i) Dumfriesshire O.H. (1970): The mean hunting success for 1966-71 was 78 finds/100 days. It declined from 1966 (85) to 1970 (62), before increasing to 84 finds/100 days (ie. above the mean) in 1971.

(ii) Kendal & District O.H. (1969): The mean hunting success for 1966-71 was 40 finds/100 days. It declined from 1966 (49) to 1969 (19), before increasing steadily to 1971, when it was above the mean at 61 finds/100 days.

(iii) Border Counties O.H. (1966): The mean hunting success for 1966-71 was 71 finds/100 days. It increased steadily from 1966 (51) to 1971 (96) over the final period of available records.

(iv) Hawkstone O.H. (post 1975): The mean hunting success for 1966-71 was 66 finds/100 days. It decreased from 1966 (80) to 1970 (44) and then, after increasing briefly in 1971 to 67 finds/100 days, it plunged to 38 in 1974 and 33 in 1975. The historian of the Hawkstone noted in 1973 that "otters continue to be difficult to find in home waters" (Greenwood, 1991). The pack was disbanded after the 1975 season. This is the latest date for the nadir of a western hunt and the possibility is discussed in Section 7.1.3.1.4 (ii) that the very high rate of killing otters carried out by the Hawkstone (the highest in Britain at 62% of the finds) may have been a factor.

(v) Pembroke & Carmarthen O.H. (1966): The mean hunting success for 1966-71 was 37 finds/100 days hunting. The number of finds/100 days progressed steadily upwards from 1966 (18) to 1971, when it reached 69.

(vi) Wye Valley O.H. This pack ceased to hunt in 1958.

(vii) Culmstock O.H. (1966): Hunting success progressed steadily upwards from 1966, when it was at its lowest at 33 finds/100 days, to 1971, when it reached 52 finds/100 days.

(viii) Dartmoor O.H. (1967): Hunting success was at its lowest over 1966 and 1967, when it was only 13 finds/100 days for each of these years. However, from 1967 it progressed steadily upwards to reach 33 finds/100 days in 1971.

The mean date of the decline nadirs of the seven western hunts was 1968.4 (see Table 9). This correlates with and follows the date (1966) when the use of aldrin/dieldrin in sheep-dip was banned. This would be expected to have its greatest effect on the western side of the country with its very large sheep population (see Section 7.1.3.2.2).

(b) Eastern Hunts:

(i) Northern Counties O.H.: The mean hunting success for 1966-71 was 30 finds/100 days hunting. There was a steady decrease from 65 finds/100 days in 1966 to 12 in 1971, ie. hunting success was still declining at the end of the last five years of available data.

(ii) Buckingham O.H.: The mean hunting success for 1966-71 was 28 finds/100 days hunting. Again there was a steady decrease from 47 finds/100 days in 1966 to 16 in 1971, ie. hunting success was still declining at the end of the last five years of available data.

(iii) Crowhurst O.H.: This pack ceased to hunt in 1958.

(iv) Courtenay Tracy O.H.: The mean hunting success for 1966-71 was 36 finds/100 days hunting. As with the Northern Counties and the Buckingham packs, there was a steady decrease in hunting success from 1966 (44) to 1971 (30 finds/100 days). Information from the history of this hunt ('Polestar', 1983) shows that hunting success continued to decline up to and including 1976, but in 1977 14 otters were found in 32 days hunting (44 finds/100 days) again. The members considered that this represented a slight but definite upward trend. Hunting, of course, ceased after that year because of legal protection in 1978. However, information from the spraint surveys, which provide a much more accurate picture over the whole southern area, do not support this optimism but suggest that the local population decline went on for a few years longer (see Section 7.1.3.2.7).

(v) Eastern Counties O.H.: The mean hunting success for 1966-71 was 38 finds/100 days hunting. It declined from 32 finds/100 days in 1966 to 26 finds/100 days in 1968, after being as high as 97 in 1953. However, after that, their hunt history (Downing, 1988) shows that they made great efforts to maintain a viable level of hunting success by concentrating more and more on the best rivers (Chanin & Jefferies, 1978) and by hunting for fewer days (down from 61 in 1953 to 32 in 1976). Consequently, their success ranged between 34 (1972) and 53 (1974) finds/100 days over the period 1969-77 and no clear pattern of decline and recovery can be discerned. However, as with the Courtenay Tracy O.H., spraint surveys again suggest that the local population decline continued for several years longer. Indeed, in this eastern area, the decline would have almost certainly continued to the point of population extinction if captive-bred otters had not been released here (see Sections 7.1.3.2.7 and 7.2).

Thus, all the available information for the eastern hunts suggests that the nadir did not occur for them until some time after hunting had ceased in 1977.

7.1.3.2.7 The timing of the nadirs of the declining eastern populations

Unlike those of the west, the records of hunting success of the eastern packs of otter hounds gave no indication of an early nadir in the population declines of the otters of the eastern side of the country (see Section 7.1.3.2.6). This source of information ended with legal protection in 1978. This is not a problem, however, because, owing to factors discussed in Section 7.1.3.2.8, the declines of the eastern otter populations 'bottomed out' long after those of the west and within the period of structured otter surveys using spraints and footprints. Thus, the three national surveys provided the percentage of occupied sites for the north-east, south-east, southern and eastern regional areas of the eastern side for the periods 1977-79, 1984-86 and 1991-94. These regional areas are as mapped in Figure 41. The results of these surveys show that, unlike the rapidly rising site occupation of the area populations of the western side throughout these three survey periods, site occupation in the eastern regional areas was very similar in the years 1977-79 and 1984-86 before increasing markedly by 1991-94 (see Table 10). Thus, the overall numbers of occupied sites for these three survey periods was 332, 639, and 1,062 respectively in the west and 45, 36 and 154 in

Figure 41. The seven regional areas of England delineated for purposes of analysis of the timing of the decline nadir in various parts of the country. Also shown are the alternate 50km survey squares (red). The 50km survey squares included in each area are listed below with their Ordnance Survey letters and whether they are in the north-west or south-east quarters of the 100km squares of the National Grid.

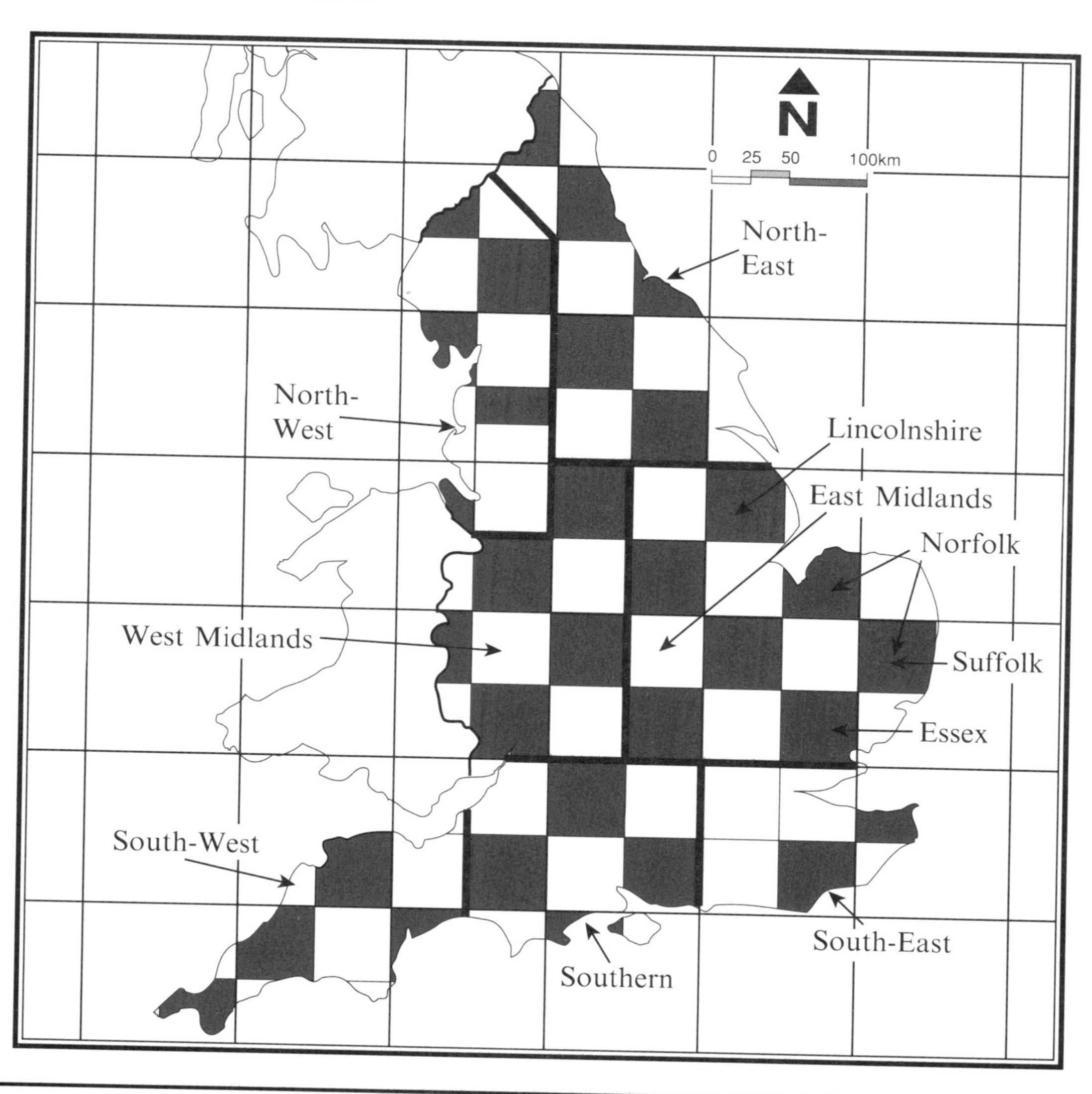

Area delineated for analysis	50km survey squares included
North-East	NT s/e; NZ n/w; NZ s/e; SE n/w; SE s/e. NZ s/e not surveyed in 1977-79.
North-West	NY n/w; NY s/e; SD n/w; SD s/e; SJ n/w.
Eastern (East Anglia & East Midlands)	SK s/e; TL n/w; TL s/e; SP s/e; TF n/w; TF s/e; TM n/w. TL s/e not surveyed in 1977-79.
West Midlands	SP n/w; SJ s/e; SO s/e; SO n/w; SK n/w. SK n/w not surveyed in 1977-79.
South-East	TQ s/e; TR n/w. TR n/w not surveyed in 1977-79.
Southern	ST s/e; SU n/w; SU s/e; SZ n/w.
South-West	SW s/e; SX n/w; SS s/e; SY n/w.

the east. This distribution of occupied sites between the three date periods is significantly different on the eastern and western sides (3 x 2 χ^2 test: $\chi^2 = 26.32$; 2 df; $p<0.001$). This finding supports the conclusion that, whereas the western populations were recovering rapidly throughout the 18 years of the national surveys from 1977 to 1994, the decline curves of the eastern population were "bumping along the bottom" (with very similar percentage site occupation in both 1977-79 and 1984-86) for about ten years before showing more marked signs of recovery by the time of the 1991-94 survey. Within this ten year period, the north-east and south-east regional areas showed nadirs in 1978 and 1979 respectively, whereas that of the southern regional area occurred in 1984 (their surveying dates at the time of the 1977-79 and 1984-86 surveys). The changes in percentage site occupation for the eastern regional area population, on the other hand, differed to those of all the above three areas. It showed a much lower number of occupied sites in 1984-86 than in 1977-79, indicating a further marked decline rather than a levelling-out in the loss of occupied sites.

These dates for the nadirs of the regional areas of the eastern side of the country are listed in Table 11 and converted into the corresponding Table of nadirs for hunt territories in Table 9. The mean date for the nadirs of the eastern hunt territories is 1979.7. Again, this correlates with and follows the date (1975) when all use of aldrin/dieldrin in seed dressings for autumn and spring-sown cereals was banned. This would be expected to have its greatest effect on the eastern (arable) side of the country with its major wheat bulb fly problem (see Figure 38).

The continuing decline in the eastern regional area, reported above, is complicated by the releasing of captive-bred otters into Norfolk and Suffolk from 1983 (Jefferies *et al*, 1986). However, there are more data for these two counties than for any other areas because of a series of county bridge surveys carried out by Essex University, the Otter Trust and the Suffolk Naturalists' Society from 1969 to 1989. Fortunately, these were carried out between the national surveys, which means that a much more complete picture of events can be constructed here.

The first survey of Norfolk was carried out by Macdonald and Mason (1976), who surveyed 233 bridge sites for signs of otters over 1974-75. (40 or 17.17% of sites were occupied.) This was followed by the Otter Trust's Norfolk bridge surveys of 1980-81 (Clayton & Jackson, 1981; 8.03% of 473 sites occupied) and 1988-89 (Jessop & Macguire, 1990; 13.95% of 473 sites occupied). In addition, the national surveys of 159 600m sites in Norfolk in 1977-79, 1984-86 and 1991-94 indicated 16 (10.06%), 2 (1.26%) and 24 (15.09%) sites occupied respectively (see Table 10). These results are shown plotted in sequence in Figure 42 and show a steep decline with a sharply defined nadir before an apparently equally marked recovery in 1988-89 and 1991-94. However, closer analysis shows that during the 1984-86 survey only one of Norfolk's two positive full survey sites was of wild-born stock (that on the River Wissey) whilst the other derived from the Otter Trust's release on the River Thet in 1984 (Ridding & Smith, 1988). Thus, the Norfolk population had declined to virtual extinction on the surveyed rivers by 1984-86. Examination of the situation in the county, river by river (Table 12), shows that nearly all of the apparent recovery by 1991-94 consisted of released captive-bred otters and their dispersing progeny. Only the River Wissey appeared to have a very small persistent wild-bred population where they had been protected within the Stanford Battleground (see Section 5.5).

A pioneer survey of Suffolk was carried out by West (1975) for the Suffolk Naturalists' Society from 1969 to 1972. Sixty (or 30%) of the 200 sites with record cards showed positive signs of otters. These occurred in 14 of the 30 10km x 10km squares surveyed and nine of the main river systems (see Table 12). Thus, the status of the otter in Suffolk at that date still appeared reasonably secure. However, despite the conservation strategy developed for the species in that county (Cranbrook *et al*, 1976) with the River Deben as the "County Otter Reserve", the population declined even more rapidly than that of Norfolk and by the time of the 1977-79 national survey only one (1.7%) of the 60 Suffolk sites showed positive signs. Release of captive-bred otters started in Suffolk in 1983 (River Black Bourn; Jefferies *et al*, 1986) and, when the Otter Trust's survey of Suffolk was carried out in 1984 (Wayre, 1985), no spraints from the wild-bred population could be found. Signs

Table 10. The numbers of sites surveyed three times and the numbers and percentages of these found to be positive in the 1977-79, 1984-86 and 1991-94 otter surveys of England (data from Lenton *et al* (1980), Strachan *et al* (1990) and this report). Those sites surveyed three times in Wales** at these dates (data from Crawford *et al* (1979), Andrews & Crawford (1986) and Andrews *et al* (1993)) and those surveyed twice in Dumfries & Galloway* (data from Green & Green (1980, 1987)) are given also (see text).

The survey data for England have been grouped into seven regional areas (see Figure 41) and also two major geographical zones comprising the Eastern (including the south) and the Western (including Wales) Sides or halves of the country. These correlate with major agricultural regions (see Figure 38).

Note that the proportions of the positive results in the three dated periods are markedly different in the two halves of the country. These indicate differences in the dates at which the population decline reached a nadir in the two halves of southern Britain.

Regional area (see Figure 41)	No. of sites surveyed	Number and percentage found positive		
		1977-79	1984-86	1991-94
EASTERN SIDE				
North-East	360	18 (5.0%)	20 (5.6%)	68 (18.9%)
South-East	134	0 (0.0%)	2 (1.5%)	3 (2.2%)
Southern	327	7 (2.1%)	6 (1.8%)	27 (8.3%)
Totals (3 areas)	**821**	**25 (3.0%)**	**28 (3.4%)**	**98 (11.9%)**
East Midlands	417	2 (0.5%)	4 (1.0%)	20 (4.8%)
Lincolnshire	142	1 (0.7%)	0 (0.0%)	2 (1.4%)
Norfolk	159	16 (10.1%)	2 (1.3%)	24 (15.1%)
Suffolk	60	1 (1.7%)	2 (3.3%)	10 (16.7%)
Totals (Eastern area)	**778**	**20 (2.6%)**	**8 (1.0%)**	**56 (7.2%)**
Eastern Side (overall)	**1,599**	**45 (2.8%)**	**36 (2.3%)**	**154 (9.6%)**
WESTERN SIDE				
Dumfries & Galloway*	414	341 (82.4%)	360 (87.0%)	Not available yet
Wales**	1,008	207 (20.5%)	393 (39.0%)	529 (52.5%)
North-West	382	9 (2.4%)	34 (8.9%)	97 (25.4%)
West Midlands	542	25 (4.6%)	45 (8.3%)	167 (30.8%)
South-West	417	91 (21.8%)	167 (40.0%)	269 (64.5%)
Totals (England)	**1,341**	**125 (9.3%)**	**246 (18.3%)**	**533 (39.7%)**
Western Side (overall) (England + Wales only)	**2,349**	**332 (14.1%)**	**639 (27.2%)**	**1,062 (45.2%)**

were only found where otters had been introduced. This was confirmed by the 1984-86 national survey, which showed that both of the two positives (3.3% of the 60 sites) were derived from captive-bred released stock, as was the case for all ten occupied sites (16.7%) in the present 1991-94 survey (see Tables 10 and 12). Thus, the Suffolk otter population had declined to virtual extinction by the early 1980s (Figure 42). The recovery was apparently all due to introduced stock.

Further examination of the other sub-areas within the eastern regional area (see Table 10) shows that the number of positive sites in Lincolnshire (50km x 50km square TF n/w) remained very low (ie. with little or no recovery from 0 to 2 positives) and the small increase in Essex (not included in Table 10 because it was not surveyed in 1977-79) from 0/107 in 1984-86 to 3/107 in 1991-94 (see Figure 13) may have resulted from dispersing otters from the releases in Suffolk (see Section 5.5). Only the East Midlands sub-area (survey squares SK s/e, SP s/e and TL n/w) appears to have exhibited some signs of recovery from four occupied sites in 1984-86 to 20 in 1991-94. However, this occurred on two rivers only. Thus, occupied sites on the River Nene increased from 0 to 2 to 5 and those on the River Great Ouse from 1 to 2 to 3 in the surveys of 1977-79, 1984-86 and 1991-94 respectively. About half of the apparent recovery was due to newly occupied sites along the River Cam and its tributaries, and is thought to consist of dispersing second and third generation otters from the releases on the rivers Black Bourn and Thet (see Section 5.5). Thus, apart from the East Midlands sub-area, the whole of the eastern regional area has shown no nadir and no recovery at all and at best would have remained as a few very small populations and individuals if the Otter Trust releases had not reinforced and then linked them. Why has there been no recovery in this whole regional area (or indeed in the south-east regional area; see Table 10)? We think that the reason must be that fragmentation of the population into very small, isolated and in the long-term non-viable fractions (see Jefferies, 1989a) prevented it happening. The losses from these small populations (indeed, some may be only one animal) owing to road and fyke net mortality may have been sufficient to continuously remove all of the 'surplus' animals so that none was available for dispersion and so colonisation of adjacent areas. Without reinforcement from released stock most of these population fragments would have petered out eventually, like that of the well-documented Glaven/Stiffkey/north Norfolk coastal marshes population studied by Weir (1984). (See also Jefferies, 1988b; Keymer *et al*, 1988; Spalton & Cripps, 1989, Wells *et al*, 1989.) The lack of recovery could not have been due to continued pollution by organochlorine insecticides and PCBs in the east, because the initial test otter releases and those that followed on were all successful, with breeding and subsequent colonisation of adjacent areas (see Section 7.2). Otters still occupy the release sites over 12 years afterwards.

7.1.3.2.8 The relationship between changes in hunting success and the timing of recovery and the scores for intensities of arable cultivation and game preservation

The mean dates for the western and eastern nadirs of the population decline and recovery curves are similar to those dates at which bans on dieldrin use became effective. This could be seen separately for the west and the east when separate bans caused considerable western and eastern area reductions in rates of contamination (see Sections 7.1.3.2.6 and 7.1.3.2.7). However, the link between an area timing of the onset of recovery and the intensity of cultivation, and so dieldrin use (ie. cause and effect), may be considered more certain if there is a significant correlation between the two. The individual hunt or regional areas with early dates for nadirs should correlate with low-intensity cultivation (ie. mostly pasture and livestock), whereas hunt or regional areas with late dates for nadirs should correlate with high-intensity cultivation (ie. mostly arable areas). Intermediate dates should occur in those areas with both forms of farming in similar amounts.

In order to examine more closely this relationship between the severity of the decline and the timing of recovery of otters, on the one hand, and the intensity and form of the main agriculture of an area, on the other, a method of scoring variation in the relative intensity of cultivation was required. First, the outlines of the hunt territories or regional areas were traced over the map of the distribution of similar levels of cultivation (three bands: less than

Table 11. The year of the nadir of the otter population decline before recovery began in each of the regional areas of the Eastern and Western Sides of southern Britain. The scores for the intensity of cultivation (high: arable; low: stock and grassland) in each of these regional areas is given for purposes of comparison and correlation (see Section 7.1.3.2.8). The dates of the nadirs for the regional areas of the Western Side of southern Britain were obtained from the data for the hunt territories included within them (see Table 9), as surveying had not started at the time that the declines of the Western populations 'bottomed out'. Note that the regional areas of the Eastern Side and those of the Western Side of southern Britain used in this analysis had similar total areas.
*Notes: † The dates of the nadirs for these regional areas were taken as the years in which they were surveyed within the national survey dates of 1977-79 and 1984-86. * There was no nadir in the decline of the otter population of the East Midlands and East Anglia, but a continuing progression towards extinction. This was only relieved by the release of captive-bred stock. ** This mean does not include the date 1986.*

Regional area (see Figure 41)	Sub-area or County	Area of Regional area or County (km²)	Score for Intensity of Cultivation	Year of nadir before recovery	Notes
EASTERN SIDE					
North-East		20,100	51.54	1978†	
South-East		10,400	57.31	1979†	
Southern		13,450	52.64	1984†	
Eastern		32,600	74.85	post 1986* No nadir	
Sub-areas of Eastern regional area	East Midlands	13,850	67.87	1978	
	Lincolnshire	5,910	80.00	post 1986* No nadir	
	Norfolk	5,370	80.00	post 1986* No nadir	
	Suffolk	3,800	80.00	post 1986* No nadir	
	Essex	3,670	80.00	post 1986* No nadir	
Eastern Side (overall)		**76,550**	**62.44**	**Mean ** 1980.3**	
WESTERN SIDE					
North-West		19,800	22.65	1969	Date of nadir from hunt territories
West Midlands		19,950	43.51	1966 post 1975	
South-West		12,750	25.06	1966 1967	
Wales		20,800	10.00	1966 1966 post 1975	
Dumfries & Galloway County		6,500	10.00	1970	Not included in totals
Western Side (overall) (England & Wales only)		**73,300**	**25.16**	**Mean 1968.7**	

Table 12. The changes in otter occupation of the main rivers of Norfolk and Suffolk from 1969 to 1992. The results of four county surveys are interspersed between those of the three national surveys. The total numbers of positive sites (full survey sites + spot checks) are given where otters were found to be present, except for the Suffolk survey of 1969-72 in which West (1975) estimated the number of otters present and whether that population was viable or endangered. The columns between those for survey results contain data on any changes in number of signs, extinctions or releases occurring between survey dates. The coloured bars indicate whether the otters shown to be present are considered to be part of the original stock (red) or derived from captive-bred and released stock (green) or both. It can be readily observed that the wild native stock decreased and became almost completely extinct by the time of the 1984-86 national survey, in which their signs were only found on one river. The steadily increasing importance of the captive-bred stock from 1983 onwards is again easily discernible. The abbreviation 'NS' indicates 'not surveyed' and the bracketed numbers indicate references, ie. (1) West (1975), (2) Macdonald & Mason (1976), (3) Lenton *et al* (1980), (4) Clayton & Jackson (1981), (5) Weir (1984), (6) Jessop (1985), (7) Jefferies (1988b), (8) Keymer *et al* (1988), (9) Wells *et al* (1989), (10) Jessop & Macguire (1990), (11) Strachan *et al* (1990), (12) this report, (13), (14), (15) Wayre (1985, 1986, 1993a).

River	Suffolk County Survey 1969-72 (1)	Norfolk County Survey 1974-75 (2)		National 1977-79 (1978) (3)		Norfolk County Survey 1980-81 (4)		National 1984-86 (1985) (11)		Norfolk County Survey 1988-89 (10)		National 1991-94 (12)	
NORFOLK													
Ant		1 positive (2)		NS		No signs (4)		NS	Release 1988	2 positives (10)	Release 1990	NS	
Babingley		1 positive (2)		6 positives (3, 12)		1 positive (4)		No signs (11, 12)		No signs (10)		No signs (12)	Fresh spraint (12) 1994
Bas		No signs (2)		NS		1 positive (4)		NS		No signs (10)		NS	
Blackwater		No signs (2)		No signs (3, 12)		No signs (4)		No signs (11, 12)		No signs (10)		No signs (12)	
Bure & Thurne		12 positives (2)		NS		No signs (4)		NS	Glaven release of 1987 moved to Bure	7 positives (10)	Release on Ant 1988/1990	NS	
Burn & Marshes		1 positive (2)		3 positives (3, 12)		No signs (4)		No signs (11, 12)		No signs (10)		No signs (12)	
Chet		No signs (2)		No signs (3, 12)		No signs (4)		No signs (11, 12)		No signs (10)		No signs (12)	
Cut-off Channel		No signs (2)		NS		1 positive (4)		NS		No signs (10)		NS	
Glaven & Marshes		1 positive (2)		NS		6 positives (4)	3 deaths 1984, 1986 (7,8,9)	NS	Release 1987	6 positives (10)	Release 1991/1992	NS	
Great Ouse & Drains		No signs (2)		No signs (3, 12)		No signs (4)		No signs (11, 12)		No signs (10)		No signs (12)	
Heacham		No signs (2)		No signs (3, 12)		No signs (4)		No signs (11, 12)		No signs (10)		No signs (12)	
Little Ouse		2 positives (2)		No signs (3, 12)		No signs (4)	Release on tributaries 1983, 1984	1 positive (11, 12) poss. from Black Bourn release	Dispersing from Black Bourn & Thet	8 positives (10)	Dispersing from Black Bourn & Thet	4 positives (12) Signs throughout river	
Muck Fleet		No signs (2)		NS		1 positive (4)		NS		No signs (10)		NS	
Nar		No signs (2)		No signs (3, 12)		No signs (4)		No signs (11, 12)		No signs (10)		No signs (12)	
Scarrow Beck		3 positives (2)		NS		No signs (4)		NS		No signs (10)		NS	
Stiffkey & Marshes		3 positives (2)		2 positives (3, 12)		2 positives (4)	3 deaths 1984, 1986 (7, 8, 9)	No signs (11, 12)	Release on Glaven 1987	No signs (10)	Release on Glaven 1991/92	1 positive (12)	
Tas		No signs (2)		No signs (3, 12)		No signs (4)		No signs (11, 12)	Possible movement from Wensum	1 positive (10)		No signs (12)	

Table 12 continued

River	Suffolk County Survey 1969-72 (1)	Norfolk County Survey 1974-75 (2)		National 1977-79 (1978) (3)		Norfolk County Survey 1980-81 (4)		National 1984-86 (1985) (11)		Norfolk County Survey 1988-89 (10)		National 1991-94 (12)	
Tiffey		No signs (2)		NS		No signs (4)		NS		No signs (10)		NS	
Thet		No signs (2)		No signs (3, 12)		1 positive (4)	Release 1984	1 positive (11, 12) From release	Released otters breeding	12 positives (10)		6 positives (12) Signs throughout system	
Tud		1 positive (2)		NS		No signs (4)		NS	Possible movement from Wensum	1 positive (10)		NS	
Waveney		No signs (2)		No signs (3, 12)		No signs (4)	Release 1984	1 positive (11, 12) From release	Still marking 1987	No signs (10)	Release 1992	16 positives (12)	
Wensum		9 positives (2)		6 positives (3, 12)		13 positives (4)		No signs (11, 12)	Possible movement from Little Ouse via Wissey	26 positives (10)		3 positives (12)	
Wissey		3 positives (2)		5 positives (3, 12)		12 positives (4)	Signs greatly diminished 1985 (11, 12)	2 positives (11, 12) Signs much diminished	Possible movement from Little Ouse	2 positives (10)	Release 1992	6 positives (12)	
Yare		3 positives (2)		NS		No signs (4)		NS		No signs (10)	Release 1989/1992	NS	
SUFFOLK													
Alde	4 otters, viable (1)			No signs (3, 12)				No signs (11, 12)			Spread from Minsmere (12)	2 positives (12)	Breeding 1992 Cubs seen (15)
Black Bourn	Extinct before 1969 (1, 6)			NS			Release 1983	Bred in 1984, 1985 Otter Trust Survey (13, 14)	Occupied & breeding 1983 onwards		Occupied & breeding 1983 onwards	Otter seen (12) during survey	
Blyth	3 otters, endangered (1)		Disappeared 1975 (6)	No signs (3, 12)				No signs (11, 12)				No signs (12) but unconfirmed sighting	Possible itinerant from Minsmere (12)
Deben	3 otters, viable (1)			No signs (3, 12)			Signs 1982 (11) Sparse signs 1984 (5)	No signs (11, 12)			Possibly from Minsmere	1 positive (12) (footprints)	
Dunwich	Present before 1976 (3)		Disappeared 1977 (3)	No signs (3, 12)				No signs (11, 12)			Spread from Minsmere (12)	Occupied (12) Spraints found	
Easton, Covehithe, Benacre	3 otters, viable (1)			No signs (3, 12)				No signs (11, 12)				No signs (12)	
Gipping	Last record 1955 (6) Extinct 1969 (1)			No signs (3, 12)				No signs (11, 12)				No signs (12)	
Hundred	2 otters, endangered (1)			1 positive (3, 12)			Sparse signs 1984 (5)	No signs (11, 12)			Spread from Minsmere (12)	Occupied (12)	
Lark	Extinct before 1969 (1, 6)			NS				No signs Otter Trust Survey (13)				NS	
Minsmere	2 otters, viable (1)		Disappeared 1977 (3, 6)	No signs (3, 12)			Release 1985/87	2 positives (11, 12) From release			Occupied & breeding 1985 onwards	Occupied (12) Spraints found	1991 Breeding (12) Cubs seen
Stour	4 otters, endangered (1)		Disappeared 1978 (6)	NS				No signs (11, 12)			Spread from Black Bourn (12)	2 positives (12) First noted in 1993	
Waveney	10 otters, endangered (1)			No signs (3, 12)		No signs (4)	Release 1984	1 positive (11, 12) From release		No signs (10)	Release 1992	16 positives (12)	

20%; 20-60%; over 60% of farmland cultivated; see Figure 38) and the proportion of each hunt territory or regional area falling into each of the three bands of cultivation intensity was counted using a transparent square grid. Then relative scores of 10, 40 and 80 were attached to each of the above three levels of cultivation intensity. By multiplying each of these cultivation intensity scores by the proportion of the territory or regional area including them, an overall score for each of these areas could be obtained (eg. Northern Counties O.H.: 0.0476 x 10 + 0.8730 x 40 + 0.0794 x 80 = overall score of 41.75). Using this method, the maximum score for an area with a high intensity of cultivation would be 80 and the minimum score for a low intensity of cultivation would be 10. In practice, most large areas would be a mosaic of small pieces of arable, urban land, nature reserves, fens and grassland with stock, all of which would have different cultivation scores. However, the higher the overall score of a large area, the greater the amount of arable, and the lower the score, the greater the amount of pasture with livestock (particularly sheep, as they form 57% of the mammalian livestock reared), ie. high scoring areas would have been using organochlorine seed dressings and the low scoring areas would have been using organochlorine sheep-dips when these materials were causing serious wildlife mortalities in the 1950s to 1970s. The area cultivation scores for the 13 hunt territories and the nine main regional areas of England, Wales and southern Scotland are shown in Tables 9 and 11.

If the above scores for intensity of cultivation in each of the hunt territories (for which we have data on 11 areas rather than for nine Regional areas) are ranked in descending order and divided into three approximately equal-sized groups (see column 2 of Table 13), then the past level of hunting success and its changes within each group can be compared with the decreasing amount of local arable and so the amount of past dieldrin use for seed dressings. Examination of the data in column 3 of Table 13 shows that, before dieldrin use started in 1956, there was no relationship between hunting success (in the period 1950-55) and the amount of arable in the hunting area. However, by 1966-71, with dieldrin in use for seed dressings on a variety of crops grown in the cultivated fields, the three groups show a series with an inverse relationship between hunting success and the arable score (see column 4), ie. as the amount of arable increased then hunting success decreased. Also, there is a positive relationship between the amount of arable and the percentage decrease in hunting success between 1950-55 and 1966-71 (see column 5), ie. as the amount of local arable increased there was a greater loss in hunting success compared to that of the pre-dieldrin years. If, for the purposes of statistical analysis, the hunts are divided into those in high and low arable areas (with a significant difference in cultivation score; see column 6), then, whereas there is no significant difference in hunting success in the two groups before dieldrin use started (see column 7), there is a significant 76.1% difference afterwards (see column 8). Again, the percentage decrease in hunting success before and after the crash is also significantly different in high and low scoring areas (see column 9). All of these significant differences provide strong pointers to a change in agricultural practice (and, by its timing, the use of dieldrin in particular) as being the main causal factor in the otter population crash of 1957.

It should be noted that dieldrin was also used in sheep-dip in the low scoring grazing areas of the west and did indeed cause a decline in the otter population of those areas (see Table 13). However, the amount of dieldrin used in sheep-dip was smaller than that used for arable purposes in the east. Also, its use in sheep-dip was banned from 1966.

A further correlation can be made between the amounts of arable (with their related higher and longer dieldrin usage) and their overall effect on the form of the local otter decline and recovery curves. Thus, comparison of the individual estimated dates for the nadirs of these curves with the scores for cultivation in each hunt territory shows a linear relationship between the two (Figure 43). This relationship is significant ($r = +0.8655$; 10 df; $p<0.001$; equation: $y = 64.37 + 0.2615x$, where y = date of nadir (1978 = 78) and x = score for intensity of cultivation). The correlation coefficient is reduced ($r = +0.8364$; 10 df; $p<0.001$) if a $\log_{10}$ transformation is used for the cultivation score. (Compare with the relationship between the game preservation scores and the nadir dates below.)

Figure 42. The changes in the otter populations of the East Anglian counties of Suffolk and Norfolk, as indicated by percentage site occupation. The data are derived from two county surveys of Suffolk (1969-72 and 1984), three county surveys of Norfolk (1974-75, 1980-81 and 1988-89) and the three national surveys (1977-79, 1984-86 and 1991-94). The total information shows that these two county populations did not 'bottom out' and recover as did those of other areas of England, but continued downwards to the point of virtual extinction. That of Suffolk would have been extinguished before 1984 and that of Norfolk by 1986 if they had not been reinforced by the captive-bred otters released by the Otter Trust starting in 1983.

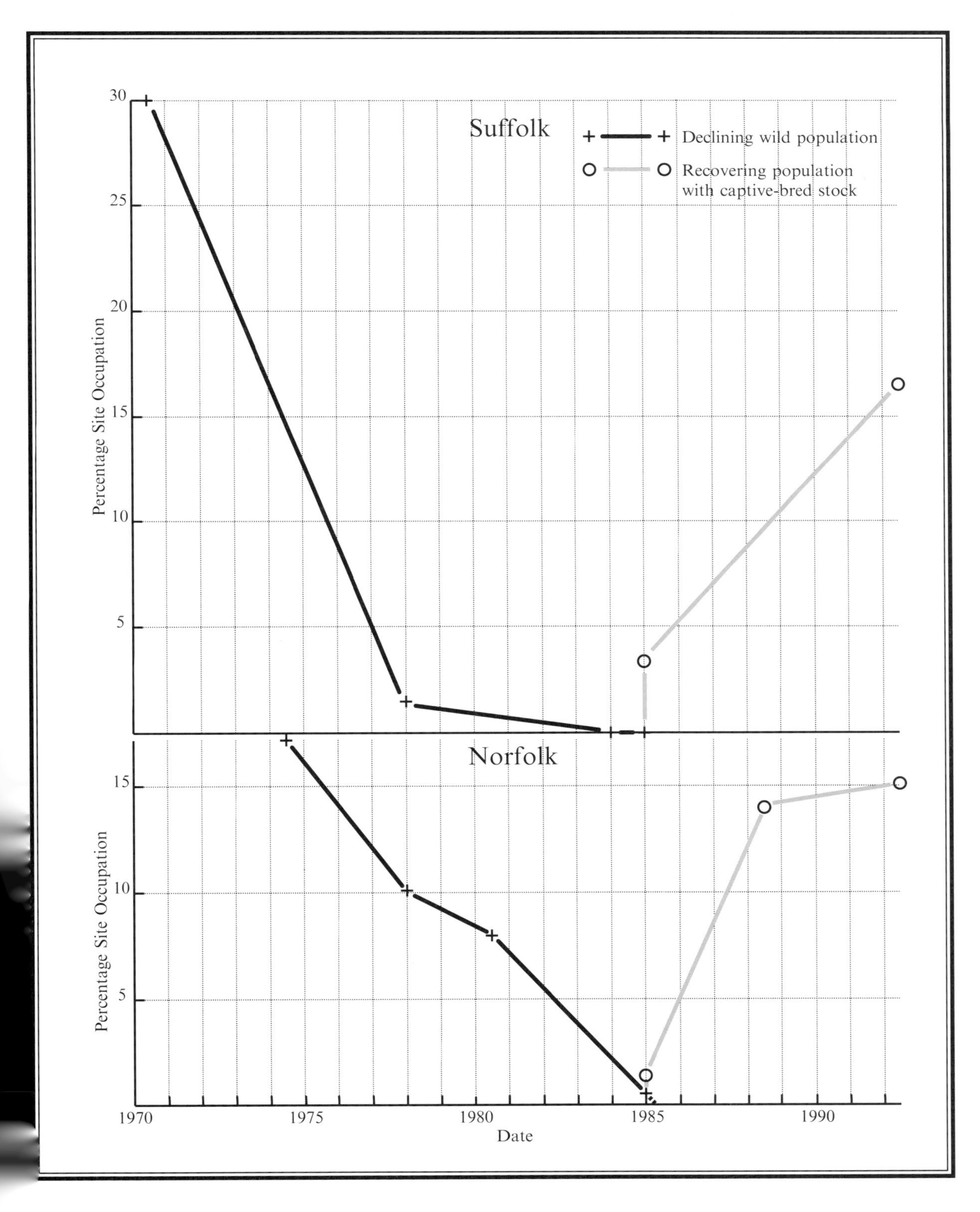

A similar linear relationship to the above exists between the amounts of arable (cultivation scores) and the estimated nadir dates for the decline and recovery curves of the nine regional areas listed in Table 11, if these are used instead of hunt territories ($r = +0.8722$; 7 df; $p<0.01$; equation: $y = 63.98 + 0.2768x$).

These significant overall correlations between the extent of area arable and the forms and nadirs of the local decline curves (ie. using a different approach and data to that of changing hunting success) provide further strong evidence that the use of dieldrin in agriculture and its banning were the major causes of decline and recovery of the otter population in Britain.

If an examination by regression analysis is made of the relationship between the nadirs of the decline and recovery curves for each hunt territory and the local scores for intensity of game preservation (see Table 9), then again a significant correlation is found between the two (Figure 43; $r = +0.8944$; 10 df; $p<0.001$; equation: $y = 54.73 + 15.075x$, where y = date of nadir (1978 = 78) and x = the logarithm of the score for intensity of game preservation). However, here the correlation coefficient is highest if a $\log_{10}$ transformation is used for the game preservation score, compared to that obtained with the normal, untransformed data ($r = +0.8661$; 10 df; $p<0.001$). Thus, the form of the relationship is not linear but curved, as shown in Figure 43. The significance of this relationship is not an anomaly due to there being a similar ranking of the hunt territories into east and west on the amount of arable and of dieldrin usage as on gamekeeper density. This can be seen from the fact that the hunt territories were not ranked in exactly the same way for these two factors (compare Tables 8b and 13). For example, Culmstock, Dartmoor, Border Counties and Dumfriesshire O.H. are ranked 5, 6, 8 and 10 out of 11 for the cultivation score and 9, 11, 5, and 6 out of 11 for the game preservation score. In addition, the form of the relationship differs (ie. one is linear and the other curved).

The effect of persecution was most marked and showed a group gradient between hunting success and game preservation before the crash in 1957 (see Table 8b). However, the killing of otters by gamekeepers, water bailiffs and the hunts, either indirectly or directly, continued after 1957 and did not stop when the population was reduced to a low level. Thus, this significant relationship confirms that, although a contributory cause rather than the main cause, persecution still had a measurable overall effect on the form of the decline and recovery curves.

7.1.4 Conclusions regarding decline and recovery and the causal factors

Information in the above Section 7.1.3 leaves little doubt that the crash of the otter population, in Britain at least, was due to the two major factors set forward by Jefferies (1989b). These were (1) a long history of persecution, which stressed the population by reducing its size and the proportion of breeding aged adults, and (2) the introduction of the highly toxic and persistent cyclodiene organochlorine insecticides as seed dressings, sprays and sheep-dips in 1956. The latter were the immediate cause of the crash by producing both a high mortality of breeding adults and a reduction in the breeding success of the survivors. The supporting basis for this conclusion can be set out as follows.

(1) Persecution

(i) The otter population crash showed a marked eastern bias, as does the intensity of game preservation (see Sections 7.1.2 and 7.1.3.1.5).

(ii) Eastern hunts killed a higher percentage of otters found and hunted than did those of the west (see Section 7.1.3.1.4 (iii)).

(iii) The Hawkstone O.H. had the highest killing rate of any of the western hunts and the local otter population curve showed the latest date for the nadir of any of those hunts (see Section 7.1.3.1.4 (ii)).

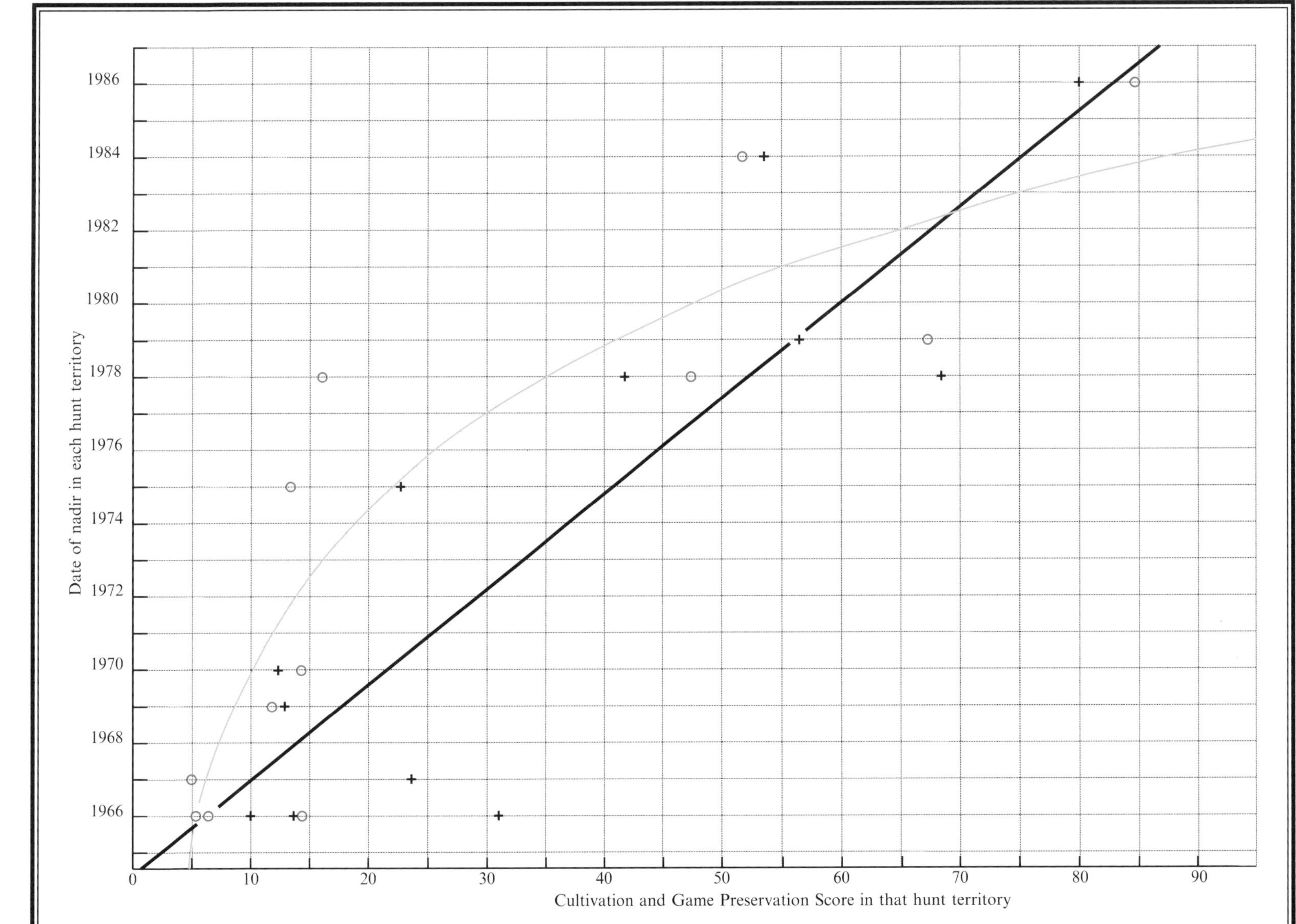

Figure 43.

The estimated dates of the nadirs of the otter population declines for each hunt territory plotted against the scores for (a) intensity of cultivation (+) and (b) intensity of game preservation (O) within those territories (see Sections 7.1.3.1.5 & 7.1.3.2.8 and Table 9). The calculated regression lines for (a) (black line) and (b) (red line) are also shown. The otter populations of eastern Britain, ie. those inhabiting areas with the highest intensities of arable cultivation and game preservation, were the last to start their recovery (see Figure 40).

Table 13. The relationships between the degree of intensity of cultivation and hunting success in the 11 hunt territories before and after the population crash. Column (2) shows the calculated scores for intensity of cultivation for each of the hunt territories ranked in descending order of amount of arable. Columns (3), (4) and (5) show the group rankings for hunting success (finds/100 days) before and after the otter crash in 1957 and its percentage decrease between 1950-55 and 1966-71. These data are shown divided into two groups with high and low intensity of cultivation in columns (6), (7), (8) and (9). Note that there is no significant difference in hunting success in areas with high and low amounts of arable in the period before the population crash (column 7). However, after the introduction of dieldrin in large concentrations as seed dressings in 1956, the difference in hunting success between areas with high and low amounts of arable is statistically significant (column 8). The difference in the percentage decrease in hunting success between 1950-55 and 1966-71 in these two areas is significant also (column 9).

Hunt territory (1)	Score for intensity of cultivation (2)	Hunting success 1950-55 (3)	Hunting success 1966-71 (4)	Percentage decrease in hunting success (5)	Score for intensity of cultivation (6)	Hunting success 1950-55 (7)	Hunting success 1966-71 (8)	Percentage decrease in hunting success (9)
Eastern Counties	80.0	72	38	47.22				
Buckingham	68.5	70	28	60.00				
Courtenay Tracy	53.6	76	36	52.63				
Mean ± S.E.	**67.37 ± 7.64**	**72.67 ± 1.76**	**34.00 ± 3.05**	**53.28 ± 3.70**	49.70 ± 8.93	70.50 ± 4.66	33.17 ± 3.10	53.03 ± 2.87
Northern Counties	41.7	55	30	45.45				
Culmstock	30.9	88	44	50.00				
Dartmoor	23.5	62	23	62.90				
Hawkstone	22.8	121	66	45.45				
Mean ± S.E.	**29.72 ± 4.39**	**81.50 ± 14.96**	**40.75 ± 9.48**	**50.95 ± 4.13**	14.34 ± 2.21	83.60 ± 10.54	58.40 ± 8.36	28.51 ± 9.88
Border Counties	13.9	71	71	0.00				
Kendal & District	12.9	59	40	32.20				
Dumfriesshire	12.1	89	78	12.36				
Pembroke & Carmarthen	10.0	78	37	52.56				
					Significance of difference			
					t = 3.5121	t = 1.2105	t = 3.0447	t = 2.5905

(iv) Hunting intensity, in terms of number of days spent hunting, was significantly gher in the east than in the west (see Section 7.1.3.1.4 (iv)).

(v) The western hunts showed a significantly higher success rate prior to the crash in)57 than did the hunts of the east. This suggests that the eastern otter population was ready at a lower density than was that of the west, owing to earlier persecution where ame preservation was at its highest intensity (see Section 7.1.3.1.5).

(vi) The effect of removing the persecution factor can be seen by the significantly creased catches of otters after the 1914-18 World War when gamekeeping and hunting rtually ceased for four years (see Section 7.1.3.1.3).

(vii) The Eastern Counties O.H. was removing a greater number of otters from Norfolk nd Suffolk in the 1950s to 1970s than the Otter Trust released to rebuild the population ccessfully in the 1980s and 1990s. (see Section 7.1.3.1.4 (i)).

(viii) There is a significant correlation between the intensity of game preservation area y area and the dates of the nadirs of the area decline and recovery curves. The earliest adirs occurred in areas of low game preservation (see Section 7.1.3.2.8).

(2) Organochlorine insecticides

(i) The otter population crash started within 18 months of the introduction of the highly oxic cyclodiene organochlorine insecticides, aldrin, dieldrin and heptachlor. It started multaneously over England, Wales and southern Scotland, indicating the sudden country-ide introduction of a new, possibly man-made, environmental factor (see Sections 7.1.2 nd 7.1.3.2).

(ii) The effect was greatest in the south and east, which coincided with the region aving the highest area usage of these materials for seed dressings (see Sections 7.1.2 and .1.3.2).

(iii) The populations of the two predatory birds, the peregrine falcon and the parrowhawk, crashed at exactly the same time and with the same area of greatest effect. hese two declines are independently considered to have been caused by dieldrin use (see ection 7.1.3.2.5).

(iv) Sudden changes in the rates of decline and the start of recovery shown by the vestern and eastern otter population curves occurred at dates which were correlated with the ntroduction of each of the bans on organochlorine usage (see Sections 7.1.3.2.4, 7.1.3.2.6 nd 7.1.3.2.7).

(v) There is a significant linear relationship between the intensity of arable cultivation f an area (and so dieldrin usage) and the date of the nadir in the decline curve for that area. his is earliest where the amount of cultivation is low. So these dates showing when decline hanged to recovery were earlier in the west than in the east. Thus, recovery is now rogressing from the west to the east, as was the case for the above two predatory birds (see ection 7.1.3.2.8).

(vi) Relating hunting success to intensity of arable cultivation showed that, whereas efore dieldrin use started in 1956, there was no significant correlation between otter unting success and the amount of arable in an area, after 1956 there was a significant nverse relationship between the two. As arable cultivation (and dieldrin use) increased, then unting success decreased (see Section 7.1.3.2.8).

(vii) The part played by PCBs in the decline itself was apparently negligible. Very little PCB was found in British mammals, including otters, in the 1960s and changes in the rate of decline and timing of recovery do not correlate with the few PCB bans (see Section 7.3.1.2). PCBs have the potential to affect the rate of recovery, though whether they have done so is still uncertain (see Section 7.3.1.2.4).

Further reinforcement for the above conclusions regarding persecution and organochlorines is given by the fact that only two major actions relating to otter conservation had been taken in Britain by the late 1970s. The first was the granting of legal protection for the otter in England in 1978 and the second was the series of bans placed on the uses of the organochlorine insecticides starting in 1962. The first brought the direct organised hunting of otters with hounds to an end and hopefully reduced the numbers of otters killed during game preservation. The second brought an end to the large wildlife kills of the late 1950s and early 1960s and eventually a lower level of pollution in British waterways. These two actions were sufficient to cause the cessation of the decline and the initiation of recovery in Britain. Thus, we have here a classic case of hypothesis followed by research and action, followed by recovery confirming the hypothesis, as set out by Jefferies and Mitchell-Jones (1993) when writing about the design of recovery plans. Unfortunately, as they noted, these do not form an orderly progression with any mammals of conservation importance, as management needs are too urgent to wait for scientific proofs and action has to be taken on the hypothesis.

One noticeable factor in this story is the remarkable speed at which the otter population reacted to the presence of the cyclodiene organochlorine insecticides in the food chain. The population crash started within 18 months of the first major use of these compounds in 1956. Also, when the first ban was instituted in 1962 there was a marked change in the rate of population decline after only one year. However, as the years passed and presumably environmental and animal contamination increased, then the reaction time lengthened. It was 2.4 years before there was a measurable response by the western population to the ban on dieldrin use in sheep-dips in 1966 and 4.7 years before there was a response by the eastern population to the final 1975 ban on all dieldrin use in cereal seed dressings. Another corresponding factor is that the early nadirs of the western declines were clearly marked, with a rapid change from decline to recovery, whereas those of the eastern population, which had received high dieldrin levels for many years, were shallow, with the population curve 'bumping along the bottom' for up to ten years before there were signs of an upturn and recovery. Presumably, again, this was due to the greater level of environmental contamination in the east in the 1970s and 1980s than in the west in the 1960s.

7.1.5 Aspects of recovery

7.1.5.1 The pattern of recovery

The three otter surveys of Wales (Crawford *et al*, 1979; Andrews & Crawford, 1986; Andrews *et al*, 1993) and the three otter surveys of England (Lenton *et al*, 1980; Strachan *et al*, 1990; this report) show a much higher site occupation in the west of southern Britain than in the east. This correlates with past agricultural usage of dieldrin and the timing of bans on its sale (see Sections 7.3.1.1.1 and 7.3.1.1.5). Also, there is a progression throughout the three surveys of England suggesting a wave of recovery from west to east (see Figures 2, 3a and b and Section 7.1.3.2.5). This could have occurred by dispersal eastwards of animals bred in the high-density areas of the west or by breeding within the low-density areas, as pollution levels decreased, or both.

Examination of the changes in percentage site occupation in particular 50km squares from 1977-79 through 1984-86 to 1991-94 (Figures 2, 3a and b) shows the presence of a distinct 'edge effect'. Thus, in the north, squares NT s/e, NY n/w, NY s/e, SD n/w, SE n/w (and presumably NZ s/e) have similar low-level populations in 1977-79 (Figure 3a). By 1984-86 these had increased in squares NY n/w and SD n/w, whereas they had remained similar and low in the others. However, by 1991-94, all these squares showed an increase but with a decreasing progression from the north. Thus, the mean percentage site occupation was 49.8% for squares NT s/e and NY n/w, 27.0% for squares NZ n/w, NY s/e and SD n/w, and 10.2% for squares NZ s/e, SE n/w and SD s/e. It would appear then that most of the recovery in the northernmost counties of England was due to a dispersal movement southwards, originally from lowland Scotland across the border and then from animals bred in northern English squares. There could have been no local breeding to produce 7.32% from 0% in square SD s/e, for example. Only square NZ n/w in the north retained a percentage site occupation of 17 and above throughout the 14-year period and so could have produced most of its increase locally. The otter population of lowland Scotland remained stronger than that of northern England (Green & Green, 1987) and the decline was not so marked (Chanin & Jefferies, 1978).

On the Welsh borders, only square SO n/w appears to have retained a sizeable and increasing population throughout 1977-79 to 1991-94. The increase was probably due to local breeding. Squares SJ n/w, SJ s/e and SO s/e, on the other hand, had 0% site occupation in 1977-79, with a small presence in 1984-86 increasing to a mean of 28.1% by 1991. This was almost certainly the result of dispersal of otters eastwards from Wales and the borders, as the population could not have increased locally by that amount in just seven years. The small presence in square SP n/w in 1991-94 probably represents the 'spearhead' of that advance.

In the south-west, square ST s/e remained with a very low percentage site occupation in 1977-79 and 1984-86 but showed a sudden increase to 17.4% by 1991-94. Again this was almost certainly due to dispersal from the nearby high-density population of Devon and Cornwall to the west.

Where populations have been lifted artificially above a critical level, local breeding may improve population status considerably, as has occurred in the squares of East Anglia (TF s/e, TL n/w, TM n/w, TL s/e). Here there is no nearby strong population to provide young otters to disperse into the area, but released captive-bred otters (see Section 7.2.5.2) are known to have bred (Jefferies & Hanson, 1988a, 1991).

Where populations were below a critical level in 1977-79 and 1984-86 and had no chance of 'reinforcement' by dispersal from stronger populations, their rate of increase has been extremely slow (eg. in squares SU s/e, TR n/w and TQ s/e).

In conclusion, most of the recent improvements in the northern and western populations of England, outside Devon and Cornwall, have been due to dispersal from adjacent stronger populations in lowland Scotland, Wales and the West Country, rather than to local breeding. However, a small amount of the noted improvements countrywide could have been generated locally by breeding, as has been demonstrated by the populations of East Anglia and the south-east. Obviously, the waterways of central and eastern England are now able to sustain a breeding population of the species.

The resulting map (Figure 2) now shows some degree of otter presence in all the 50km squares surveyed, even in the Midlands where resident otters have been absent for many years (Jefferies, 1991b).

7.1.5.2 Rate of expansion

Research has shown that another large mustelid, the badger, has a particularly low rate of recolonisation of ground cleared of the species in the past by gamekeepers, organochlorine insecticides and the Ministry of Agriculture (Kruuk & Macdonald, 1985; Cresswell *et al*, 1990; Reason *et al*, 1993). Recolonisation by the otter too appears to be surprisingly slow (Section 6.2) considering its mobility and capacity for wandering between watersheds (Jefferies, 1989c) and across upland areas several miles from waterways (Batten, 1920).

It is known that an adult male otter can swim at the rate of 1.7 to 2.0 km/hour and can cover a distance of 14.45km in a single night by such continuous swimming (Green *et al*, 1984). Speeds when travelling over land could be even higher. This means that it is technically feasible for an otter to complete a journey across England in a little over a fortnight. Thus, a journey of 180km, the distance between Wales and Cambridgeshire, could be completed in 12 nights, or 230km, the distance between Devon and Kent, could be completed in

16 nights. However, this does not appear to happen, and the east of England remains sparsely populated, with an even lower density population in the Midlands, ie. dispersing and recolonising otters do not appear to make such long journeys.

The way in which an otter, or group of otters, actually recolonises 'new' ground can be seen from examination of the behaviour of release groups. Jefferies *et al* (1986) showed by radio-tracking that the known home range develops slowly, with bursts of rapid expansion along the main river followed by periods of consolidation and accumulation of knowledge of the area and its local ditches, woods, streams and potential holt sites. Thus, the overall expansion rate is low, only 388m per day from nights 1 to 40. After that it decreases further to only 20m per day after day 163. The mean expansion rate over the first 150 days is 257m per day and, at this rate, an expansion of range of 50km would take 195 days or 6.5 months.

In practice, the expansion rate is very much slower than 50km in 6.5 months. This is because of the time taken for development and establishment of territories, adult mortality at a young age (Section 7.1.3) and a slow breeding rate. The age at first breeding is 17 months for males and 22 months for females (Jefferies, 1988a). The mean number of cubs per litter is only 1.78 on Shetland (Conroy, 1993) and the cubs stay with the mother for 10 to 12 months (Wayre, 1979). Conroy (1993), using data from Shetland, has calculated that, with present high breeding-aged adult and cub mortality, a female will produce only 1.2 female and 1.2 male cubs of breeding age in her lifetime, ie. there are few cubs which are surplus to maintaining the local population.

An indication of the rate of advance of a 'rolling front' of an established breeding otter presence can be gauged by examination of Figures 2, 3a and b. One example is given by 50km x 50km square SJ s/e, in which percentage site occupation increased from 0 to 29.7% in the 14 years from 1977-79 to 1991-94. Also, 29.7% is close to the 27.5% site occupation shown by the adjacent square SO n/w in 1977-79 before it increased to 79.1% by 1991-94. Thus, a 'rolling front' of equal site occupation had moved forward into previously 'empty' country only one square's width (or 50km) in 14 years.

This is a rate of only 3.6km per year and equates with the prediction of several decades before there is an established and considerable breeding presence across the width of England (see Section 6.2.2).

As noted above, the apparently very slow rate of movement of the 'rolling front' may be largely due to the very low intrinsic rate of natural population increase of the otter. High accidental mortality, such as that on the road, may well exacerbate this situation by reducing the length of the breeding life. However, recolonisation may be no slower than would be expected in a large mustelid (compare with the badger). There are no indications at present of contributions from any extrinsic factors such as from the known PCB contamination of the environment.

7.1.5.3 Possible genetic changes in a recovered population

The above very slow rate of recolonisation of 'empty' areas by the otter suggests that when England and Wales were fully populated by this species, with all territories occupied, then there may have been a high degree of genetic stability in the system.

A long period of such stability may have allowed the development of east/west and north/south clines in the genetic constitution of the various area populations, not all of which may have been discernible in the phenotype. Physical separation of the otters of Ireland from those of Britain by the Irish Sea has resulted in a separate subspecies *Lutra lutra roensis* with a darker pelage (Dadd, 1970) and other morphological differences (Lynch *et al*, 1996) when compared to the British animal *L.l. lutra*. These may have developed after a period of only around 8,000 years' separation. The Irish Sea was a narrow channel or had a land bridge before the sea-level rose (Yalden, 1982).

Such genotypic and phenotypic variation is of considerable scientific interest (Jefferies & Mitchell-Jones, 1993). Unfortunately, the severe decline starting in 1957 has resulted in the near extinction of the otter population of the eastern side of England. This factor, when coupled with the compensatory eastward expansion of the stronger population of western England and Wales will almost certainly have obliterated most of the evidence of any original east/west genotypic clines over southern Britain. The otter population of southern Britain will now have a strong western bias to its genetic constitution. However, the presence of any north/south genotypic or phenotypic clines may still be discernible in the future, though undoubtedly altered by the southward expansion of the otters of southern Scotland into northern England.

This destruction of any, as yet unstudied, clines is particularly unfortunate as they may have been developing since Britain was repopulated by otters from the Continent with the amelioration of the climate following the last Ice Age some 13,500 years ago (Yalden, 1982).

7.1.5.4 End point

The form of the recovery curve (see Section 6.2.1) and the predicted dates for various stages in the recovery of the otter population of England (see Section 6.2.2) show that 'complete' recovery (at whatever level that is taken to be) is still a long time off. This is worrying. The possibilities of new threats, such as disease (see Section 7.3.7) or something as yet unforeseen, increasing mortality above a critical level and causing a further population crash before the revived population is strong enough to ensure sufficient survivors for continuation are unfortunately still too high. The only way known to speed up recovery and increase area population size rapidly is by releasing captive-bred or rehabilitated animals (see Section 7.2). This form of conservation management should not be dismissed or abandoned too early.

With regard to the final population or end point achieved, it is most unlikely that the pre-crash population of England in 1956 was as numerous as the contemporary population of Ireland, for example (Jefferies, 1989b). Chapman and Chapman (1982) found that overall percentage site occupation was as high as 92% in Ireland in a survey conducted over 1980-81. This is probably one of the highest overall levels in a western European country, with individual area site occupation as high as 100%. Indeed, the otter population density of England in 1956 was almost certainly already greatly below that of Scotland or Wales, let alone approaching that of Ireland. The only comparative information for such a suggestion is that provided by the 1952-54 otter survey of River Board Areas (RBAs) by Stephens (1957) (see Section 7.1.1). She listed five (71.4%) of the seven RBAs of Wales (including Severn) in the "very numerous" category. In contrast, only six (22.2%) of the 27 RBAs of England (including Severn again) were grouped in this category, and in several areas the otter was considered to be "scarce". On the other hand, the otter was considered to be "very numerous" over all areas of Scotland. Using Stephens's grades of 4 ("very numerous"), 3 ("numerous"), 2 ("fairly small population") and 1 ("scarce"), the mean grades for Scotland, Wales and England were 4.00, 3.57 and 2.78 respectively.

However, there is no reason why the otter population of Britain, and indeed England, should not be higher and more generally distributed in the future than it was just before the crash in 1957. The British peregrine falcon population crashed at exactly the same time and over the same area as that of the otter and this is considered to be due to the same cause, organochlorine insecticides (Jefferies & Prestt, 1966; Chanin & Jefferies, 1978; Ratcliffe, 1993). By 1963 the situation was at its worst, with the number of peregrines down to about 360 pairs or 44% of the 1939 level. Voluntary bans on the use of organochlorines for spring-sown cereal seed dressings in 1962 (see Section 7.3.1.1.5) resulted in a marked recovery, and by 1971 numbers were at 54% of the 1939 level. By 1975 the figure had risen to 60%, and by 1981 to 90% of the pre-war population. By 1985 it had exceeded it (Ratcliffe, 1972, 1984; Batten *et al*, 1990). The difference is due to enhanced and enforced legal protection and a greater public esteem in a more environmentally conscious era. The otter too is now regarded with much public affection (Serpell, 1991) and as a figurehead for freshwater conservation. It is no longer looked on as just another pest. It has full legal protection so can no longer be hunted with hounds or shot. Bye-laws have been enacted to prevent its drowning in fyke nets (see Section 7.3.3.2.1) and there are extensive programmes to reduce freshwater pollution from all sources. Thus, the English otter population could indeed spring back with an enhanced population size if given time to recover without any further environmental or biological catastrophes.

7.2 Releases of otters into England

7.2.1 Historical necessity

It may be difficult to appreciate today, but by the mid-1970s the increasing reports of the local disappearance of the otter (Lloyd, 1962; Anonymous, 1969; Macdonald & Mason, 1976; Worsnip, 1976) and their confirmation countrywide by Chanin and Jefferies in 1978 suggested to the conservationists of the time that the decline would continue to the point where the species would be lost from England other than perhaps the West Country. Also, the apparent rate of the decline suggested that this may be soon. The Joint Otter Group (JOG) set up in 1976 to examine the available information considered this a strong likelihood and agreed that all measures should be taken to save the population (O'Connor, 1977, 1979). One of the measures recommended for investigation was that of releasing otters, following a suggestion put forward at the time by Philip Wayre. One of us (DJJ) was requested by JOG to carry out the necessary research with the Otter Trust. This research has been reported in detail elsewhere (see references in Section 7.2.3). The first experimental trial of the developed technique for the release of captive-bred stock took place in Suffolk in July 1983 (Jefferies *et al*, 1986; Wayre, 1989). There is an obvious need to set down what has been done for future historians of the species and also to measure its success, as it appears to have had a considerable effect in changing the course and timing of recovery in the east.

7.2.2 The objectives of releasing otters

There are several conservation reasons for releasing otters. These are:

(i) to repopulate a large area from which the otter has become extinct. This would be classed as a reintroduction (Linn *et al*, 1979).

(ii) to strengthen an existing population when this has become fragmented into non-viable fractions. This would be classed as restocking (Linn *et al*, 1979). This technique enables at least some of the local genetic material to be retained and perpetuated, when otherwise it would be lost with the eventual deaths of the few remaining isolated otters (Jefferies, 1988b; Jefferies & Mitchell-Jones, 1993).

(iii) to confirm that otters are now absent from a test river, not because it is polluted, but because there are no otters in the area to populate it (Jefferies, 1989b). Experimental release groups can be used as an invaluable practical test or 'probe' to confirm that the original cause of the decline had been determined correctly, had been rectified and had ameliorated by the time of the release (see Caughley, 1994).

(iv) to help colonise new areas in the van of the main recovery population by the 'seeding' of animals (particularly females) in suitable areas in the path of the expanding population. This would provide mates for the first immigrants (mostly young males), which would otherwise pass on as itinerants. This technique could greatly improve the present slow rate of expansion of the area inhabited by a settled, stable, territorial and breeding otter population and so the rate of population recovery.

(v) to hasten the end point of the recovered population. While a population is abnormally small, unevenly distributed and recovering as slowly as that of the southern British otter population, it is vulnerable to any new negative factor such as a disease epizootic. The faster the population recovers, the better.

Although, initially, the release techniques were devised for the extreme case, that one needed to repopulate completely large areas of Britain from which otters had become extinct (see paragraph (i) above), in practice the only releases carried out have been restocking or population strengthening exercises (see paragraph (ii)). The first releases also served as 'probes' (see paragraph (iii)). More recently the aim has been to hasten recovery (see paragraph (v)). The technique described in paragraph (iv) was tried experimentally by the Otter Trust on the rivers Wylye, By Brook and Stour in Wiltshire and Dorset in 1989 to 1991.

7.2.3 Derivation of the otters and patterns of release used in England

There are three main sources of otters for release.

(i) *Capture of wild otters:* This can only be done if native populations are available which are strong enough to withstand the loss of the individuals taken. Such a procedure has not been considered prudent in Britain.

(ii) *Captive-breeding:* The advantage of captive-breeding of animals of conservation importance is that the few animals used for the initial breeding stock can be selected from the best area for the genetic constitution required and then maintained over a long and safe productive breeding life (Jefferies & Mitchell-Jones, 1993). Also, there is more control over the age of the animal at release. Philip Wayre has been breeding *Lutra lutra* from British native stock at the Norfolk Wildlife Park and Otter Trust since 1970 (Wayre, 1989).

(iii) *Rehabilitated wild stock:* Injured adult otters (eg. from road traffic accidents) can be legally taken into captivity and given veterinary attention and convalescence. Similarly, dependent cubs (eg. when the mother has been killed) can be legally reared to independence in captivity. Both are required to be released when no longer incapacitated. This technique of 'salvaging' wild otters for future release has been pioneered by Green and Green (1991) of The Vincent Wildlife Trust.

There have been three different forms or patterns of introduction and two derivations of the otters used by the two programmes releasing otters into eastern and southern England in the 1980s and 1990s.

(i) *Multiple catchments:* The Otter Trust and the Nature Conservancy Council released groups of three otters (two females and one male) into gaps in the known distribution of wild otters as shown by survey. This work, starting in 1983, concentrated on the East Anglian counties of Norfolk and Suffolk (Jefferies *et al*, 1986; see Appendix 13a) and used the Otter Trust's captive-bred stock which contained a large proportion of East Anglian genetic material in its constitution. This originated from otters caught during the early part of the campaign to eradicate the coypu in these counties before legal restrictions on catching the former species (Wayre, 1989; Jessop & Cheyne, 1993). These otters were released at an age of 18 months in order that breeding should occur within a short period after release whilst the animals were likely to be still together in the same area (Jefferies & Wayre, 1984; Jefferies *et al*, 1986). Their progeny should then provide further otters for recolonisation of the surrounding area. It was believed that the expansion of these new foci by breeding and dispersion should eventually enable the fragmented wild population to coalesce with these inserts and so the remaining wild stock. Thus, in the terminology of the Working Group on Introductions (Linn *et al*, 1979), this is not a reintroduction but a programme of 'restocking' and strengthening of an existing population. The background research to the project and the reasoning behind the method used have been reported by Jefferies and Mitchell-Jones (1982), Jefferies (1984), Jefferies and Wayre (1984) and Jefferies *et al* (1985, 1986). This programme of releases has been continued by the Otter Trust, who have extended it to include central southern England. It is ongoing.

(ii) *Single catchment:* The Vincent Wildlife Trust has released groups of rehabilitated otters into North Yorkshire. Here the work has concentrated on a single large release area, the River Derwent catchment and nearby River Esk, with relatively few release sites being used (see Appendix 13b). The stock originates largely from Scotland, though a few English animals have been included. With this form of release the idea is to saturate one catchment with otters. The released animals themselves will then disperse from the release sites into surrounding empty areas, tributaries and catchments when numbers exceed the optimal carrying capacity of the original release rivers (Green & Green, 1992, in press; Woodroffe, 1993, 1994). This programme started in 1990 and continued till 1993. Again, it may be termed 'restocking' or strengthening, as a very small wild population existed in the area prior to releases (Strachan *et al*, 1990). In 1994 the programme was extended to cover other areas and river systems (J. Green, pers. comm.) (see Appendix 13b).

(iii) *Placing in the path of the expanding 'front':* As reported in Section 7.2.2 (iv) above, the Otter Trust placed four experimental groups of otters in the path of the expanding and recolonising population 'front' in Wiltshire and Dorset in 1989 to 1991 (see Appendix 13a and Section 5.7). This was a small-scale experiment and only four male and five female captive-bred animals were released.

7.2.4 Procedure and monitoring

All releases reported in Section 7.2.3 relied on the 'soft' release system developed for otters by Jefferies *et al* (1986) and detailed, with precautions, by Jefferies and Wayre (in press). The animals are maintained and acclimatised in a pen on the bank of the chosen release river for at least three weeks prior to release. After release, food is provided within the pen for a further 12-day period. However, there has been no experience of captive-bred or rehabilitated otters being unable to catch fish and feed themselves well within that period.

Fish from the selected release area were analysed for organochlorine insecticides, PCBs and heavy metals before the first release (Jefferies & Freestone, 1985), but, as there are difficulties in evaluating the threat of 'cocktails' of pollutants and the first releases showed that eastern rivers were capable of supporting otters again by the 1980s, this has been discontinued by the Otter Trust.

The success of the first two release groups of captive-bred otters was monitored by radio-tracking, using a harness which disintegrated after 50 days leaving the animals free of any encumbrance (Mitchell-Jones *et al*, 1984; Jefferies *et al*, 1986). After radio loss, these and successive groups were monitored by spraint survey, footprints and observation (Jefferies *et al*, 1986; Ridding & Smith, 1988; Jessop & Cheyne, 1993). The rehabilitated otters were again monitored, first by radio-tracking, using a similar harness, and then by spraint survey and observation (Green & Green, in press). All bodies of otters retrieved from the release areas were autopsied to determine cause of death (Jefferies & Hanson, 1988a, 1991; Jessop & Cheyne, 1993; Green & Green, in press). All released otters have been tagged by implanted microchips with unique numbers since 1991.

7.2.5 Release statistics

7.2.5.1 The numbers released

The Otter Trust has released 55 captive-bred otters (24 males/31 females) into English rivers between 1983 and the end of 1993 (see Appendix 13a). 37 (67%) of these (17 males/20 females) were released into the East Anglian counties of Norfolk (27) and Suffolk (10), as the Otter Trust's breeding stock has had a considerable input of East Anglian genetic material (Jessop & Cheyne, 1993). A further 18 captive-bred otters (7 males/11 females) were released to rivers of central southern England (see Appendix 13a). A total of 16 release sites has been used on the same number of rivers.

The Vincent Wildlife Trust has released 25 rehabilitated otters into England between 1990 and 1993 (see Appendix 13b). 21 (84%) of these were released into the River Derwent catchment and the remainder to the River Esk. A total of 11 release sites was used on the two systems. In all, 80 adult otters have been released by the two programmes up to the end of 1993. The releases of 1994 have brought this total to 105. The latest animals have been released largely in Lincolnshire and the East Midlands (P. Wayre and J. Green, pers. comm.).

7.2.5.2 Confirmed breeding successes

There is no doubt that most of the released groups of otters, both captive-bred and rehabilitated, have bred successfully. This has been ascertained from dated sightings of groups of otters (mother with cubs) in the release areas, which have been obtained from many independent observers, and the finding of small footprints alongside adult ones during survey. The first known cubs usually appear just over one year after the release date in both programmes (Jefferies & Hanson, 1988a; Green & Green, in press). Where this could be ascertained more accurately at the Minsmere release (Otter Trust release 04), cubs were born 14.5 and 17 months after release when the females were 2 years 10.3 months and 3 years 0.6 months old (Jefferies & Hanson, 1988a).

One landowner living at the site of the River Thet release (Otter Trust release 02) has recorded each of the litters born in most years since the release date of 16th July 1984 (Ridding & Smith, 1988; Wayre, 1994). Otters are known to be capable of producing cubs at approximately 12-month intervals (Jefferies & Hanson, 1991), though the interval is usually 15 to 18 months (Wayre, 1988). These frequent sightings of otter cubs on the Thet are occurring on a river where no signs of otters had been seen since before 1974 (see Table 12). The records have been listed by P. Wayre and published annually in the Otter Trust's Journal (eg. Wayre, 1994). Where sites have been subjected to many hours of observation by interested landowners and conservation officers, the lists of successful breedings are probably almost complete. Known successful breeding occurred on the first Otter Trust release site (River Black Bourn) in 1984, 1985, 1986 (twice, ie. two females with cubs), 1987, 1988, 1989 and 1993, and on the second Otter Trust site (River Thet) in 1985, 1986 (twice), 1987 (three females with cubs), 1988 (twice), 1991, 1992 (twice), 1993 and 1994 (twice) (R. Shuter, pers. comm.). It is almost certain that some of the otter cubs seen at the earliest release sites in recent years represent the third generation. Thus, a female with two well-grown cubs was seen near the River Black Bourn on 8th April 1994 (K. Nicholas, pers. comm.). At the date of birth of these cubs (ie. late 1993), the original two females which were released on 5th July 1983 would have been 13 years old. Even if they had survived that long, this would have been a considerable age for a fit and breeding female (see Jefferies, 1988a). It should be noted also that three females were observed breeding on the River Thet in 1987. The two marked gaps in both the Black Bourn and Thet series may have been when the original females died and the second generation of females took over the local breeding holts.

Wayre (1994) has added all the various records of certain breeding at Otter Trust release sites, and these came to a total of 36 by the end of 1993. Using similar data, Green and Green (in press) of The Vincent Wildlife Trust concluded that successful breeding had occurred on the Derwent catchment on 8 to 11 occasions. Both are obviously minimal totals.

7.2.5.3 Casualties among the released otters

There have been small numbers of casualties in both release programmes. These have amounted to 11 (or 13.7%) of the 80 released. However, death was often not until many months after release and after breeding (Jefferies & Hanson, 1988a).

(i) The Otter Trust: Casualties have occurred in both the initial release generation and their progeny. Thus, road traffic accidents caused the deaths of the released males in the Otter Trust releases (see Appendix 13) 04 (Minsmere; Jefferies & Hanson, 1988a), 10 (River Yare; Jessop & Cheyne, 1993) and 16 (River Stort; Jessop & Cheyne, 1993; Mason & Macdonald, 1993) and one of the released females in release 23 (River Itchen; Wayre, 1994). Two of the male progeny from release 04 were also killed on the roads (Minsmere; Jefferies & Hanson, 1988a, 1991). One further male was killed in the pre-release pen during fighting with the two females (River Glaven; Jessop & Cheyne, 1993). The released male from release 03 (River Waveney; Jessop & Cheyne, 1993) disappeared from the river during a period of eel fyke net fishing and the released female from release 07 (Catfield; Jessop &

Cheyne, 1993) was found dead in the River Ant, which again is fyke-netted. These are both considered to have been fyke net casualties (Jessop & Cheyne, 1993).

(ii) The Vincent Wildlife Trust: There have been four known losses from the VWT release and rehabilitation programme in the River Derwent catchment. Two released males were drowned in a single illegally set eel fyke net in 1992 (Woodroffe, 1993; Green & Green, in press). This was one of only nine illegally set nets known to have been deployed in the river during the release programme and demonstrates again the efficiency of these traps for catching otters, even when the latter are at low density (see Jefferies *et al*, 1993). A third otter, a released female, was killed on the road in 1992, 18 months after its release (Woodroffe, 1993; Green & Green, in press). A fourth otter was found dead recently alongside a pheasant release pen (Green & Green, in press). The cause of death of this animal is uncertain at present.

Thus, out of 13 known casualties, seven (54%) were caused by road traffic accidents, four (31%) were known or likely fyke net casualties, and a further two (15%) were or may have been due to other traumas. These are common causes of deaths among wild otters and are to be expected. There have been no known casualties among the released otters or their progeny which have been due to pollutants. All retrieved bodies of Otter Trust casualties have been autopsied and analysed for toxic pollutants. Some of these have been published (Jefferies & Hanson, 1988a, 1991; Mason & Macdonald, 1993).

7.2.6 Measuring the success of the two release programmes

The only real criterion for the success of any release programme is the measurable long-term presence of breeding populations occupying the chosen release sites and showing some expansion at its conclusion over the region as a whole. There is no doubt that most of the released otters of both programmes are breeding successfully (see Section 7.2.5.2) and that nearly all of the chosen release sites (eg. River Black Bourn, River Thet, Minsmere) are still occupied (some after 12 years) with no further additions other than four replacement males (see Appendix 13). Only the releases on the River Glaven on the north Norfolk coast appear to have moved from the chosen release site, showing a recent preference for the River Bure and possibly the Wensum systems after remaining for a period of three years up to March 1994 (R. Monteath, pers. comm.). Presumably the continuous long-term occupation of release sites elsewhere up to the time of survey is being maintained after the death of the original release stock. Further, examination of the map of England showing the rivers now occupied through natural recolonisation and those considered to be occupied by released otters and their progeny (Figure 6) shows that the two release programmes have succeeded in occupying the large areas attempted, to the extent that this is demonstrable during a national spraint survey.

That the above recovery is due to the otter releases themselves is shown by several indisputable factors.

(i) That the two main release areas, Derwent and East Anglia, have not become occupied by movement and dispersal from their adjacent territory is shown by the dearth of otters in the areas to the west of them. Also they stand out as occupied whilst the area between them, largely Lincolnshire, and the area to the south of East Anglia, largely Kent and Sussex, neither of which have had release programmes devoted to them, stand out as still very poorly occupied by otters.

(ii) The losses of otters from East Anglian rivers, followed by their recolonisation one after another after local releases, are clearly demonstrated by the data in Table 12. These recolonisations follow a logical sequence from the release points. The River Little Ouse was colonised throughout from the expanding populations on the Rivers Black Bourn and Thet. From the Little Ouse the otters could have moved westwards and entered the River Cam system in Cambridgeshire, which suddenly became occupied around 1992 (see Section 5.5).

The Minsmere release has provided a nucleus from which the Dunwich and Hundred rivers, the Alde and possibly the Blyth have been colonised. Also, the release on the Black Bourn is a possible source of the otters colonising the River Stour on the Suffolk border with Essex since 1993 (see Section 5.5).

(iii) The timing of the changeover from rapid and continuous decline to a resurgence of the otter population of Suffolk from 0.0% of survey sites occupied by wild otters in 1984-86 to 16.7% occupied by wild-born and captive-bred otters in 1991-94 dates exactly to the start of otter releases there in 1983. Similarly, that of Norfolk expanded from 0.6% of sites occupied by wild otters in 1984-86 to 15.1% occupation after releases started there in 1984 (see Strachan *et al*, 1990 and Section 5.5, Table 10 and Figure 42 in this report).

(iv) Some further support for the success of the release programmes is provided by other data analyses already reported. Thus, in Section 6.1.7 we noted a very high spraint density in East Anglia coupled with what is still a relatively low rate of site occupation. As there are factors correlating spraint density with otter density, it was suggested that there are very high densities of otters in certain productive colonised areas of East Anglia. This would be in accord with the presence of breeding centres (ie. release sites) from which young otters were dispersing to other low-density or empty catchments.

(v) In Section 6.2.2 we reported that, though the otter population of East Anglia was declining rapidly towards extinction between the first and second surveys, after releases and between the second and third surveys the percentage increase in site occupation in Anglian Region was 552% from a starting figure of 1.28% occupation. This is much larger than the percentage increases for Southern, Severn-Trent and North West Regions (40, 69 and 244%), which started with higher site occupation (2.07, 2.29 and 2.78% respectively). This rate of increase could only occur if extra otters were being fed in and these were surviving and breeding. Similarly, the only other Region which has received releases, Yorkshire, showed a percentage increase of 25% on a starting percentage occupation of 1.77% between the first and second surveys. However, the percentage increase was very much higher, at 400% on a starting percentage occupation of 2.21%, between the second and third surveys and after releases had started. Whereas the former increase is below average and below the regression line in Figure 23, the latter, following releases, is considerably above it.

In addition, the small-scale experimental placing of four groups in the path of the expanding 'front' (reported in Sections 5.7, 7.2.2 (iv) and 7.2.3) has shown some signs of success. The sudden increase in the rate of recovery in parts of the Wessex Region between the national surveys of 1984-86 and 1991-94 appears to have been due to the addition of this stock (see Section 5.7).

Thus, both programmes could be said to have succeeded in accomplishing their aims, which in the case of the Derwent release included the successful rehabilitation of incapacitated otters.

7.2.7 Conclusions

The reason for starting research on the release programmes was that of the rapid decline and potential extinction of the native otter population. Information that the 'corner might have been turned' with some incipient recovery was not available until after the 1984-86 survey, by which time the first four otter releases had been effected. Consequently, the urgency has decreased and the principal objectives of the programme revised to strengthening the eastern populations, retaining some of the genetic constitution of these populations (which otherwise would have been lost altogether) and hastening recovery (see Section 7.2.2). It should be remembered that calculations have shown that full recovery is still several decades off (see Section 6.2). However, there have been many invaluable 'spin-offs' from the release work:

(i) The undoubtedly successful results of both release programmes show that conservationists now have available a technique for rebuilding otter populations rapidly from almost nothing, should further problems arise. Indeed, as noted by Caughley (1994), the Otter Trust releases represent one of only five captive breeding programmes actually successfully returning an endangered species to the wild.

(ii) Further, the VWT work pioneers a reliable method of safely rehabilitating otters. This is required by the Wildlife and Countryside Act 1981.

(iii) One of the useful 'spin-offs' has been the information provided by the radio-tracking studies (Jefferies *et al*, 1986) on how otters use rivers in lowland England. This would have been difficult to obtain otherwise because the density of otters in the south is too low to make trapping one for radio-tracking a feasible proposition.

(iv) Another of the most striking results of the strengthening programme, particularly in East Anglia, has been the finding of the considerable coverage and the production of a measurable (by survey) and visible otter population after such a relatively small number of animals has been released (ie. only 37 for Norfolk and Suffolk). This is far below the number predicted to be necessary by detractors of the programme. Perhaps this should not be surprising if a simple calculation of the potential for multiplication is made. Thus, if one allows that cubs are born when the female reaches an age of two years old and that two cubs, one male and one female, can be born each year until death occurs at nine years old (see Jefferies, 1988a), then it can be calculated that, if all subsequent mortality is 'natural' at nine years old, the two released otters (one male and one female) have the potential to become 282 after only ten years. In fact, 20 females have been released to Norfolk and Suffolk in the ten years of the release programme to 1993, though some have only been released for a few years to date. Such theoretical numbers as those calculated are, of course, improbable because of the high possibility of death long before nine years old. However, where the fortunes of the released females have been followed closely by local enthusiasts (eg. rivers Black Bourn and Thet) breeding has been recorded for many consecutive years (see Section 7.2.5.2) and long after the suggested length of breeding life in the Shetlands of only 2.4 years (Conroy, 1993). Such a large number of cubs must go somewhere, so the high rate of recolonisation of many Norfolk and Suffolk rivers and into Cambridgeshire and Essex is not surprising. Another feature of the release programme is the relatively small number of casualties and bodies found considering the potential numbers of animals involved. This again may be an indication that, with the large number of empty but suitable sites available, any density-dependent mortality could be reduced. Starvation is unlikely and long-distance dispersal unnecessary, so road mortality may be lower than elsewhere. The results of monitoring suggest that the survival and productivity of each breeding female may well be much higher in East Anglia than in Shetland (Conroy, 1993). The higher rate of recolonisation by captive-bred animals in the east compared to that shown by the 'rolling front' from Wales and the south-west of England in the west may again support this suggestion.

(v) The above information on the large effect of placing a mere 37 otters into East Anglia over ten years provides useful insight into the potential effect of removing 123 adult otters, again over only ten years, as occurred when, the Eastern Counties O.H. killed this number in Norfolk and Suffolk out of a small and rapidly dwindling population between 1957 and 1966 (Downing, 1988). The present data place this mortality in its correct perspective and add weight to the concern expressed with regard to persecution mortality in Section 7.1.3.1.4.

(vi) Finally, the initial releases acted as 'probes'. These confirmed the suggestion, provided by data on the feral mink population which was spreading eastwards, that contamination of the aquatic environment had ameliorated sufficiently to support mustelids again and that the reason the rivers were empty was because there were no otters locally to occupy them.

It could be said against releasing otters that putting extra animals into a population hanges the genetic make-up of that population and breaks up any long-term clines. Iowever, in the present case, the original genetic make-up and east/west clines were already lestroyed when the eastern population was almost completely lost after 1957, followed by a trong eastward expansion of the original western population. In addition, the use of releases alvages and perpetuates what genetic material is left in the east, even if diluted, rather than etting it be lost by default through continued decline and eventual extinction. The use of aptive-bred otters containing local genetic material provides an even stronger genetic basis or using this technique. Also, far from encouraging inbreeding, as has been suggested, the elease groups, each made up of unrelated otters, greatly increases the genetic pool available nd prevents inbreeding between the very few and isolated small groups of wild otters nown to be present. The future of the release programmes in a conservation strategy is iscussed in Section 7.4.

.3 Factors influencing the otter population

A range of factors which could influence the size and distribution of the otter opulation of Britain was listed by Chanin and Jefferies (1978). Some of the most important f these (ie. persecution and pollution by organochlorines) now have controls (ie. legal rotection and bans on dieldrin uses) which appear to have succeeded in that the British opulation is showing a marked recovery. When the Joint Otter Group reviewed the ituation in the late 1970s (O'Connor *et al*, 1977, 1979), it noted that otter numbers had pparently decreased to the point where every individual was important to recruitment to the opulation and that action should be taken to reduce all unnecessary deaths. A start has been nade on reducing accidental mortality by the introduction of guards for fyke net mouths. Iowever, other factors are still present (eg. road mortality) and some have the potential to ncrease in importance in the future (eg. disease and further pollution with man-made mole-ules). This Section reviews and updates the information on the various factors considered o affect the otter at both population and individual levels and lists attempts to ameliorate heir adverse effects.

.3.1 Pollution

Pollution, in one form or another, is considered to be one of the major causes of loss of he otter in the countries of Western Europe (Chanin & Jefferies, 1978; Mason & Macdonald, 986). This is not surprising considering that the otter is a species living at very low density nd feeding in the actual medium, water, which drains, channels and concentrates the ollutants from areas of human habitation and from industrial and agricultural land.

The pollution it has had to contend with has changed over the years. In the nineteenth-entury industrial revolution most of the waterways of Britain were relatively unpolluted, ut there were other areas with high levels of organic waste or very highly toxic industrial vaste in which no fish could live. The twentieth century has seen the advent of the new nan-made molecules, eg. PCBs and DDT, for which no enzyme systems had been evolved o break them down. Today, freshwater pollution consists of a vast range of materials orming a 'cocktail' which may be selectively broken down or accumulated in the tissues of he animals forming the aquatic food chain. Evaluating the toxicity of this 'cocktail' to an ndividual of one species on the basis of a few standard pollutant analyses is very difficult nd becoming ever more complex. Toxicity trials on otters for comparative purposes are ery few for obvious reasons. Also, the effect on one individual animal depends on the hanging rates of breakdown and elimination and of repair and tolerance within the tissues f that individual and may not be of great importance to the population. Whether or not a opulation is affected depends on how many breeding-age individuals are killed relative to he rate at which they are replaced in that population. Thus, a population can increase lespite considerable levels of pollution in its environment. The analogy of the slowly filling ath, with taps running but with the drain plug out, works for both individuals and opulation. For the individual, the running taps represent the incoming pollutant while the

Table 14. The estimated area usage of the organochlorine insecticides, (a) aldrin/dieldrin, (b) DDT/BHC and (c) heptachlor, in terms of acres treated annually. The data are derived from Tables published in the Report by the Advisory Committee on Poisonous Substances used i Agriculture and Food Storage (Cook, 1964) and were collected for the years 1962/63. These data are divided into usage in four regions based on the Ministry of Agriculture, Fisheries & Food Advisory Regions at that time. These are (1) East: the East Midlands, Eastern and South East (Wye) Advisory Regions, (2) North: the Northern, Yorkshire and Lancashire Advisory Regions, (3) West: the West Midlands Region and the whole of Wales and (4) South: the South East (Reading) and South West (Bristol and Starcross) Advisory Regions. These four regions are approximately equivalent to the eigh areas of Table 10 and Figure 41 used in this report, in the following combinations: (1) East (Eastern + South-East), (2) North (North-East - North-West), (3) West (West Midlands + Wales) and (4) South (South + South-West).

Note that the acreages of the three tables (a, b & c) below are not strictly additive, as many crops may be treated with seed dressings against soil pests at drilling and also later sprayed or dusted with different insecticides to kill foliage pests.

Crop	East	North	West	South
(a) Estimated annual area usage of aldrin and dieldrin in 1962/63: Acres treated				
Cereals (Wheat, Barley, Oats)	150,800	94,400	20,000	30,700
Potatoes, Sugar-beet and Brassicas	248,300	50,500	43,900	26,000
Other crops	39,550	6,471	1,622	980
Total acreage treated with aldrin/dieldrin	**438,650**	**151,371**	**65,522**	**57,680**
(b) Estimated annual area usage of DDT and BHC in 1962/63: Acres treated				
Cereals (Wheat, Barley, Oats)	1,171,700	321,600	180,200	450,300
Potatoes, Sugar-beet and Brassicas	98,500	14,100	8,200	14,400
Other crops	194,900	29,630	96,000	122,500
Total acreage treated with DDT/BHC	**1,465,100**	**365,330**	**284,400**	**587,200**
(c) Estimated annual area usage of heptachlor in 1962/63: Acres treated				
Cereals (Winter Wheat)	56,000			
Sugar-beet seed	143,000			
Total acreage treated with heptachlor	**199,000**			

drain represents the breakdown rate, repair rate and tissue tolerance. If the former runs too fast for the latter, the bath slowly fills, pollutants accumulate and death occurs. For the population, the running taps represent replacement of breeding adults while the drain represents their death. Only a slight reduction in the flow from the taps or increase in the drain-hole size results in an empty bath or no population.

7.3.1.1 The organochlorine insecticides and their metabolites

7.3.1.1.1 The history of usage

A synthetic organic molecule with insecticidal properties, DDT, was first discovered in 1939 and was manufactured in large quantities in Britain and the USA during the second half of the 1939-45 World War. It was used mainly to control insect vectors of human diseases. From 1945, DDT and the new synthetic organochlorine gamma benzene hexachloride (or lindane) were used increasingly for agricultural pest control. The much more toxic (to vertebrates) organochlorines, aldrin, dieldrin and heptachlor were available to commercial growers after 1955. Aldrin, dieldrin and heptachlor were used as seed dressings on wheat (both autumn/winter and spring-sown), barley, oats and sugar beet. They were attached to the seeds with a 'sticker' in order to control the soil pests, wheat bulb fly (particularly prevalent in eastern England; see Figure 38), frit fly, wireworms and leatherjackets. However, much of the dressing was knocked off in the drill and fell on the field surface (Jefferies *et al*, 1973). Crops of potatoes and edible brassicas (eg. cabbage) were also treated. The organochlorines were also applied as sprays and as aldrinated fertiliser. Aldrin dosage in some fertiliser mixtures was as high as 3.6 to 4.1 kg of active ingredient per acre (0.405 ha) (Cook, 1964). As arable farming had polarised towards the east by the 1950s–1960s (see Figure 38), most of these crops were grown in that half of Britain, particularly the south-east, and very large total acreages were treated here every year from 1956 (see Table 14 for usage in 1962/63).

Dieldrin was also available as a veterinary product for use in sheep-dips to control fly strike (see Appendix 16). It was used extensively because of its persistence which provided a long period of larvicidal activity (12-20 weeks). Dipping (obligatory by law) was only necessary once rather than the usual twice a year. Cook (1964) noted that "the majority of farmers are using those single dips which contain dieldrin". They were designed to give bath concentrations of 0.02 to 0.05% of active ingredient and their disposal resulted in pollution of local waterways. A mean of 2.4 mg/kg^{-1} wet weight was found in home-produced mutton kidney fat by 1964 (Cook, 1964). Most of this veterinary use was in the west because of the polarisation of British agriculture.

7.3.1.1.2 Residues of organochlorine insecticides in English otters

One of us (DJJ) considered over three decades ago that the otter was a potential victim of the organochlorine insecticides, as the species occurred at very low densities, inhabited the rivers draining agricultural land and was a fish-feeder like the grey heron. The latter species accumulated high concentrations of dieldrin and showed population effects (Prestt, 1970; Cooke *et al*, 1982).

Consequently, many otter bodies were collected, autopsied and analysed. The first of these was found dead in 1962. Jefferies *et al* (1974) reported that, of 31 otters examined between 1963 and 1973, 81% showed measurable residues of dieldrin with liver wet weight concentrations up to 13.95 mg/kg^{-1} (a lethal level). The series was continued up to 1992 (Jefferies & Hanson, 1987, 1988b). Only a few of these analyses have been published so far and the data are to be analysed and reported separately. Further chemical analyses of English otters have been carried out by Essex University (Mason *et al*, 1986a; Mason, 1988). Some already published residue levels for English otters are listed in Appendix 14.

These analytical results are usually expressed either as mg/kg^{-1} wet weight of tissue or as mg/kg^{-1} in extractable lipid. pp^1-DDE is one of the commonest of the organochlorine residues found in the organs of otters and other wildlife. It is the main metabolite of pp^1-DDT produced in living animals and, as it is persistent, it passes along food chains. Although pp^1-TDE (pp^1-DDD) is present in small amounts as a constituent of technical DDT, the main source of its presence in otter and other tissue is through post-mortem breakdown of the parent compound pp^1-DDT (Jefferies & Walker, 1966; Jefferies, 1972). Aldrin and heptachlor are rapidly metabolised to their epoxides, dieldrin (HEOD) and heptachlor epoxide, in living tissues and it is these compounds which are usually found at analysis. One of the most frequently used of the organochlorine insecticides, gamma BHC (lindane), is seldom found as a residue because of its rapid breakdown in the body after death, even if this is maintained in deep freeze at -20° C (French & Jefferies, 1968). Consequently, its effects have been largely misattributed.

7.3.1.1.3 Effects on wildlife: birds

Research on the British populations of the peregrine falcon and sparrowhawk at the Nature Conservancy, Monks Wood, showed that these species both exhibited the classic response of a wildlife predator at the top of the food chain to the persistent organochlorine insecticides absorbed in its prey (Newton, 1986; Ratcliffe, 1993). Ratcliffe (1970) showed that reproductive abnormalities started to occur soon after 1945, with eggshell thinning and breaking. This correlated with the start of the use of DDT at the end of the 1939-45 World War. However, this reduction in reproductive success did not immediately result in a population decline. Declines followed soon after 1956 with the start of the use of the more toxic aldrin and dieldrin as cereal seed dressings. At that time, thousands of wild birds, particularly wood-pigeons were found dead in the arable fields of eastern England. The population crashes of the peregrine falcon and sparrowhawk, which were most severe in the south and south-east of England, appear to have been due to a high mortality of breeding-aged adults following the consumption of a few highly contaminated prey (Jefferies & Prestt, 1966). Recovery of both species followed rapidly on the voluntary and mandatory bans (see Section 7.3.1.1.5). This recovery occurred from the north and west and was virtually complete in the peregrine by 1985 (Crick & Ratcliffe, 1995) and the sparrowhawk by 1990 (Newton & Haas, 1984; Stroud & Glue, 1991) (see Section 7.1.3.2.5).

Kingfishers and herons were also found dead with high levels of dieldrin in their bodies (Cooke *et al*, 1982; Haas & Cooke, 1983). In fact the heron was carrying higher organochlorine residues than any other species of wild bird in Britain in the 1960s (Prestt, 1970). These frequently reached lethal levels. Also, virtually every egg laid had an abnormally thin shell (Prestt, 1970). The heron population took many more years to recover from the 1962-63 severe winter (when it showed a decrease of 41% on normal numbers) than was the case for the other severe winters this century (Reynolds, 1974).

7.3.1.1.4 Effects on wildlife: mammals

The data on mammals affected by the organochlorine insecticides are very much fewer than those on birds. Mammals have fewer interested observers, they are less visible, and may also die unnoticed underground. The most obvious casualty was the fox, and many of these were reported by the hunts. Some 1,300 foxes are known to have died in eastern counties of England in just five months over the winter of 1959/60 (Taylor & Blackmore, 1961; Thompson & Southern, 1964). Many of these died in convulsions and disease was suspected. Blackmore (1963) later showed experimentally that this mortality was undoubtedly linked with the organochlorine seed dressings, particularly dieldrin, to which this species is extremely sensitive. Any level over 1 mg/kg^{-1} wet weight in the tissues could be considered a lethal concentration. Jefferies (1969) autopsied and analysed the bodies of 17 badgers found dead in south-eastern England and concluded that six certainly and six others very probably died of dieldrin poisoning. The source of the dieldrin for both fox and badger was considered to be dead and dying wood-pigeons. This bird had been found dead with up to 41 mg/kg^{-1} wet weight dieldrin in the muscle. Three to six dead wood-pigeons would kill a fox (Blackmore, 1963) and ten would kill a badger (Jefferies, 1969).

Besides foxes and badgers, the small mammals, field mice, bank voles and field voles, inhabiting fields drilled with dressed wheat were known to become contaminated with dieldrin and many field mice received lethal doses (Jefferies *et al*, 1973; Jefferies & French, 1976). Pipistrelle bats sampled in East Anglia too were found to be carrying one third of the lethal level of organochlorine insecticides as 'background' residues, with just under the lethal level after hibernation (Jefferies, 1972). However, in none of these species was it considered that the effects on their populations would be more than local, ie. the bodies examined were not considered at the time to be indicators of national population crashes like those of the sparrowhawk and peregrine (see Section 7.3.1.1.3).

The otter provides the best example of the population of a predatory mammal crashing because of the presence of the toxic cyclodiene organochlorine insecticides in its diet. This report provides much more circumstantial evidence linking the two events. The close correlation of timing of decline and recovery and area of greatest effect now makes that link almost certain and has been covered already (see Section 7.1.4). The full history of the otter's decline has taken longer to work out than has that of the peregrine and sparrowhawk. The explanation for the much more rapid recovery of these predatory birds (see Section 7.3.1.1.3) compared to that of the otter lies partly in (i) the greater mobility of the birds, which were not largely restricted to linear environments (eg. rivers and streams), and (ii) these birds were not so affected by dieldrin sheep-dips in the west as was the otter (but also see Section 7.1.5.2).

7.3.1.1.5 Bans on organochlorine insecticide uses

The widespread, visible and very numerous wildlife mortalities of the late 1950s and early 1960s caused great concern. Many gamebirds, particularly pheasants, were found dead, as well as wood-pigeons. The cause was unknown to begin with and the fox deaths, for example, were attributed to a form of encephalitis (Blackmore, 1963). However, the new seed dressings and fertilisers were soon under suspicion. A joint committee of the Royal Society for the Protection of Birds, the British Trust for Ornithology and the Game Research Association (Cramp & Conder, 1960; Cramp *et al*, 1962) was set up in order to list the casualties, their numbers and distribution, and the Nature Conservancy formed a small research group (which included one of the authors, DJJ) in 1960 to study the effects of the new chemical farming on Britain's wildlife and to advise central government. The results of this research (reviewed by Moore, 1965, 1987; Cooke *et al*, 1982; Sheail, 1985), confirming cause and effect, and public concern and anxiety for human welfare eventually brought bans on all the uses of the organochlorine insecticides. Unfortunately, these came piecemeal over the 20 years from 1962 to 1983. This long drawn-out removal of these substances had a considerable effect on the extent of the decline of the otter population and almost allowed its extinction in the east (see Section 7.1.3.2.7).

Three reports by the Advisory Committee on Poisonous Substances used in Agriculture (Cook, 1964; Advisory Committee, 1967; Wilson, 1969) considered the undoubted agricultural value of the organochlorines and what would be lost if their uses were banned in Britain. The increasing scientific basis for the concern about wildlife caused them to recommend progressive bans, which were acted upon after consideration by the Minister for Agriculture. These bans had a rapid effect upon the status and recovery of the declining predatory bird populations (Newton, 1986; Ratcliffe, 1993) and much more slowly upon that of the otter (Jefferies, 1989b; this report, see Section 7.1). The bans and the dates when they became effective were as follows.

1962: Most of the deaths of seed-eating birds were found to have occurred in cereal growing areas in spring. Consequently, it was agreed in summer 1961 that a voluntary ban should be imposed on the use of aldrin, dieldrin and heptachlor seed dressings on spring-sown cereals and that in autumn they should be restricted to cereals in districts where there was a real danger from wheat bulb fly (ie. over much of the south-east; see Figure 38). This voluntary agreement came into effect in spring 1962 and was renewed each year afterwards. Cook (1964) noted that it "appears to have been well honoured in practice" and Moore (1965) agreed that the arrangement had worked well, with "the huge casualties which were such a shocking feature of the English countryside from 1956 to 1961 having been greatly reduced". The extremely large quantities of organochlorine insecticides which must have been in use from 1956 to 1961 can be envisaged when it is realised that the millions of acres treated and listed in Table 14 are those where these substances were still applied after the voluntary ban had come into effect (ie. 1962/63). This ban would have had its greatest effect in the arable areas of the north-east, midlands, south and south-east (ie. outside the East

Anglian wheat bulb fly area; see Figure 38). The effect on the otter population was greater (see Figure 39) than was envisaged in 1978 (Chanin & Jefferies, 1978).

1966: A ban on the use of aldrin and dieldrin in sheep-dip was imposed at the end of 1965. It came into effect on 1st January 1966. The effect of this ban would have been greatest in the west (where most of the sheep were reared).

1975: A ban was imposed on the use of aldrin and dieldrin in all seed dressings in 1973. This was a mandatory ban. However, in order to give time for the using-up of the stock, the withdrawal of dressings was delayed until the end of 1974. 1975 was the first year with no aldrin/dieldrin seed dressings. This ban made a considerable contribution towards the recovery of otters in the east (see Section 7.1.3.2.7).

1981: All agricultural use of aldrin and dieldrin was finally banned in 1981. A ban on the use of DDT followed in 1982.

1983: By 1983 all regular uses of the persistent organochlorine insecticides had ceased, though timber treatments continued for a few years longer.

7.3.1.2 Polychlorinated biphenyls

7.3.1.2.1 History, uses and bans

During the early days of chemical analysis of wildlife tissues for organochlorine insecticides and metabolites, it was noted that an additional series of peaks sometimes appeared on the chromatograms. It was not until 1966 that Jensen identified a series of such peaks in pike and white-tailed eagles in Sweden as corresponding to polychlorinated biphenyl (PCB) compounds. Holmes *et al* (1967) then showed that the compounds with long retention times occurring in British specimens were also PCBs, and since 1966 these have been quantified in addition to the organochlorine insecticide residues. Later work showed that PCBs were worldwide contaminants (Risebrough *et al*, 1968).

PCBs are produced and marketed under a number of commercial names, eg. Aroclor. These are actually mixtures of over a hundred different PCB congeners. It is the concentration of these mixtures or, rather the constituent PCB congeners remaining after metabolism and passage through a food chain, which are quantified and listed with the organochlorine residues (see Appendix 14). In recent years individual PCB congeners have been quantified, adding to the complexity. Aroclor 1254 (biphenyl with 54% w/w of chlorine) was the commercial PCB mixture with the 'fingerprint' closest to that found in British wildlife specimens in the 1960s and 1970s (Jefferies & Parslow, 1976). These mixtures have many industrial uses such as constituents of protective coatings, plasticisers, sealers, adhesives and printing inks. In liquid form they are used as hydraulic fluids, in thermostats, cutting oils, grinding fluids and they are incorporated into electrical apparatus, such as transformers, as a dielectric.

After the finding of large amounts of PCBs (particularly Aroclor 1254) in seabirds dying with PCB lesions in the Irish Sea in the large 'wreck' of 1969 (Holdgate, 1971; Parslow & Jefferies, 1973), the industry quickly responded and Monsanto met with Nature Conservancy staff at Monks Wood to agree a ban on the use of their PCBs in open systems in 1970 (Sheail, 1985). Bans on all further use of PCBs in the closed systems of new equipment (eg. transformers) were brought in during 1986 and H.M. Government has agreed to phase out and destroy all PCBs already in use in such equipment by the end of 1999 at the latest (Anonymous, 1990).

7.3.1.2.2 Toxicity and effects on terrestrial wildlife

The toxicity of commercial PCB mixtures is very complex because the many congeners are not equally toxic. Their toxicity may also depend on whether they are present singly (when effects may be absent) or in combination with certain other congeners

(Kihlstrom *et al*, 1992). When tested on finches in the laboratory, the lethal toxicity of a commercial PCB mixture such as Aroclor 1254 was found to be 1/13 of the toxicity of DDT (Prestt *et al*, 1970) and DDT has only 1/6 to 1/14 of the lethal toxicity of dieldrin (De Witt *et al*, 1960). A British grey heron (from Derbyshire) analysed in the late 1960s had accumulated 900 mg/kg^{-1} wet weight PCB in the liver (Prestt *et al*, 1970) (this would be several thousand mg/kg^{-1} lipid weight), whereas only 10 to 12 mg/kg^{-1} wet weight dieldrin could have been lethal. Thus, overall avian lethal toxicity would appear to be low compared to that of dieldrin.

There are few data on lethal toxicity of PCBs to mammals, though it would appear that mammals can generally metabolise PCB congeners more rapidly than can birds and consequently they have lower residues. Thus, birds feeding on mammals have only 40% of the PCB levels found in birds feeding on birds (Prestt *et al*, 1970) (see also Section 7.3.1.2.4 (iv)). No laboratory tests on the comparative lethal toxicities of PCB and dieldrin to otters have been made to our knowledge, for obvious reasons. However, no certain PCB-caused otter or other wild mammal casualties are known in Britain, though this is not the case for dieldrin (Blackmore, 1963; Jefferies, 1969; Jefferies *et al*, 1973, 1974). It is known that dieldrin is very toxic and rapidly lethal to mammals. Foxes have suffered convulsions and died while running, and laboratory tests have shown that only 1 mg/kg^{-1} wet weight in the tissues of a fox is indicative of dieldrin poisoning (Blackmore, 1963). This is less than the PCB loads carried by many otters today.

PCBs are known to be capable of producing many detrimental sublethal effects on birds and mammals in the laboratory and these appear to be linked to a basic effect on thyroid activity (Jefferies, 1975; Jefferies & Parslow, 1976; Brouwer *et al*, 1989). Such laboratory tests have shown that the American mink is very sensitive to PCBs and that these reduce reproductive success (Aulerich & Ringer, 1977; Bleavins *et al*, 1980; Kihlstrom *et al*, 1992). This sensitivity has been extrapolated to the otter by some authors (Mason & Macdonald, 1993), but, as Leonards *et al* (1994) have pointed out, this extrapolation is highly speculative. The otter may be more like the ferret, which is known to be much less sensitive to PCBs than is the mink (Leonards *et al*, 1994). Because of this doubtful extrapolation Mason and Macdonald (1993) use a level of 50 mg/kg^{-1} of PCB in tissue lipid as that likely to be associated with reproductive failure in the otter. However, a liver lipid concentration of 56.7 mg/kg^{-1} PCB was found in a trapped otter from Islay in 1985 which was pregnant with three healthy embryos and even higher residues have been found in lactating animals which have recently given birth (D.J. Jefferies and H.M. Hanson, personal data). Kruuk *et al* (1993b) have other similar examples.

7.3.1.2.3 Residues in otters and their evaluation

Some published PCB residues in otters are given in Appendix 14. Evaluation of a PCB residue is very difficult because the concentration given is that of the total congeners remaining after metabolisation and breakdown in the environment. This cannot be related easily back to a concentration of the original mixture found in a laboratory animal, as each congener has a different toxicity. The few published concentrations given in Appendix 14 may appear to show a bias, with higher levels in the east (as suggested by Mason and Macdonald (1993) (see Section 7.3.1.2.4), but this is partly due to the inclusion of a few old and sick animals (such as 1 and 9) in the eastern group with abnormally low amounts of lipid left in the organs. This greatly magnifies the organ concentration when this is given in terms of mg/kg^{-1} lipid (Jefferies & Hanson, 1987) (cf. lipid and wet weight concentrations of otters 9 and 12 in Appendix 14). Such animals should not be used for monitoring purposes and the actual lipid level in the organ should always be stated (see Appendix 14). For example, otters with low organ lipid have been found with 244.7 and 128.5 mg/kg^{-1} PCB in liver lipid in Devon and Dyfed, Wales, in 1988 and 1987 respectively (D.J. Jefferies and H.M. Hanson, personal data). These show that high lipid PCB levels are not the prerogative of the east, where the otter population is low. In addition, Kruuk *et al* (1993b) showed a mean level of around 150 mg/kg^{-1} PCB in liver lipid in otters from Shetland, where the population is thriving (Green & Green, 1980, 1987).

7.3.1.2.4 The effect of the PCBs on the British otter population

After an exhaustive review, Smit *et al* (1994) concluded that "available data is insufficient to either accept or reject the hypothesis that PCBs constituted a major causal factor in the decline of the European otter". However, examination of the situation here suggests that there is little doubt that the PCBs were not a major causal factor in the British otter population crash starting in 1957. This can be seen from the following facts.

(i) PCBs were first produced and used in 1930 and, while they probably became more available to wildlife after the 1939-45 World War, there is no reason to suppose that use suddenly increased to produce an exceptional effect on peregrine falcons, sparrowhawks and otters in 1957 (Prestt *et al*, 1970).

(ii) The timing of the stages in the recovery of the otter do not correlate with any of the few bans on the uses of PCBs, whereas they do correlate with those on the uses of the organochlorine insecticides.

(iii) PCBs do not have the lethal toxicity of dieldrin (see Section 7.3.1.2.2) and it is the killing of breeding-aged adults which causes sudden population crashes rather than small reductions in breeding success. For example, the sparrowhawk and peregrine falcon were producing thin-shelled eggs in Britain from 1945, when DDT was marketed, but the populations did not crash until dieldrin was used and caused high mortality.

(iv) Levels of PCBs in otters in the 1960s and early 1970s were at trace level and too low for accurate quantification (see Jefferies, 1992b). This was due to the fact that the input rate was still low and mammals had a greater capability of metabolisation of PCBs than did birds (see Section 7.3.1.2.2). This can be seen from the lack of PCB residues at that time in badgers (Jefferies, 1969), bats (Jefferies, 1972), field mice and bank voles (Jefferies & French, 1976). It was not due to lack of analytical capability, as shown by the 559 bird specimens analysed for PCBs between 1966 and 1968 (Prestt *et al*, 1970). Indeed, all specimens were analysed for PCBs at Monks Wood from 1966.

(v) The otter, peregrine falcon and sparrowhawk all showed a marked area bias in their population declines and recoveries. However, Prestt *et al* (1970) could show no area bias in the distribution of PCBs in 559 specimens from British birds, including sedentary species. The eastern area bias claimed for PCB residues in otters and used as the basis for the hypothesis that PCBs are a causal factor for the low population there (Mason & Macdonald, 1993) is by no means clear cut. It could be due to the inclusion of otters with abnormally low lipid reserves (see Section 7.3.1.2.3). The anomaly of the PCB-sensitive mink population expanding rapidly into East Anglia (Strachan & Jefferies, 1993; this report), where PCB levels are said to be too high for inhabitation by otters (Mason & Macdonald, 1993), should be noted too.

(vi) Ornithologists using much larger databases on the predatory birds, with many hundreds of analyses, could find no definite connection of the PCBs to their population declines (Newton & Haas, 1984; Newton, 1986; Ratcliffe, 1993). It would be very odd if the sudden declines, occurring at exactly the same dates and concentrated in the same areas, had different causes in otters and predatory birds. Use of Occam's razor would suggest that the main causal factors were likely to be similar in the two cases.

Recovery of the otter population is taking place all over Britain, despite the ubiquitous PCB contamination. However, it is very difficult to know whether, and if so by how much, the later higher concentrations of PCBs might have affected the rate of recovery of the otter population in the 1970s to 1990s. It is impossible to estimate whether it could have taken place more rapidly without the high PCB levels. Their toxicity to otters and effects on the British otter population might have been exaggerated because of lack of pertinent data, but they are present in the aquatic environment in considerable quantities and have known effects on mustelids, at least in the laboratory. Despite all the doubts expressed in Sections 7.3.1.2.2 and 7.3.1.2.3, they remain an unknown hazard which cannot be ignored.

Smit *et al* (1994) hypothesise that local otter populations (eg. in Shetland) may be able to endure high PCB concentrations if other stresses are low, but not if these increase. Such a hypothesis would fit the cases of the Irish Sea guillemot 'wreck' of 1969 (Holdgate, 1971) and possibly the phocine distemper epizootic in common seals in 1988 (Thompson & Hall, 1993).

7.3.1.3 Oestrogenic compounds

The suggestion that DDT might have oestrogenic effects was first put forward by Burlington and Lindeman (1950) to explain the inhibition of testicular growth and secondary sexual characters in DDT-dosed cockerels. Their hypothesis was that since the molecular configuration of the DDT molecule is similar to that of the synthetic oestrogen diethylstilboestrol, it might elicit the same response. Bitman *et al* (1968), using laboratory rats and hens, showed that the ortho para isomer of DDT was much more active than the para para isomer in this respect. Some mammalian oestrogenic activity is also present in low-chlorine polychlorinated biphenyl (PCB) mixtures (Jefferies, 1975).

The recent finding of both male and female organs in roach caught near a sewage effluent outfall in a south-eastern British river has rekindled interest in the abilities of some man-made molecules to mimic oestrogen (Tyler, 1995). The 'feminisation' of freshwater fish in some areas suggests that these materials may be released in industrial sewage. Various compounds have been suspected besides the organochlorine insecticides and PCBs. These include (i) industrial plasticisers, (ii) nonylphenol, a breakdown product of the alkylphenolethoxylates (APEOs) (used in detergents, particularly in the textile industry, paints and herbicides), and (iii) polycyclic aromatic hydrocarbons (PAHs) (arising from the burning of natural fuels such as oils) (Tyler, 1995). If these oestrogenic mimics are persistent and can bio-accumulate in tissues with passage through the food chains, as can the organochlorine insecticides and PCBs, then obviously both the otter and man are at risk. This field requires much more research to evaluate the wildlife and human hazards.

7.3.1.4 Heavy metals

Concentrations of mercury, cadmium, zinc and copper found in the livers and kidneys of individual English otters and published in the literature are given in Appendix 15. These levels can be compared with those in Welsh otters from the previous decade (Jefferies, 1992b; also see Andrews *et al*, 1993). Metal concentrations are usually expressed on a dry weight basis. The mean multiplication factors for obtaining wet weight for these four individual otters were x 0.2973 for liver and x 0.2214 for kidneys. Zinc and copper are essential elements and are normally regulated in the mammalian body. Mercury and cadmium, on the other hand, have no metabolic function.

The cadmium concentration could have arisen from water draining riparian arable fields or from roadside drains. Rain-water sprayed from roads is contaminated with cadmium from impurities in oil and tyres (Lagerwerff & Specht, 1970). Also, cadmium is a known contaminant of phosphate and super phosphate fertilisers (Schroeder & Balassa, 1963) and has been found in field mice and bank voles from arable fields (Jefferies & French, 1976).

Mercury has been used in agricultural and industrial fungicides for many decades. Thus, field mice and bank voles are known to have accumulated seriously high levels of mercury (a mean of 0.39 mg/kg^{-1} wet weight in whole bodies of field mice in the 14 days after drilling) from consuming cereal seeds dressed with mercury-based fungicides (Jefferies *et al*, 1973; Jefferies & French, 1976). Water draining from arable fields would contain mercury, particularly during the period shortly after drilling. It has been found that there is serious mercury contamination of waterways draining to the Wash in East Anglia because of its use in fungicides employed for horticultural bulb-dipping (Moriarty & French, 1977). Once in the water and the freshwater invertebrate fauna it can be accumulated by fish (Jefferies & Freestone, 1985; Spalton & Cripps, 1989) and could then be passed on to otters.

Elevated mercury levels in some coastal areas may be from natural mercury from sea spray. Sea-water and sea fish have naturally high concentrations of most heavy metals.

Both cadmium and mercury are highly toxic and can produce nephrotoxicity (Nicholson *et al*, 1983) as well as impairment of nerve functions with death following high doses. Mercury levels are usually higher than cadmium levels in otters (see Appendix 15) and, although numbers are very small, they appear to have been higher in Wales (liver dry weight concentrations of 18.04 and 12.37 mg/kg^{-1}; Jefferies, 1992b) than in East Anglia. However, the Welsh otter population was already recovering in 1975 when these were measured. Tissue mercury concentrations are very difficult to evaluate, as the toxicity of mercury varies with its organic form. Ethyl, phenyl and methyl mercury have all been used as fungicides on seeds, but only elemental mercury is measured and reported at analysis. Very few studies examine its organic form before discussing its toxicity. In addition, as aldrin/dieldrin and organomercury were used together in large quantities as seed dressings, the effect of the latter is very difficult to separate from the effect of the former.

One of the Welsh otters had a liver cadmium concentration of 34.12 mg/kg^{-1} dry weight (Jefferies, 1992b), which makes the reported English concentrations (Appendix 15 and Mason *et al* (1986b)) appear small. The sublethal effects of this level are unknown, but it was not lethal as the animal died as a road casualty.

However, whatever the toxicity of the heavy metal concentrations found, these materials are not harmless. Cadmium and mercury form part of the 'cocktail' of contaminants consumed by otters on a daily basis and will undoubtedly contribute to the toxicity of the overall load.

7.3.1.5 Acidification

Acid precipitation results in acidification of waterways, which results in the decline and eventual loss of the fish biomass in the affected area. Also among the marked effects of conifer afforestation of the uplands is an exacerbation of this stream acidification (Nilsson *et al*, 1982; Stroud *et al*, 1987). Another problem of acidification is the mobilisation of metals such as aluminium. Elevated aluminium levels are toxic to fish. Obviously, a severe reduction or the complete absence of fish is likely to restrict the distribution of the otter. This problem is considered to be one of the reasons for the loss of otters from many areas of Sweden. Populations only thrive in eutrophic waters (M. Olsson & F. Sandegren in Mason & Macdonald, 1986). Also, Mason and Macdonald (1987) found that otters were not resident on streams draining conifer plantations in upland Wales where the pH fell periodically below a value of 5.5, whereas they were resident on tributaries draining open moorland.

7.3.1.6 Organic pollution

The widespread use of nitrate fertilisers on arable land has led to a steady increase in the level of soluble nitrate in many rivers (again largely in the east). The average level doubled in the River Thames and trebled in the River Lee between 1940 and 1980 (Mellanby, 1980). Levels as high as 30 mg/litre were found by 1980 (Mellanby, 1981). This is a problem for potable waters because, although safe for adult humans and animals, the nitrate is transformed into nitrite by micro-organisms in the gut of young animals. The nitrite combines with haemoglobin in the blood to form methaemoglobin, which reduces oxygen transport. Mellanby (1981) noted that no serious cases of methaemoglobinanaemia had been reported in babies at that date, but obviously there is a potential problem for the young of aquatic mammals, such as the otter, water vole and water shrew. No research has been carried out on this point.

The occasional incident involving the release of untreated sewage or farm slurry causes almost complete deoxygenation of the water owing to removal of oxygen during oxidation of complex organic compounds. In flowing water, a 'plug' of deoxygenated water moves downstream killing all the fish which have not escaped into side streams and tributaries. The effect on the otter is solely one of removal of its food supply. Fortunately, such adverse effects are only temporary and the river may be restocked with fish or recolonised naturally from its tributaries.

7.3.2 Persecution

7.3.2.1 Direct and Indirect persecution

Howes (1976) and Jefferies (1989b) have suggested that much of the nineteenth century otter population decline was due to the killing of otters by gamekeepers and water bailiffs for the purposes of game preservation. This was both direct (ie. the intentional taking of otters by shooting and riparian trapping) and indirect persecution (by the widespread setting of traps and snares to take all predators indiscriminately). Unlike hunting with hounds, where accurate tallies were kept and some conservation measures employed, it is difficult to estimate the numbers of otters taken for game preservation purposes. However, this appears to have been sufficiently large to affect the numbers taken by the packs of otter hounds, both in the nineteenth century and up to the decline in the 1950s ('Polestar', 1983; and see Section 7.1.3.1.5). Jefferies (1989b) considered that the extent of this combined persecution from 'keepers and organised hunting was one of the major causal factors in the decline, in that the population was already small and stressed (through too high a removal of breeding adults) before the widespread use of dieldrin precipitated the sudden crash observed in 1957. By 1964, the otter hunters brought in a rule of not killing the hunted otter whenever this could be avoided. However, killing by 'keepers continued. As many as 13 are known to have been trapped in one locality in the north-west during a 12-week period in 1964 (J. Paisley, in Woodroffe, 1994) and this is just one known example. When legal protection for the otters of England and Wales came into force in 1978 (see Section 7.3.2.2), the organised packs of otter hounds were disbanded but the killing of otters during predator control still occurred. Bodies of otters killed in snares have been received in the last ten years in the VWT survey of causes of otter mortality (Jefferies & Hanson, 1987). This may be indirect persecution, as with the taking of pine martens (see Jefferies & Critchley, 1994), but the result is the same. Unfortunately, direct persecution may still continue, as has been reported for other legally protected predators (Pickford, 1994).

The effect of this direct and indirect persecution mortality on slowing the rate of population recovery is unknown. It could have a considerable effect when a population is fragmented into isolated groups with very few individuals surviving (see Section 7.1.3.2.7) or when otters are recent recolonists of a previously unoccupied area.

With regard to the next decade, many of today's anglers and angling clubs have never had otters on their stocked rivers. Also, there has been a considerable growth of freshwater fish-farms for trout and carp in many areas since the otter disappeared from most catchments in England. Both could form further potential sources of human antagonism to the increasing otter population, although the take of the otter is small in comparison with the many miles of river inhabited and most problems at fish-farms can be eliminated by electric fences and/or guard dogs. It is hoped that some of these potential difficulties can be ameliorated by education.

7.3.2.2 Legal protection

The Nature Conservancy Council obtained legal protection (prohibiting killing and taking) for the otters of England and Wales in 1978 when the species was added to Schedule 1 of the Conservation of Wild Creatures and Wild Plants Act 1975. More complete protection was afforded by its later inclusion in Schedules 5 and 6 of the Wildlife and Countryside Act 1981. This covered the otters of England, Wales and Scotland and came into force in 1982. Under this Act it is an offence intentionally to kill, injure or take an otter except under licence or intentionally to damage, destroy or obstruct access to its holt or disturb it whilst occupying that holt. Also, under Section 11 (2), a person is guilty of an offence if he sets in position an article (any trap) which is of such a nature and so placed as to be calculated to cause bodily injury to any wild animal included in Schedule 6 which comes into contact therewith. However, it is a defence for a person to show that the article was set in position for the purpose of killing or taking any wild animal which could be lawfully killed or taken by those means and that he took all reasonable precautions to prevent injury thereby to any wild animals included in that Schedule. This would appear to include snares and unguarded fyke nets, though this point could only be decided in a court of law.

7.3.3 Accidental mortality

7.3.3.1 Road mortality

An estimation of the number of badgers in Britain (250,000: Cresswell *et al*, 1990) and their productivity, together with information on the percentage occurrence of various causes of mortality, allows the calculation of an astonishing figure of 47,500 deaths of badgers annually owing to road traffic accidents (Harris, 1989). Also, recent analysis of the data on the causes of death of pine martens in England and Wales shows that both the numbers of road deaths and their percentage of total mortality increases each decade (Strachan *et al*, in press). This is more likely to be due to increasing road traffic density than to increasing marten numbers. There is now mounting evidence that suggests that road mortality is becoming one of the major causes of mortality in British otters too.

During his study of the small population of otters of the north Norfolk coast, Weir (1984) reported that at least four otters were killed by vehicles in the late 1950s and early 1960s and four were killed between 1969 and 1975. There were definite 'black spots' at presumed frequent crossing places. Thus, two deaths occurred at Cley sluice and two at Hempstead Mill. This population ceased to exist when the last otter died in April 1986 (Spalton & Cripps, 1989).

Twelve out of the 18 otter bodies (67%) found in Devon between April 1982 and September 1989 had been killed in road traffic accidents (Marshall, 1991).

Of 12 coastal otters from Scotland and Wales (1981-1988) which were found to be infected with gastrointestinal parasites at autopsy, nine were killed on the road (75%) (Jefferies *et al*, 1990).

Weber (1991) examined 29 coastal otters from Shetland and nine others from mainland Scotland (Aberdeenshire, Invernessshire) for gastrointestinal parasites between 1983 and 1987. He noted that "most of the autopsied individuals were road kills".

Twelve of the 23 bodies of otters from England, Scotland and Wales which were analysed for organochlorine insecticides by Mason *et al* (1986a) were killed in road traffic accidents (52%) over 1982-1985. The percentage was higher (68% plus a further 4% probable casualties) in the 75 otters examined by Jefferies and Hanson (1988b) over 1983-1988.

The highest known cause of death of otters released by the Otter Trust and The Vincent Wildlife Trust has been road traffic accidents (see Section 7.2.5.3). Thus, seven out of the 13 known casualties (54%) were killed on the roads (Jefferies & Hanson, 1988a, 1991; Jessop & Cheyne, 1993; Green & Green, in press). Jessop and Cheyne (1993) also reported the deaths of two other otters on the roads of the south-east of England in 1992, one near Royston on a catchment adjacent to the River Lea and one on the River Rib near Buntingford. Both are from a very low-density population likely to have been derived from the East Anglian releases via the River Cam.

There have been over 50 recorded road kills in the expanding otter population of Devon and Cornwall alone in the last seven years (1987-1994) (H. Marshall, M.R. Lane, H. Hanson, pers. comm.). Many of these road accidents occur where otters are faced with greatly increased current speeds under bridges or with a blocked or flooded culvert on a minor tributary, stream or ditch. They cross the road as the only alternative. Others may happen where a road crossing is unavoidable, as during a migration between two river catchments (Jefferies, 1989c). Of course, it can be said that the bodies of road casualties of any species will be more easily found than those from most other causes of death and so their percentage of otter mortality could be artificially high. However, they remain our largest number and percentage of bodies and known causes of death and thus have major importance. Also, there are situations where road accidents have had the potential to eliminate small and isolated otter populations such as those present in eastern counties (eg. the released population at Minsmere in 1987; Jefferies & Hanson, 1988a). Thus, the subject of otter road mortality is being studied in greater detail by D.J. Jefferies, J. & R. Green and H.M. Hanson and will be reported later. D.J. Jefferies would value any further records and descriptions of circumstances of road mortality in otters.

7.3.3.2 Drowning in underwater traps

7.3.3.2.1 Eel fyke nets

Unfortunately, the modern eel fyke nets, set in long chains in rivers and estuaries, have proved to be very effective at catching and drowning otters, and the number of such casualties is large. The British recorded casualties (76 drownings, plus two rescued alive, between 1973 and 1993) are fully reported and discussed individually by Jefferies *et al* (1984, 1993), Jefferies (1985, 1990b), Strachan *et al* (1990) and Woodroffe (1993). The true number killed is likely to be very much higher than this, as there will be a natural tendency not to report the deaths of a legally protected species occurring in this way. Dead otters found in the sea off Shetland and in Norfolk rivers may be unreported creel and fyke net casualties respectively (Jefferies *et al*, 1993; Jessop & Cheyne, 1993).

Regrettably, of the 78 known eel fyke captures in Britain since 1973, 25 (23 casualties + 2 survivors) or 32% came from the small and severely reduced population of England. Analysis of the published individual records provides even greater concern, as none came from the strongly recovering population of the south-west, but all came from isolated fragments of the population in the north-west, Midlands and east coast, where loss of individual otters could have been critically important. Thus, in the north-west, six otters were drowned and one survived in fyke nets set near to the small breeding population at Leighton Moss (Lancashire) between 1979 and 1984. One of these was a lactating female (Jefferies *et al*, 1984; Strachan *et al*, 1990). This was a heavy mortality on a small population which still survives (see Section 5.1). A further female and three cubs were drowned in a tarn near Kendal in 1982 (Jefferies *et al*, 1984). One cub from the very small and isolated but still surviving population fragment in the east Midlands, that on the River Nene (Northamptonshire), was drowned in 1981 (Jefferies, 1990b).

With regard to the small and most threatened east coast populations, two animals were caught (1988 and 1989; Jefferies, 1990b) from the isolated population fragments in South Humberside and on the Lincolnshire coast near Skegness. Nine further otters are known to have drowned in Norfolk and Suffolk rivers, broads and lakes between 1973 and 1984. These came from the very small wild East Anglian population which was declining rapidly at that time prior to restocking (see Section 7.1.3.2.7) (Jefferies *et al*, 1984; Jefferies, 1985). The most recent (1993) casualties have been two males from the restocking project carried out by the VWT on the River Derwent system in North Yorkshire (Woodroffe, 1993) (see Section 7.2.5.3).

Eel fyke nets are used all year round for catching yellow eels and between September and November for catching silver eels on migration to the sea (Gibson, 1984). However, an analysis of the dates of catching otters shows that 76% of casualties occur between September and November with a peak in October (Jefferies, 1990b; Jefferies *et al,* 1993). This suggests that otters may be particularly interested in fishing for silver eels when these are available.

Eel fyke nets have an undoubted attraction to otters when they contain eels in the cod end. They appear to have the capacity to catch otters even where the latter are at a very low population density (Green & Green, 1981; Jefferies, 1985). Thus, this exacerbates the already severe problem of having the greatest net density (along the east coast of England; Jefferies *et al*, 1988) where the otter populations are smallest and most threatened.

Research carried out and reported by a committee of otter conservationists, fisheries scientists and an eel-net manufacturer hosted by The Vincent Wildlife Trust has produced a fyke net entrance guard which allows acceptable catches of eels (Jefferies *et al*, 1988). These have been manufactured for the VWT and are in use in most NRA Regions where there are bye-laws making their fitting mandatory. Unfortunately, the Anglian Region, where the problem is at its greatest, does not at present have such a bye-law, although guards are distributed to licensed fyke netsmen. However, the two most recent net casualties in North Yorkshire were caused by an unlicensed netsman, but a prosecution failed on a technicality (Woodroffe, 1993).

7.3.3.2.2 Creels

Twelves (1983), Jefferies *et al* (1984, 1993) and Jefferies (1990b) list 66 known deaths of otters at sea in marine crustacean (edible crab and lobster) creels between 1975 and 1992. All of this recorded mortality is from Scotland, particularly the Northern and Western Isles, and has occurred in the modern two-chamber or 'parlour' creels. These authors suggest that the true mortality between 1975 and 1992 may have been at least 230 for Orkney, Shetland, Skye and South Uist alone.

Although none of this marine mortality was around the English coast, this was almost certainly due to the recent low population in this country. In the past there have been many records of otters drowning in crab and lobster pots (presumably the old single funnel, hemispherical basket-work traps) set off the rocky coasts of Cornwall (Clark, 1906) and Devon (Cox, 1947). Nine otters were recorded drowned by these authors alone. It seems likely then that, with the present recovering population of the south-west, marine mortalities will occur again and perhaps in larger numbers with the modern two-chamber creels which make escape more difficult (Twelves, 1983). The problem of creel mortality is exacerbated because 79% of the casualties are breeding-age females (Jefferies *et al*, 1993) because only females and young males are small enough to enter the funnels (Twelves, 1983). Jefferies *et al* (1984, 1993) have suggested creel modifications which might reduce this mortality, but none have been tested or developed as yet.

7.3.3.2.3 Other fishing gear

Otters are known to have drowned in other forms of freshwater fishing gear in the past. Thus, bodies have been reported in the box traps set for perch in Lake Windermere (1950) and pike (1960s) and crayfish (1969) traps set in the River Itchen in Hampshire (Jefferies *et al*, 1984). So far only one otter casualty is known to have been caused by the new, almost transparent, monofilament drift nets. This animal was killed by strangulation after having put its head through one of the meshes (Isle of Harris; Jefferies *et al*, 1989). It appears to have been a casualty of coastal net debris rather than active fishing at sea. It seems likely that, with a recovering otter population, particularly in the south-west, and the present large amount of fishing debris both floating and on shore (Bourne, 1977), such mortalities will occur with increasing frequency on English coasts in the future. Actively fished monofilament nets drowned 536 auks, mostly razorbills, in sheltered bays off the Cornish coast in 1988 (Flumm, 1988), and otters again appear to be greatly at risk.

7.3.4 Fish biomass

Kruuk *et al* (1993a) have suggested that otter populations inhabiting freshwater systems are food-limited and that the utilisation of streams and rivers by otters is strongly correlated with their fish biomass. This implies that other parameters of habitat quality, such as vegetational cover and forms of human disturbance, may not be important unless they affect fish biomass. On the other hand, the frequency of organic and industrial pollution incidents, causing oxygen deficiency and high fish toxicity respectively, would be of importance because of their direct effect on fish presence and availability. This research was carried out in Scotland, where there is a high incidence of seasonal non-violent deaths among otters. These are attributed to periods of food shortage (Kruuk *et al*, 1987; Conroy, 1993). Thus, as a further aid to interpretation of the present distribution of otters in England and in order to examine whether this factor was important in determining future range expansion in the south of Britain, data on fish biomass, on managed fisheries and on the frequency of pollution incidents was sought for each Region from the NRA. A digest of this information follows, but the last factor is discussed further in Sections 5 and 7.3.1.

European Community (EC) designated salmonid fisheries should exceed a sampling biomass of 15g/m^2, while designated cyprinid fisheries should exceed a sampling biomass of 20g/m^2. These are the targets for fisheries management. However, calculation of an accurate figure for the fish biomass of an entire river system is virtually impossible and extrapolation from sample electro-fishing counts can result in a figure 10 to 100 times too high or too low. The problem is further compounded by the anadromous and catadromous nature of the salmonid species and eels, which may result in a large proportion of the fish biomass on some rivers migrating into and out of the river at various times of the year. Nevertheless, each Region sets biomass targets for its rivers for use in monitoring. Stocks and recruitment levels are assessed using sample electro-fishing.

(i) North West Region: The rivers here are principally game fisheries. NRA sampling of salmonid fry and parr throughout the Region has shown that recruitment is ‘healthy’ (ie. substantial and up to requirements for continuity) on the upper reaches of all river systems and that these meet biomass targets.

(ii) Northumbrian Region: The rivers of this Region are, for the most part, game fisheries. However, populations of mixed coarse fish occur along the slower and deeper stretches. NRA sampling of fry and parr throughout the Region has shown a ‘healthy’ recruitment on the upper reaches of all river systems and that these meet biomass targets.

(iii) Yorkshire Region: The River Esk provides the only game fishery in this Region. Although brown trout occur on most of the other river systems, these are predominantly mixed fisheries supporting good populations of coarse fish, especially throughout the middle and lower reaches of each system. Electro-fishing sampling has shown that they meet NRA biomass targets. Eels are recorded as numerous throughout.

(iv) Severn-Trent Region: NRA electro-fishing surveys show a good distribution of mixed coarse fish on the rivers of the Severn-Trent Region. These are primarily cyprinid fisheries and they meet designated fisheries targets. Eels are widespread, but concern about recruitment levels has led to several restocking releases of elvers in the upper Trent.

(v) Anglian Region: Most rivers support a good population of mixed coarse fish. These populations meet biomass targets, although many sites are regularly restocked by angling clubs to increase rates of recruitment. Eels are sufficiently numerous to support a commercial eel-netting industry which fishes around the coast and estuaries as well as inland.

(vi) Thames Region: A good ‘healthy’ and diverse fish fauna is recorded throughout most of the Region. However, it fails to meet the EC designated cyprinid and salmonid fisheries targets of 20g/m^2 and 15g/m^2 respectively at many sites.

(vii) Wessex Region: Fisheries on principal rivers vary in the degree that they meet biomass targets. The rivers of the Somerset Levels, together with the rivers Stour and Avon, support good mixed coarse fish populations which do meet biomass targets. Also, they have some large eel populations. However, salmonid populations are heavily influenced by the extent of stocking and these generally fall below targets. The small coastal river systems all show good populations of trout and eels.

(viii) South West Region: The rivers of this Region are classed as game fisheries. They support purely freshwater and migratory salmonids, together with locally good populations of eels and mixed coarse fish meeting designated biomass targets.

(ix) Southern Region: The chalk streams of the Region support salmonid fisheries which meet designated biomass targets. Cyprinid fisheries also meet biomass targets and support a good diverse fish fauna, with ‘healthy’ populations of eels throughout.

(x) Welsh Region: In addition to the principal fishery of migratory salmonids, brown trout and a good mixed population of coarse fish occur throughout the River Wye and meet designated biomass targets. The River Dee in England supports migratory salmonids and a good mixed coarse fishery with good numbers of eels.

It would appear that in England only in Thames and Wessex Regions do some of the rivers or stretches of them fall much below designated biomass targets for salmonid and cyprinid fisheries. This has not apparently prevented a considerable recolonisation of Wessex Region rivers, as shown by the 18.83% occupation of full survey sites in the 1991-94 otter survey. Thames Region is only recently being penetrated by the 'front' of the otter recolonisation and it remains to be seen how and if recolonisation is affected by the slightly lower fish biomass here. It seems doubtful that in southern Britain fish biomass would be limiting distribution as yet because population density is still low. In addition, with an adult male otter's home range measured at over 25km of river (Green *et al*, 1984; Jefferies *et al*, 1986) and fish biomasses as high as 20g/m^2 of cyprinids (this equates to 1.5kg (or the daily food intake of an otter) per 75m^2 (or 8.5 x 8.5 metres of surface)) there would seem to be an ample food supply for a very much higher density otter population than we have at the moment or for some decades to come. No bodies of otters showing indications of starvation have been autopsied in the series from England examined by Jefferies and Hanson (1987).

7.3.5 Habitat destruction

Coghill (1978) believed that habitat destruction was a principal reason for the disappearance of otters in Worcestershire. However, Chanin and Jefferies (1978) considered that, although there was no doubt that there had been a steady attrition in optimal otter habitat since the 1939-45 World War, there were two reasons for suggesting that this was not the cause of the original decline. Thus, (i) there were no data suggesting a sudden increase in habitat destruction in the mid-1950s, and (ii) the variation in type, extent and rate of habitat change over the whole country was so great that it was inconceivable that they could have caused a decline of otters simultaneously over virtually the whole of Britain (see Section 7.1).

Much has been written on the habitat features and conditions considered necessary for the otter to survive and breed in an area (O'Connor *et al*, 1977; Macdonald *et al*, 1978; Lenton *et al*, 1980; King & Potter, 1980; Lenton, 1982). These include considerable cover in the form of brambles, shrubs and bankside and emergent herbaceous vegetation. Mature trees, such as ash, oak, sycamore and beech, are valuable when they have suitable bankside root formations for supplying some of the need for riparian holts (Macdonald *et al*, 1978). A good supply of fish (over 200mm in length), particularly eels, is required for food, and disturbance should be minimal. Organisations, such as The Vincent Wildlife Trust through its Otter Haven Project (Lenton, 1982), carried out much useful work for otter conservation by providing such features in order to ensure that they were present where otters were known or expected. Also, habitats like these were sought out in East Anglia to provide the first localities for releasing captive-bred otters, eg. the River Black Bourn in Suffolk used by the Otter Trust in 1983 (Jefferies *et al*, 1986; Wayre, 1989).

It is true that in lowland Britain such habitats would be considered the optimum. Their presence or creation where otters are few and are being released or they are recolonising naturally ensures, as far as this is possible, that (i) they will stay and breed in that chosen conservation area and (ii) their productivity is likely to be at its highest. However, there are many indications that, when otter populations are high, many sites are inhabited or colonised which we would have considered suboptimal or even completely unsuitable a few years ago. Thus, it is not essential to have such high quality habitats available everywhere for the otter to have eventually a much more extended coverage of England than it has at the present. This can be observed from historical data regarding otters in England in the nineteenth and early twentieth centuries and in Ireland today.

Richard Jefferies, writing between 1877 and 1882 (Jefferies, 1883), knew the otter to nhabit the River Thames in Greater London at least as far down as the seaward side of 'eddington Lock. Also, otters were known to use the Grand Union Canal in the centre of ondon at that time. One was killed in St John's Wood in 1863 (Harting, 1865). Between he First and Second World Wars otters were seen in the Thames at Westminster Bridge 1922), at Hammersmith (1923) and at Syon Marsh near Kew Gardens (1937: Fitter, 1945). 'he waters of the Thames and the canal were far from unpolluted at these times, but the ollution would have been organic rather than with man-made molecules inimical to life.

Similarly, the drains and dykes of the Cambridgeshire and south Lincolnshire fens do ot look to us today to be ideal otter habitat, with their long straight channels running hrough flat arable farmland and their raised banks almost devoid of marginal cover. Iowever, Millais (1904-06) considered the slow-moving rivers of Cambridgeshire and Iuntingdonshire, such as the Great Ouse, very good otter country because of their abundant eels". This description could be extended to the fen drains between the rivers Velland, Nene and Great Ouse as well. According to local farmers these were certainly well opulated by otters before the population crash, despite their lack of cover. There are few old ublished records of otters on the fens because it was largely the rivers which were hunted, but Ailler and Skertchly (1878) noted that they occurred in Burwell Fen near Ely, and Barclay 1938) recorded that they "occasionally bred in the fens". More recently, we have photo-raphic evidence of an otter which was shot around 1955 in the Great Raveley Drain, near Voodwalton Fen (H. Hanson, pers. comm.), and one of us (DJJ) received the body of an tter road casualty killed at Ramsey St Mary's in Huntingdonshire on 21st May 1965. This vas a female, probably nursing young, found where the B1040 road runs alongside the New)yke. Thus, otters are well able to survive and breed where cover is minimal provided there s a good food supply.

With regard to breeding holts (ie. those chosen and used by the female to give birth nd rear her cubs), these are usually found to be the most secure and secluded of all holts Harper, 1981; Green *et al*, 1984; Jefferies, 1987). However, it is known that the females an produce and rear their cubs in very disturbed and insecure conditions when and where tter populations are at high densities (see Section 7.3.6). Thus, one litter of cubs was reared n a holt below one of the busy main jetties at the Sullom Voe Oil Terminal in Shetland Berry, 1985).

The Otter Trust's choice of their 1991 release sites for captive-bred otters on the River tort and at nearby Amwell in Hertfordshire were particularly sharply criticised by some onservationists because they were considered unsuitable and atypical of otter habitat Wayre, 1993b). The area had long stretches of bank strengthened by piling and many noored boats, with much disturbance. However, these critics may have been misguided. Examination of the photographic plates in the report of the 1980-81 otter survey of Ireland Chapman & Chapman, 1982) shows that otters can and do inhabit apparently very poor abitat when populations are large and at high densities. Thus, Plate 3 shows the newly analised banks of the River Clare in Galway with no bankside cover and piles of debris and redgings, but well marked by otters. Also, spraints were found on the concrete blocks and ubble at the edge of a highly disturbed and used car park at Limerick on the Shannon stuary. Again, there was no cover at all (Plates 6 and 7 of Chapman & Chapman, 1982). The Otter Trust's released animals have stayed and bred in Hertfordshire in the Stort area Wayre, 1993b). However, while otters are still at low density, as they are in England, they an move very easily and take up habitation elsewhere. This can include movements etween watersheds (Jefferies, 1989c). For example, captive-bred otters released on the River Glaven in north Norfolk in 1987 apparently moved into the adjacent River Bure ystem two years later (Jessop & Cheyne, 1993).

At very high densities, on the other hand, otters will need to be able to tolerate an pparently much lower quality of environment than the optimum described above. Examination of Figure 21a shows that otters will inhabit suboptimal habitat, including plands, as their population increases. Adults consume only 1.4 to 1.5kg of food each day Wayre, 1979) but inhabit very large home ranges (male: 39.1km; female: 16.0 to 22.4km of waterway; Green *et al*, 1984). Also, they are catholic feeders and will eat many ot animals besides large fish (Weir & Banister, 1972; 1977; Mason & Macdonald, 19 Chanin, 1981; Wise *et al*, 1981; Breathnach & Fairley, 1993). They will even eat carr (Cuthbert, 1973; Hewson, 1995). Thus, it would not be difficult for them to find suffici food somewhere within most 20 to 40km stretches of waterway anywhere in south Britain today. It should be remembered too that they can and do swim 11.5km (7.2 mi Jefferies *et al*, 1986) in a night, so can feed in one area and then move many kilome away where cover may be available for daytime shelter.

As noted by Jefferies (1989b), the two factors likely to have been the major causes the crash of the otter population starting in 1957 were (i) pollution by man-made toxins (dieldrin and organomercuries) and (ii) persecution by hunting and gamekeeping practi With the effects of these two factors reduced, there are no habitat reasons why the recover otter population (see Section 6.2) should not inhabit much of England and reach h population densities once again, given the present climate of interest in the wellbeing of species. Indeed, we are now seeing for the first time otters moving into areas where considered the habitat to be of too low a quality to support them in the first two surveys (Section 3). This suggests that the increasing population pressure has reached a stage i few areas when it is now forcing some otters into suboptimal habitats again.

7.3.6 Disturbance

Increasing disturbance was listed among the likely problems for the declining o population when the Joint Otter Group reviewed the situation in 1977 (O'Connor *et* 1977). However, Chanin and Jefferies (1978) showed that likely sources of disturbar including boating and angling, had increased steadily in recent decades, so concluded disturbance was not a major factor in the sudden crash of the otter population in Brit starting in 1957-58. Also, there is no correlation between the number of rod licences iss to anglers in each Water Authority Region and the present state of the otter population that Region (Strachan *et al*, 1990). Consequently, conservationists have become less c cerned over disturbance as more information has been gained (Chanin, 1993). Indeed, in present 1991-94 survey, apparent tolerance of disturbed areas was noted by frequent reco and occupied sites indicating that otters were travelling through built-up areas. Thus, t are now using rivers passing through cities and towns with human populations in excess 100,000. Such cities and towns which otters are known to have traversed recently or in past include Cambridge (River Cam), Canterbury (River Stour), Carlisle (River Ede Darlington (River Tees), Exeter (River Exe), Gloucester (River Severn), Lancaster (Ri Lune), Maidstone (River Medway), Newcastle (River Tyne), Norwich (River Yare/Ri Wensum), Oxford (River Thames/River Cherwell), Plymouth (Tamar estuary), Pres (River Ribble), Worcester (River Severn) and York (River Ouse). However, as they w often recorded owing to their marking of sites, these otters could have been exploring ma (see below).

Jefferies (1987) has reviewed the few available data on the effects of individ sources of disturbance on the otter, using reported observations and information from rac tracked otters in Perthshire (Green *et al*, 1984) and Suffolk (Jefferies *et al*, 1986). Jeffe pointed out that otters use intelligent means of getting around a source of disturbance rat than retreating from it. This they achieve by travelling through dense cover or by swimm under water so that their presence remains undiscovered. Thus, where cover continues to sufficient, then the effects of disturbance may be ameliorated or eliminated.

On the other hand, perhaps we should not dismiss disturbance as a problem entirely is notable that otters are largely nocturnal on mainland Britain, particularly in the so (Jefferies *et al*, 1986) and when at low population density (Jenkins, 1980). Those rad tracked in Perthshire and Suffolk usually emerged within an hour of dusk and emerge times changed with dusk times (Green *et al*, 1984; Jefferies *et al*, 1986). This behavi may be a long-term reaction to continuous daytime disturbance. Also, Jefferies (1987) ma one caveat regarding the apparently low effect of disturbance on observed otters, that ma of the data from radio-tracking and spraints are derived from males; there is much l information on females, particularly on females with cubs. Much of the site-marking with spraints is carried out by the males (Green *et al*, 1984; Hillegaart *et al*, 1985). In addition, female resting sites were invariably found to be situated in the most secure habitat available (Green *et al*, 1984). Female couches (ie. above ground daytime resting sites) were covered with dense scrub, while male couches were often exposed. The finding by these authors, that female otters used twice as many underground holts (68%) as surface couches (32%) whereas males showed the reverse trend (couches 67%; holts 33%), strongly suggests that the female has a higher sensitivity to disturbance than the male.

Despite the often quoted breeding holt below one of the busy jetties at the Sullom Voe Oil Terminal on Shetland (Berry, 1985), the holts that we know to have been chosen by the female to give birth and rear her cubs are the most secure and/or secluded of all holts. Green *et al* (1984) and Harper (1981) also make reference to the added security of breeding holts. There is, in addition, a reduction or cessation of spraint-marking around a breeding area (Östman *et al*, 1985), suggesting that security is actually enhanced at this time. Jefferies (1987) provides examples of the marked reaction of females with cubs at times of disturbance, including evacuation of the area.

Further, some effect of disturbance might be expected in the otter, as there is evidence for it in Britain's other large nocturnal mustelids, the badger and the pine marten. With these species there are more data available for analysis.

The badger, like the otter, appears to be able to get habituated to background noise such as traffic, but unusual or sudden disturbance has a considerable and significant effect, such as late emergence (Humphries, 1958; Neal, 1986; Jenkinson, 1991). Scott (1960) and Barker (in Neal, 1986) both found that the amount of cover around the sett was a most important factor and that, when good cover was present badgers would tolerate greater disturbance than when it was absent. This would support what little is known about otters. Aaris-Sorensen (1987) attributed the long-term trend of badger decline around Copenhagen to unintentional human disturbance.

A study of the localities selected for inhabitation by pine martens living at very low density in England and Wales again showed a significant tendency towards those with the lowest disturbance (Strachan *et al,* in press).

More data are required on the effects of various forms of disturbance on otter behaviour. In the meantime, it seems best to reduce disturbance and to increase cover wherever possible where populations of breeding otters are resident or expected. Strachan *et al* (1990) have pointed out that, if disturbance increases and cover is decreased, then it is possible that the otter may retreat from the disturbed area and feed elsewhere, thus effectively reducing the area available for inhabitation.

In conclusion, it seems likely from the available evidence that otters select low disturbance situations when possible, but that more and more disturbance has to be tolerated as the population increases in density with the consequent competition for food and holt sites. Thus, at low densities, as in England, otters have a wide choice of holt sites, including breeding holt sites, and, with an ample food supply, can afford to feed only at night, so avoiding disturbance. At high densities, as in Shetland, on the other hand, otters may be forced to feed through the daylight period, particularly if the food supply is small and limiting (see Section 7.3.4), and even breed under the busy oil terminal jetty at Sullom Voe.

7.3.7 Diseases

Keymer (1993a) has reviewed the literature of the diseases of wild and captive *L. lutra*. Urolithiasis, pneumonia, cataracts, neoplasms, salmonellosis, pseudotuberculosis, tuberculosis and canine distemper have all been recorded. Botulism, leptospirosis and Aleutian disease have been suspected. However, as Keymer remarks, little is known about otter diseases and most descriptions of the above usually refer to single animals or only very small samples. The reports also span many decades and none of these diseases are believed to have been the cause of population changes.

Probably the diseases of most concern to conservationists, with regard to their ability to produce adverse population effects in the otter, are canine or mink distemper and Aleutian disease. Both are contagious and lethal to mustelids and could pass from species to species, eg. from mink or ferret to otter. Phocine distemper and bovine tuberculosis are further possible problems. All form a potentially very serious hazard.

7.3.7.1 Contagious diseases

7.3.7.1.1 Canine and mink distemper

Otters are susceptible to canine distemper (Harris, 1968; Keymer, 1993a) and presumably to the related mink distemper, which is known to have the potential to cause very high mortality among farmed mink (Kennedy, 1951). Canine distemper has been recently confirmed in two captive otters in Germany (Keymer, 1993a), and Harting (in Millais, 1904-06) reported two deaths in captive otters in Britain in the nineteenth century, the disease having been contracted from dogs. Nineteenth-century zoologists and dog owners would have been well aware of the symptoms and problems of canine distemper. The potential for this disease to cause an epizootic in otters is now probably much slighter than that of Aleutian disease (see Section 7.3.7.1.2). Most dogs are now vaccinated and since 1982 hounds are no longer used to hunt otters. Thus, the likelihood of close contact between sick dogs and otters is very much reduced. Also, there are fewer mink escaping from fur farms now and many captive mink are vaccinated too.

7.3.7.1.2 Aleutian disease

Aleutian disease is caused by a parvovirus and is an immune deficiency disease which is predominantly an infection of captive mink (Porter, 1986; Keymer, 1993a). It was named from the strain of mink in which it was first discovered in 1956 in the USA. The first confirmed record of the disease in Britain did not occur until 1990, presumably from imported animals. Although originally considered doubtful, it is now known that otters and feral mink will share the same holt at the same time (Green *et al*, 1986), which must considerably increase the chances of contagion. There is also the strong possibility that otters bite and kill mink (see Sections 6.3 and 7.3.8.1).

This problem has become exacerbated recently because an outbreak of Aleutian disease was confirmed in domestic ferrets in 1990 and still continues. Unfortunately, there is considerable contact between ferrets from various parts of the country at ferret shows and races, so there has been ample opportunity for the disease to have become widely dispersed in only five years. Most of the affected animals have been from the Wessex area, but simultaneous cases have been recorded in Berkshire, Lancashire, London and Oxfordshire also (McKay, 1993). The numbers recorded as positive for the disease were 26, 14, 6 and 19 in 1990, 1991, 1992 and 1993 respectively (M. Oxenham, in McKay, 1993). The actual number infected may be very much higher than this, since many dead animals are not autopsied. The result has been a considerable concern among ferret owners regarding spreading the infection at shows. There is no effective treatment for this lethal disease.

From the point of view of otter conservation, our major concern is that this disease has another carrier (besides mink) by which it could get out into the field. Indeed, ferrets may prove to be a greater problem than mink. Feral mink are already established now over much of the country, but there are few modern escapes owing to the strict legislation and inspection regarding cages and fencing. This may mean that imported fur-farming diseases will stay within the confines of the captive mink population. However, the *raison d'être* of ferrets is that they should be released into the countryside for bolting rabbits; consequently very many escape and survive in the warrens (Birks, 1993). Thus, the chances of escape of this disease from the domestic to the sylvatic condition are very high.

As with the otter holts shared between mink and otter, it has been found by radio-tracking that otters frequently use rabbit burrows for both day and night-time shelters (Jefferies *et al*, 1986). There is then a high possibility of an otter and a ferret meeting underground. Although bodily contact between otter and mink or ferret is likely to be avoided by the smaller animal (though attack by the otter could occur), in the close confinement of a holt or burrow aerosol infection from droplets, breathed or coughed out, is possible. Such aerosol transmission of Aleutian disease can occur from up to a metre away (McKay, 1993), so contact is not necessary.

The finding of symptoms and lesions closely resembling those of Aleutian disease in a dying Norfolk otter in 1986 (Spalton & Cripps, 1989; Wells *et al*, 1989) lends considerable credence to the concern that the disease could enter the otter population. This was some time before the 'confirmed' presence of the disease was reported among Britain's domesticated mustelids. We believe that Aleutian disease, once within a few individual otters, could spread rapidly through the whole population by contact, by fighting and by aerosols in shared holts. The resulting epizootic could be as catastrophic as that of the outbreak of phocine distemper among Norfolk's common seals in 1988 (see Section 7.3.7.1.3). The additional potential possibility of a reduced immune response owing to contamination with organochlorine insecticides and PCBs (Friend & Trainer, 1970; Jefferies, 1975; Safe, 1984) could exacerbate the situation. The possibility of a PCB-induced reduction in the immune response to phocine distemper virus in the common seals of the North Sea during the epizootic of 1988 (see Section 7.3.7.1.3) is still under discussion (Thompson & Hall, 1993).

7.3.7.1.3 Phocine distemper and bovine tuberculosis

Another disease which caused considerable concern for the future of Britain's recovering otter population at the time was the outbreak of phocine distemper (PDV) in North Sea common seals in 1988. This disease, caused by a morbillivirus, brought about the deaths of 18,000 common seals in the North Sea, and 50% of those in the Wash were found to have died (Thompson & Hall, 1993). Scottish grey seals were affected too but to a lesser extent (Hall *et al*, 1992). As this virus is related to that causing canine distemper (CDV), to which otters are susceptible (Harris, 1968; Keymer, 1993a), there was great concern that it could jump the species gap and infect coastal otters (Jefferies, 1988b). Keymer (1993a) suggests that this is indeed a worry but is unlikely to occur in the field. Inhalation is considered to be the main route of infection of canine distemper and he considered that there would not have been close contact between otters and seals. Also, as otters are not gregarious, passage between otters was unlikely. However, otters are curious, as are other carnivores such as mink, and would examine closely any dying or moribund seals found. Distemper could be passed between otters at inhalation at contact smelling, at fighting (Jefferies & Hanson, 1991) or by aerosol inhalation when in close confinement underground in holts. Otter hunting diaries show that adult males and females were frequently taken from the same holt (eg. Greenwood, 1991). There were many diseased grey and common seals along the coasts of the Northern and Western Isles, areas which have particularly high densities of coastal-living otters. Fortunately, no examples of phocine distemper have been suspected or confirmed in British otters, either in 1988 or since.

Stephens (1957) reported that the body of a male otter from Cornwall showed lesions of "very advanced tuberculosis of the lungs". Which form of tuberculosis (avian or bovine) is unclear but the latter would seem most likely. An infection of bovine tuberculosis in an otter or otters from the south-west of England, where the disease occurs at a very high incidence in the local badgers, would seem to be a possibility. Mustelids appear to be sensitive to this disease. Also, at least one aggressive interaction between otter and badger is known (G. Liles, pers. comm.) and badgers have been found sharing part of a large holt in Scotland (Green *et al*, 1984).

In conclusion, we must emphasise that the possibility of a wildlife disease epizootic remains one of the biggest threats to the recovering otter population.

7.3.7.2 Blindness

There is considerable interest in the condition of blindness in British otters. Williams (1989, 1993) has collected 22 records of its occurrence, all between 1957 and 1980. Keymer (1993b) suggests that this blindness could be due to keratitis or cataracts. He considered the latter to be the most likely cause and suggested that these cataracts may be caused by organochlorine poisoning (see Section 7.3.1). Jefferies (in press) suggests that hyperkeratosis and corneal opacity are possible with avitaminosis A (see Jefferies, 1975) and agrees that the timing would support the hypothesis that organochlorine contamination was the most likely basal cause of this blindness. Its high incidence has apparently remained unrecorded since the bans on dieldrin and DDT use.

7.3.7.3 Parasites

A survey of parasites in 56 British coastal-living otters showed the presence for the first time of larvae and adults of the sealworm *Pseudoterranova decipiens* in the stomach, intestines and mediastinal cavity of seven of their hosts (Jefferies *et al*, 1990). The major fish host of this nematode is the cod and the more frequent mammalian hosts are the grey seal and cetaceans. Examination of 29 further coastal otters from Shetland showed *P. decipiens* in one and another anisakine nematode *Anisakis simplex* in eight others (Weber, 1991). Prevalence of *P. decipiens* in otters reached 15% on mainland Scotland and 17% on the Hebrides, but none was found in 12 English otters. Burdens up to 183 *P. decipiens* adults were found in one otter and the condition of the gut walls of some affected otters caused initial concern. Young and Lowe (1969) demonstrated an association between gastric lesions and the presence of anisakine nematodes in grey seals and harbour porpoises and Rausch (1953) considered *P. decipiens* to be pathogenic to the sea otter.

7.3.8 Problems caused by other mustelids

7.3.8.1 Feral American mink and feral ferrets

The interactions between otters and feral American mink have been covered in detail already in Section 6.3 and would appear to result in the mink being driven out or killed in areas reoccupied by otters when the latter again occur at high density. The otter may interact in the same way with the polecat, now spreading into England again from Wales (Jefferies, 1992b). However, there are two ways in which the feral mink population of Britain could provide problems for the conservation of a recovering otter population.

First, it is now known that both otters and mink will share the same resting sites during the day (Green *et al*, 1986). Consequently, when the mink scent is hunted by packs of hounds as a modern substitute for the otter, this places the latter in some danger. An example is given by the otter and mink which were flushed from the same stick pile by a mink hunt on the River Creedy in Devon (Green *et al*, 1986). Although hounds may be 'entered' only to mink, they may 'chop' any accompanying animal before they can be driven off. This must provide concern for the mink hunts as well, with regard to the provisions of the Wildlife and Countryside Act 1981. Radio-tracking has shown that male otters move almost continuously, occupying a different holt or hover each day; 37 different sites were used by one male in Perthshire (Green *et al*, 1984). Thus, with an expanding population, there is now no way of being certain that a likely resting site is not being used by an otter, with or without a mink, on many catchments across the breadth of England. Otters are now present in all ten English Regions of the National Rivers Authority (see Section 4). In addition, the presence of mink in all these Regions (Strachan & Jefferies, 1993) means that mink hunting is now carried out in many areas; there were 17 registered packs of mink hounds in 1990 (Downing, 1990). Even if direct danger to otters is obviated somehow, the passage of a pack of hounds along a riverbank causes considerable disturbance. Macdonald *et al* (1978) noted that otters abandoned a favoured site in Norfolk for several months after a pack of otter hounds passed through in the early 1970s.

Secondly, there are the considerable problems of mink diseases, which may be carried over into the field from fur-farming origins. Many of these diseases, such as Aleutian disease and mink or canine distemper, are likely to be as lethal to otters as they are to mink. The problems of such contagion are discussed in more detail in Section 7.3.7.1. The suggestion from the analysis of the data on the interaction between otters and feral mink (see Section 6.3) that otters may bite, injure, kill and possibly occasionally eat mink obviously greatly increases the chances of disease passing between the two species.

Much of what has been written about the problems of contagious diseases of mink also applies to those of feral ferrets. The infection of most concern is again that of Aleutian disease (see Section 7.3.7.1.2). Britain's small but recovering polecat population, another species of conservation importance (Birks, 1993), faces the same hazard from the diseases of mink and ferret as does the otter.

It is difficult to quantify the extent to which mink may kill and/or eat otter cubs, particularly in areas where otters are few and mink are many. Mink would certainly kill any small or young riparian mammals, and there is a period when female otters leave the cubs hidden and unattended for several minutes. Also, the eating of otter cubs by mink has been recorded (Jefferies, 1991a). However, it probably occurs too infrequently to be ranked as a serious problem.

7.3.8.2 Asian short-clawed otters

Observations and two road casualties (1986 and 1991) over a period of eight years have confirmed that there is a population of Asian short-clawed otters living wild and breeding near Oxford (Jefferies, 1990a, 1992a). The records cover a very compact area with a diameter of only 17km. The waterways concerned are the rivers Thames and Thame, the Bayswater Brook and the Oxford Canal, immediately east, north and west of Oxford. These animals are presumably originally derived from escapes or releases, as the species has become a popular exotic pet.

The native distribution of this species (*Aonyx cinerea*) covers India and eastwards to Borneo and Java (Harris, 1968). Most of this area is also inhabited by the Eurasian otter (*Lutra lutra*) and the Indian smooth-coated otter (*Lutrogale perspicillata*). Competition is avoided by largely differing diets and occupation of different sections of the same rivers. *A. cinerea* is an invertebrate specialist and *L. lutra* is a fish specialist, though of course both will take the other form of food as part of their diet (Chanin, 1985).

Lutra lutra may be moving into the upper Thames area from the west, as 2 out of 74 sites on the upper Thames were found to be positive in this 1991-94 survey (see Section 5.6). Occupation of the same area of the Thames catchment may mean that the larger *L. lutra* (7.0 to 10.1kg) will kill the smaller *A. cinerea* (2.7 to 5.4kg; Chanin, 1985) or drive it into waterways which are seldom hunted by the former.

If *A. cinerea* proves to be a carrier of any diseases to which it is relatively insensitive or largely immune, but which could be lethal to *L. lutra* (ie. similar to measles, which was introduced to the Inuits by Europeans with lethal consequences; Dubos *et al*, 1969), then this could prove to be a serious problem. However, this seems unlikely judging by the experience of the Otter Trust in keeping both species. Otherwise the only conservation problem caused by the presence of *A. cinerea* would seem to be that of potential confusion at times of survey of *L. lutra*.

On the other hand, the presence of breeding *A. cinerea* since at least 1983 (Jefferies, 1990a) provided an unplanned experiment or 'probe' testing the waters of the Thames catchment in a similar way to that of the first Otter Trust release on the River Black Bourn in Suffolk in 1983. It proved that the middle Thames waters were then sufficiently unpolluted to allow inhabitation by otters of whatever species and that *L. lutra* was absent solely because there were none left to occupy the area.

7.4 Conservation of otters, past, present and future

The decline of the British otter population, its timing and the area of greatest effect as described in this report correlate with the start of use of the highly toxic and persistent organochlorine insecticides. However, there was also a subsidiary cause, that of the persecution and killing of breeding-age adults over many decades before the decline started. Early amelioration of these two factors was essential to ensure the survival of the otter. In consequence, two major conservation actions have been completed, ie. the bans on the uses of the organochlorine insecticides from 1962 to 1983 and the provision of legal protection in 1978 and 1982. These two actions appear to have reversed the negative population trend and the pattern of recovery fits their timing. Their deployment while there was still a population large enough to recover and expand from the west and the north has ensured that recovery is well established by 1996. Where the population was too small, fragmented and approaching extinction, ie. along the eastern coastal zone of England, the insertion of groups of released otters formed the only way to get recovery started (see below). Thus, most of what was essential to be done has been completed already with some degree of success.

There are many indicators that the organochlorine pollution of the English waterways has degraded to the point where it is no longer the main factor governing otter numbers and distribution, although these materials are still present in otter tissues. Further degradation of the residues in the sediments is a matter of time and little can be done to speed it up. However, the occurrence of further pollution is in the hands of man. The otter swims and feeds in our drinking water.

From now on, the rate of recovery will be dependent upon the otter's intrinsic rate of natural increase. The female otter does not breed until two years of age and then has the capacity to produce one litter a year. However, Conroy (1993) has calculated, using data from Shetland that, with a breeding life expectancy of only 2.4 years and an average litter size of 1.78 cubs, then with cub mortality a female will produce only 1.2 male and 1.2 female cubs of breeding age within her lifetime. Even if litter sizes are larger in the south (eg. Stubbe (1977) estimates 2.4 for Germany), this is an extremely slow potential rate of increase and accounts for the very slow movement of the otter population 'front'. Anything which will reduce natural and accidental mortality (assuming deaths due to persecution and insecticidal poisoning are already reduced) would be of great importance in that it would increase the numbers of dispersing and recolonising otters. Thus, all steps should be taken to reduce accidental mortality which should be controllable. A successful start has been made in controlling one of the causes of this mortality by research on and production of guards for fyke nets by the VWT. Their use needs to be enforced by bye-law, particularly in Anglian Region, where fyke netting is carried out on a large scale and otters are particularly few. The very large problem of otter road mortality remains and schemes must be researched and tested to try to reduce this at identified 'black spots', perhaps by the use of tunnels, guiding fences, barriers and light deflectors. Natural mortality owing to starvation, as reported for Shetland, seems less likely in the south, as food is unlikely to be limiting here. Prosecutions should follow any proven cases of persecution.

Although it seems that otters can live in most aquatic situations and distribution over most of Britain should again be achievable, the final density of the otters will depend on habitat quality. High quality habitat includes a high fish biomass, low disturbance (likely to be of greater importance to females than males), low pollution and the presence of secure breeding holts and cover. For the present, the creation of more otter havens and habitat improvement should be targeted along the fringes of the naturally expanding wild otter populations, in order to provide the necessary high quality habitat for occupation by the dispersing young. This could perhaps stabilise the 'front' rapidly and speed up the production of first litters owing to itinerant animals establishing territories sooner. The value of SSSI (Site of Special Scientific Interest) status to provide further protection for known traditional breeding holts (Jefferies & Mitchell-Jones, 1989) and other forms of statutory protection for sites should be explored.

With the otter's new-found popularity and protection, this final density could be much higher than that pertaining to the pre-decline period, much as for the peregrine falcon. This may necessitate education of the present generation of waterway users and managers, such as anglers and fisheries bailiffs, so that they accept and accommodate the new increased otter populations in their areas of shared interest. Fish-farms may need to take steps to protect stock again from otters, much as they are used to doing with grey herons. Information on such practical steps will need to be made available.

The releasing of otters, both captive-bred and rehabilitated, has successfully brought about reoccupation of much of two areas of eastern England long in advance of that which could be expected from the eastward movement of the western population. By so doing, it has undoubtedly preserved the gene pool of the remaining eastern otters of those areas from complete extinction as well as incidentally acting as a useful environmental 'test probe', showing that the east will support the main body of the otter population when it reaches there. Restriction of otter releases to the counties of the eastern coastal zone was put into operation by the NCC some years ago, so that the 'natural' reoccupation from the west could take place in the Midlands and southern England and its rate observed. As a conservation tool in the present climate of wild population expansion, releases are no longer essential, having largely served their purpose. They may be phased out over the next few years if satisfactory recovery continues, though their use might yet be of value in the south-east 'corner' of England for which natural reoccupation would seem a long time off. It should be remembered that releases are the only way we know of for accelerating the rate of recovery and reoccupation of isolated, empty or sparsely inhabited areas if the need arises. Other European countries may like to take note of their undoubted success despite the often expressed concern that they would not work.

Continuation of the present monitoring scheme for otters is essential so that any indication of a faltering in the recovery and its area of effect will be quickly perceived and hopefully corrected. Comparison with the calculated recovery curve given in this report may indicate whether the recovery rate is above or below expectation. The national surveys should continue the examination of all the present sites for the calculation of percentage site occupation as before and, preferably, should include data on spraint density as an indication of otter density within occupied sites (ie. as a form of 'fine adjustment' on the technique). Provision of information on otter occupation in the non-surveyed 50km squares of England would be a useful adjunct to the national survey results. These could be completed by County Wildlife Trusts and local area otter officers as reported here.

Some monitoring for the presence of mustelid-borne viral and bacterial diseases (particularly Aleutian disease) in otters 'found dead' is needed, because these would appear to be one of the greatest potential threats to a recovering otter population. The success of the release programme and the spreading of both otter and mink populations would appear to indicate that water quality is generally no longer inimical to aquatic mustelids over quite large areas of Britain, including East Anglia. Thus, perhaps sufficient information has now been gained from the general analysis of organs and spraints for PCBs and organochlorine insecticides. With the bans on the uses of these materials already in place and difficulties in demonstrating present effects on British otters from their residue levels, perhaps more attention should be focussed on any increasing problems or area effects from acidification, nitrates, organomercury, organophosphorus sheep-dips and oestrogen mimics. Some analyses for organochlorines may still be necessary in areas where recolonisation by otters remains slow or absent, but further progress with regard to PCB residues depends on the development of a method for accurate evaluation of the residues in otters. Perhaps use of the technique of calculating toxic equivalency factors (TEFs) for the PCB congeners present in commercial mixtures for comparison with those remaining in otter tissues after environmental breakdown may be applicable (see Safe, 1990).

With regard to research, more information is required on the habitat usage and behaviour of otters living at low density in the rivers of southern England. The difficult problem of estimating numbers from survey data, such as spraint density, needs attention. Also, do present freshwater pollutant levels affect litter size and cub survival and if so by how much?

There is now a favourable 'climate' for the otter in England and it occupies a high place in the public esteem, which has not always been so. There are more people interested in otters or employed by County Trusts or NRA Regions in carrying out surveying projects or preparing the environment for future otter occupation than ever before. This augurs well for the otter's rehabilitation but even now we cannot afford to be complacent. Although recovery is well under way, there are still 78% of sites in England without otters and analysis of the present rate of recovery suggests that it will be several decades before England has a strong otter population again.

The Joint Nature Conservation Committee has recently prepared a strategy for co-ordination of conservation effort on the otter because of the large number of organisations and individuals now interested in the species.

8. ACKNOWLEDGEMENTS

We are grateful for the technical advice and information which was provided by the staff of the various departments of the NRA Regions for use in preparing this report.

This Otter Survey of England 1991-94 would not have been possible without the kind permission of all the riparian owners and tenant farmers who allowed access to the survey sites. We should also like to thank many of them for their enlightened views about otters and local knowledge of recent sightings.

We have recorded in the Regional text the tremendous amount of local conservation effort which has been directed towards restoring suitable otter habitat. For this information we are indebted to the Otters and Rivers Project officers for their help and continuing good work. Many of them organised simultaneous surveys of adjacent squares so that this report could be more comprehensive than previous surveys and give a complete 'snap-shot' picture of the status and distribution of the otter over the whole of England in the early 1990s. We should like to thank Lynn Collins, Andrew Crawford, Melanie Findlay, Duncan Glen, John Green, Paul Hoban, Penny Howell, Hilary Marshall, Graham Roberts, Graham Scholey, Helen Smith, Tim Sykes, Martin Twiss, James Williams and Gordon Woodroffe in particular.

Touring the country and living out of a camper van for 28 months would have been a very lonely pursuit, if it had not been for the tremendous support from a network of 'mammal people' who were prepared to spend time with the surveyor and provide him with excellent hospitality when needed. We thank you all.

Finally, we thank Philip Oswald for reading our text and suggesting many improvements and Terry O'Connor of the VWT, who patiently transformed our illegible handwriting and frequent changes into a typescript and disk ready for publication.

9. REFERENCES

AARIS-SØRENSEN, J. (1987). Past and present distribution of badgers in the Copenhagen area. Biological Conservation 41: 159-165.

ADVISORY COMMITTEE ON POISONOUS SUBSTANCES USED IN AGRICULTURE AND FOOD STORAGE (1967). Review of the present safety arrangements for the use of toxic chemicals in agriculture and food storage. H.M.S.O., London.

ANDREWS, E.M. & CRAWFORD, A.K. (1986). Otter Survey of Wales 1984-85. The Vincent Wildlife Trust, London.

ANDREWS, E., HOWELL, P. & JOHNSON, K. (1993). Otter Survey of Wales 1991. The Vincent Wildlife Trust, London.

ANONYMOUS (1969). The otter in Britain. Oryx 10: 16-22.

ANONYMOUS (1990). This Common Inheritance: Britain's Environmental Strategy. H.M.S.O., London.

ARNOLD, H.R. (1984). Distribution maps of the mammals of the British Isles. Institute of Terrestrial Ecology, Huntingdon.

ARNOLD, H.R. (1993). Atlas of mammals in Britain. H.M.S.O., London.

AULERICH, R.J. & RINGER, R.K. (1977). Current status of PCB toxicity to mink and effects on their reproduction. Archives of Environmental Contamination and Toxicology 6: 279-292.

BARCLAY, E.N. (1938). Mammals. In: L.F. Salzman (ed.) The Victoria History of the County of Cambridgeshire and the Isle of Ely, Vol. 1, 243-245. Oxford University Press, Oxford.

BATTEN, H.M. (1920). Habits and Characters of British Wild Animals. W. & R. Chambers, London & Edinburgh.

BATTEN, L.A., BIBBY, C.J., CLEMENT, P., ELLIOTT, G.D. & PORTER, R.F. (eds) (1990). Red data birds in Britain. T. & A.D. Poyser, London.

BELL, A.A. (1975). Dieldrin residues in the livers of kestrels and barn owls found dead in 1970-1973. Annual Report of the Institute of Terrestrial Ecology, 1974: 27-28.

BERRY, R.J. (1985). The natural history of Orkney. New Naturalist Series. Collins, London.

BIRKS, J. (1990). Feral mink and nature conservation. British Wildlife 1: 313-323.

BIRKS, J. (1993). The return of the polecat. British Wildlife 5: 16-25.

BITMAN, J., CECIL, H.C., HARRIS, S.J. & FRIES, G.F. (1968). Estrogenic activity of op[1]-DDT in the mammalian uterus and avian oviduct. Science, New York 162: 371-372.

BLACKMORE, D.K. (1963). The toxicity of some chlorinated hydrocarbon insecticides to British wild foxes (*Vulpes vulpes*). Journal of Comparative Pathology & Therapeutics 73: 391-409.

BLEAVINS, M.R., AULERICH, R.J. & RINGER, R.K. (1980). Polychlorinated biphenyls (Aroclors 1016 and 1242): Effects on survival and reproduction in mink and ferrets. Archives of Environmental Contamination and Toxicology 9: 627-635.

BOURNE, W.R.P. (1977). Nylon netting as a hazard to birds. Marine Pollution Bulletin 8: 75-76.

BREATHNACH, S. & FAIRLEY, J.S. (1993). The diet of otters *Lutra lutra* (L.) in the Clare river system. Proceedings of the Royal Irish Academy 93B (3): 151-158.

BROUWER, A., REIJNDERS, P.J.H. & KOEMAN, J.H. (1989). Polychlorinated biphenyl (PCB)-contaminated fish induces vitamin A and thyroid hormone deficiency in the common seal (*Phoca vitulina*). Aquatic Toxicology 15: 99.

BURLINGTON, H. & LINDEMAN, V.F. (1950). Effect of DDT on testes and secondary sex characters of White Leghorn cockerels. Proceedings of the Society for Experimental Biology and Medicine 74: 48-51.

BUXTON, A. (1949). A naturalist sportsman in Norfolk. The New Naturalist, a Journal of British Natural History 6: 45-48.

CAUGHLEY, G. (1994). Directions in conservation biology. Journal of Animal Ecology 63: 215-244.

CHANIN, P. (1981). The diet of the otter and its relations with the feral mink in two areas of Southwest England. Acta Theriologica 26: 83-95.

CHANIN, P. (1985). The natural history of otters. Croom Helm, Beckenham, Kent.

CHANIN, P. (1993). Otters. Whittet Books, London.

CHANIN, P.R.F. & JEFFERIES, D.J. (1978). The decline of the otter *Lutra lutra* L. in Britain: an analysis of hunting records and discussion of causes. Biological Journal of the Linnean Society 10 (3): 305-328.

CHAPMAN, P.J. & CHAPMAN, L.L. (1982). Otter Survey of Ireland 1980-81. The Vincent Wildlife Trust, London.

CLARK, J. (1906). Mammals. In: W. Page (ed.) The Victoria History of the County of Cornwall, Vol. 1, 348-352. Archibald Constable, London.

CLAYTON, C.J. & JACKSON, M.J. (1981). Norfolk Otter Survey 1980-1981. Otters, Journal of the Otter Trust, 1980, 1 (4): 16-22.

CLODE, D. & MACDONALD, D.W. (1995). Evidence for food competition between mink (*Mustela vison*) and otter (*Lutra lutra*) on Scottish islands. Journal of Zoology, London 237: 435-444.

COGHILL, I. (1978). The status of the otter (*Lutra lutra* L.) in Worcestershire. Worcestershire Nature Conservation Trust Newsletter 23: 11-13.

CONROY, J. (1993). Otter mortality and survival. In: P.A. Morris (ed.) Proceedings of the National Otter Conference, Cambridge, September 1992, 21-24. The Mammal Society, Bristol.

CONROY, J.W.H. & FRENCH, D.D. (1991). Seasonal patterns in the sprainting behaviour of otters (*Lutra lutra* L.) in Shetland. Proceedings, Fifth International Otter Colloquium, 1989. Habitat, Hankensbuttel 6: 159-166.

COOK, J.W. (Chairman) (1964). Review of the Persistent Organochlorine Pesticides: A Report by the Advisory Committee on Poisonous Substances used in Agriculture and Food Storage. H.M.S.O., London.

COOKE, A.S., BELL, A.A. & HAAS, M.B. (1982). Predatory Birds, Pesticides and Pollution. Institute of Terrestrial Ecology, Huntingdon.

COX, L.C. (1947). Otters in the sea. Field 190: 638.

CRAMP, S. & CONDER, P.J. (1960). The deaths of birds and mammals connected with toxic chemicals in the first half of 1960: Report No. 1 of the B.T.O. - R.S.P.B. Committee on Toxic Chemicals. Royal Society for the Protection of Birds, London.

CRAMP, S., CONDER, P.J. & ASH, J.S. (1962). The second report of the Joint Committee of the British Trust for Ornithology and the Royal Society for the Protection of Birds on Toxic Chemicals, in collaboration with the Game Research Association. Royal Society for the Protection of Birds, Sandy.

CRANBROOK, THE EARL OF (1977). The status of the otter (*L. lutra* L.) in Britain in 1977. Biological Journal of the Linnean Society 9 (3): 305-322.

CRANBROOK, THE EARL OF, CORNER, C. & WEST, R. (1976). A policy for the conservation of the otter in Suffolk. Unpublished typescript with the Nature Conservancy Council.

CRAWFORD, A., EVANS, D., JONES, A. & McNULTY, J. (1979). Otter Survey of Wales 1977-78. Society for the Promotion of Nature Conservation, Nettleham.

CRESSWELL, P., HARRIS, S. & JEFFERIES, D.J. (1990). The History, Distribution, Status and Habitat Requirements of the Badger in Britain. Nature Conservancy Council, Peterborough.

CRICK, H.Q.P. & RATCLIFFE, D.A. (1995). The peregrine *Falco peregrinus* breeding population of the United Kingdom in 1991. Bird Study 42: 1-19.

CUTHBERT, J.H. (1973). Some observations on scavenging of salmon *Salmo salar* carrion. Western Naturalist 2: 72-74.

DADD, M.N. (1970). Overlap of variation in British and European mammal populations. Symposia, Zoological Society of London 26: 117-125.

De WITT, J.B., MENZIE, C.M., ADOMAITIS, V.A. & REICHEL, W.L. (1960). Pesticidal residues in animal tissues. Transactions of the 25th North American Wildlife Conference: 277-285.

DOWNING, G. (1988). The hounds of spring: The history of the Eastern Counties Otter Hounds. Privately published by G. Downing, Nayland.

DOWNING, G. (1990). Cold water on otter nonsense. Shooting Times and Country Magazine, August 2-8, 1990: 44.

DUBOS, R., PINES, M., MARGENAU, H. & SNOW, C.P. (1969). Health and Disease. Time-Life, Amsterdam.

DUNSTONE, N. (1993). The mink. T. & A.D. Poyser, London.

ERLINGE, S. (1968a). Territoriality of the otter *Lutra lutra* L. Oikos 19: 81-98.

ERLINGE, S. (1968b). Food studies on captive otters (*Lutra lutra* L.). Oikos 19: 259-270.

ERLINGE, S. (1972). Interspecific relations between otter (*Lutra lutra*) and mink (*Mustela vison*) in Sweden. Oikos 23: 327-335.

FAIRLEY, J. (1984). An Irish Beast Book, 2nd edition. Blackstaff Press, Belfast.

FITTER, R.S.R. (1945). London's Natural History. New Naturalist Series. Collins, London.

FLUMM, D. (1988). Net deaths. Birds 12 (2): 9.

FRENCH, M.C. & JEFFERIES, D.J. (1968). Disappearance of gamma BHC from avian liver after death. Nature, London 219 (5150): 164-166.

FRIEND, M. & TRAINER, D.O. (1970). Polychlorinated biphenyl: interaction with duck hepatitis virus. Science, New York 170: 1314-1316.

FROST, W.E. & BROWN, M.E. (1967). The Trout. New Naturalist Series. Collins, London.

GIBSON, M. (1984). Eel fishing: An introduction. Fisheries Advisory Booklet No 2. Welsh Water, Brecon.

GOUGH, H.C. (1957). Studies on wheat bulb fly (*Leptohylemyia coarctata* (Fall.)). IV. The distribution of damage in England and Wales in 1953. Bulletin of Entomological Research 48: 447-457.

GREEN, J. & GREEN, R. (1980). Otter Survey of Scotland 1977-79. The Vincent Wildlife Trust, London.

GREEN, J. & GREEN, R. (1981). The otter (*Lutra lutra* L.) in western France. Mammal Review 11: 181-187.

GREEN, J. & GREEN, R. (1987). Otter Survey of Scotland 1984-85. The Vincent Wildlife Trust, London.

GREEN, J. & GREEN, R. (1991). Release techniques for otters: theory and practice. In: T. Thomas (ed.) The Proceedings of the Third Symposium of the British Wildlife Rehabilitation Council, 24-30. British Wildlife Rehabilitation Council, Horsham.

GREEN, J., GREEN, R. & JEFFERIES, D.J. (1984). A radio-tracking survey of otters *Lutra lutra* on a Perthshire river system. Lutra 27: 85-145.

GREEN, J., GREEN, R. & LILES, G. (1986). Interspecific use of resting sites by mink *Mustela vison* and otter *Lutra lutra*. The Vincent Wildlife Trust Report, 1985: 20-26.

GREEN, R. & GREEN, J. (1992). The rehabilitation programme for orphans and injured otters of The Vincent Wildlife Trust, Great Britain. In: C. Reuther (ed.) Otterschutz in Deutschland. Habitat, Hankensbuttel 7: 147-151.

GREEN, R. & GREEN, J. (in press). Ecological and ethological requirements of the European otter; experience of the reintroduction of the species in Britain. Cahier d' Ethologie, University of Liège.

GREENWOOD, H. (1991). The History of the Hawkstone Otter Hounds. Privately published by H. Greenwood.

GRIGOR'EV, N.D. & EGOROV, Y.E. (1969). [On the biocenotic connections of the mink with the common otter in the Bashkirian SSR.]. Sbornik trudov Nauchno-issledovatel skogo instituta zhivotnovodstva Syr'ya Pushniny 22: 26-32.

HAAS, M.B. & COOKE, A.S. (1983). Pollutants in kingfishers. Annual Report of the Institute of Terrestrial Ecology, 1982: 64-65.

HALL, A.J., POMEROY, P.P. & HARWOOD, J. (1992). The descriptive epizootiology of phocine distemper in the UK during 1988/89. Science of the Total Environment 115: 31-44.

HARPER, R.J. (1981). Sites of three otter (*Lutra lutra*) breeding holts in fresh water habitats. Journal of Zoology, London 195: 554-556.

HARRIS, C.J. (1968). Otters: A study of the recent Lutrinae. Weidenfeld & Nicholson, London.

HARRIS, S. (1989). Taking stock of Brock. BBC Wildlife 7: 460-464.

HARTING, J.E. (1865). Otters in Middlesex. Zoologist 23: 9429.

HAVINS, P.J.N. (1981). The otter in Britain. Robert Hale, London.

HEWSON, R. (1995). Use of salmonid carcasses by vertebrate scavengers. Journal of Zoology, London 235: 53-65.

HILLEGAART, V., ÖSTMAN, J. & SANDEGREN, F. (1985). Area utilisation and marking behaviour among two captive otter (*Lutra lutra* L.) pairs. Otters, Journal of the Otter Trust, 1984, 1 (8): 64-74.

HOLDGATE, M.W. (ed.) (1971). The seabird wreck of 1969 in the Irish Sea. Natural Environment Research Council, London.

HOLMES, D.C., SIMMONS, J.H. & TATTON, J. O'G. (1967). Chlorinated hydrocarbons in British wildlife. Nature, London 216: 227.

HOWES, C.A. (1976). The decline of the otter in south Yorkshire and adjacent areas. Naturalist 101: 3-12.

HUDSON, W.H. (1892). The naturalist in La Plata. Chapman & Hall, London.

HUMPHRIES, D.A. (1958). Badgers in the Cheltenham area. School Science Review 139: 416-425.

HURRELL, H.G. (1963). Pine martens. Sunday Times Publications, London.

HURRELL, H.G. (1968). Pine martens. Forestry Commission Forest Record 64. H.M.S.O., London.

JEFFERIES, D.J. (1969). Causes of badger mortality in eastern counties of England. Journal of Zoology, London 157: 429-436.

JEFFERIES, D.J. (1972). Organochlorine insecticide residues in British bats and their significance. Journal of Zoology, London 166: 245-263.

JEFFERIES, D.J. (1975). The role of the thyroid in the production of sublethal effects by organochlorine insecticides and polychlorinated biphenyls. In: F. Moriarty (ed.) Organochlorine Insecticides: Persistent Organic Pollutants, 131-230. Academic Press, London.

JEFFERIES, D.J. (1984). The otters return. Natural World 11: 10-12.

JEFFERIES, D.J. (1985). An otter casualty from Breydon Water, Norfolk. Otters, Journal of the Otter Trust, 1984, 1 (8): 23-24.

JEFFERIES, D.J. (1986). The value of Otter *Lutra lutra* surveying using spraints: an analysis of its successes and problems in Britain. Otters, Journal of the Otter Trust, 1985, 1 (9): 25-32.

JEFFERIES, D.J. (1987). The effects of angling interests on otters with particular reference to disturbance. In: P.S. Maitland & A.K. Turner (eds) Angling and wildlife in fresh waters (ITE Symposium No. 19), 23-30. Institute of Terrestrial Ecology, Grange-over-Sands.

JEFFERIES, D.J. (1988a). Dimensions and weights of some known-aged otters *Lutra lutra*, with notes on continuous growth during maturity, the largest weights achieved and a possible recent reduction in longevity. Otters, Journal of the Otter Trust, 1987, 2 (1): 10-18.

JEFFERIES, D.J. (1988b). Postscript: The demise of the otter population of the Rivers Glaven and Stiffkey and adjacent coastal marshes. In: V. Weir, The Otter, 2nd edition. The Vincent Wildlife Trust, London.

JEFFERIES, D.J. (1989a). Fragmentation of the English otter population and its effect on viability. Otters, Journal of the Otter Trust, 1988, 2 (2): 20-22.

JEFFERIES, D.J. (1989b). The changing otter population of Britain 1700-1989. Biological Journal of the Linnean Society 38 (1): 61-69.

JEFFERIES, D.J. (1989c). Otters crossing watersheds. Otters, Journal of the Otter Trust, 1988, 2 (2): 17-19.

JEFFERIES, D.J. (1990a). The Asian short-clawed otter *Amblonyx cinerea* (Illiger) living wild in Britain. Otters, Journal of the Otter Trust, 1989, 2 (3): 21-25.

JEFFERIES, D.J. (1990b). Further records of fyke net and creel deaths in British otters (*Lutra lutra*) with a discussion on the use of guards. Otters, Journal of the Otter Trust, 1989, 2 (3): 13-20.

JEFFERIES, D.J. (1991a). Predation of otters *Lutra lutra* by American mink *Mustela vison*. Otters, Journal of the Otter Trust, 1990, 2 (4): 33-35.

JEFFERIES, D.J. (1991b). The last confirmed otter *Lutra lutra* from the central midlands of England. Otters, Journal of the Otter Trust, 1990, 2 (4): 5-12.

JEFFERIES, D.J. (1992a). Another record of an Asian short-clawed otter living free in the Oxford area of England, with notes on its implications. Otters, Journal of the Otter Trust, 1991, 2 (5): 9-12.

JEFFERIES, D.J. (1992b). Polecats *Mustela putorius* and pollutants in Wales. Lutra 35: 28-39.

JEFFERIES, D.J. (in press). Blindness in British otters: further incidence, possible causes and correlation with the period of organochlorine pollution. Otters, Journal of the Otter Trust.

JEFFERIES, D.J. & CRITCHLEY, C.H. (1994). A new pine marten record for the North Yorkshire Moors: skull dimensions and confirmation of species. Naturalist 119: 145-150.

JEFFERIES, D.J. & FREESTONE, P. (1985). Chemical analysis of some coarse fish from a Suffolk river carried out as part of the preparation for the first release of captive-bred otters. Otters, Journal of the Otter Trust, 1984, 1 (8): 17-22.

JEFFERIES, D.J. & FRENCH, M.C. (1976). Mercury, cadmium, zinc, copper and organochlorine insecticide levels in small mammals trapped in a wheat field. Environmental Pollution 10: 175-182.

JEFFERIES, D.J. & HANSON, H.M. (1987). Autopsy and chemical analysis of otter bodies. The Vincent Wildlife Trust Report, 1986: 42-44.

JEFFERIES, D.J. & HANSON, H.M. (1988a). The Minsmere otter release and information gained from a detailed examination and analysis of the two casualties. Otters, Journal of the Otter Trust, 1987, 2 (1): 19-29.

JEFFERIES, D.J. & HANSON, H.M. (1988b). A second report on the autopsy and chemical analysis of British otter bodies. The Vincent Wildlife Trust Report, 1987: 34-36.

JEFFERIES, D.J. & HANSON, H.M. (1991). Evidence of fighting in a juvenile male otter road casualty bred from the otters released at Minsmere in 1985 and 1987. Otters, Journal of the Otter Trust, 1990, 2 (4): 13-24.

JEFFERIES, D.J. & MITCHELL-JONES, A.J. (1982). Preliminary research for a release programme for the European otter. Otters, Journal of the Otter Trust, 1981, 1 (5): 13-16.

JEFFERIES, D.J. & MITCHELL-JONES, A.J. (1989). Mammals. In: D.A. Ratcliffe & P.H. Oswald (eds) Guidelines for Selection of Biological SSSIs, 232-241. Nature Conservancy Council, Peterborough.

JEFFERIES, D.J. & MITCHELL-JONES, A.J. (1993). Recovery plans for British mammals of conservation importance, their design and value. Mammal Review 23: 155-166.

JEFFERIES, D.J. & PARSLOW, J.L.F. (1976). Thyroid changes in PCB-dosed guillemots and their indication of one of the mechanisms of action of these materials. Environmental Pollution 10: 293-311.

JEFFERIES, D.J. & PENDLEBURY, J.B. (1968). Population fluctuations of stoats, weasels and hedgehogs in recent years. Journal of Zoology, London 156: 513-517.

JEFFERIES, D.J. & PRESTT, I. (1966). Post-mortems of Peregrines and Lanners with particular reference to organochlorine residues. British Birds 59: 49-64.

JEFFERIES, D.J. & WALKER, C.H. (1966). Uptake of pp[1]-DDT and its post-mortem breakdown in the avian liver. Nature, London 212 (5061): 533-534.

JEFFERIES, D.J. & WAYRE, P. (1984). Re-introduction to the wild of otters bred in captivity. Otters, Journal of the Otter Trust, 1983, 1 (7): 20-22.

JEFFERIES, D.J. & WAYRE, P. (in press). Otter releases in Britain: History, Procedure and Best Practice. Joint Nature Conservation Committee, Peterborough.

JEFFERIES, D.J., CRIPPS, S.J., GORROD, L.A., HARRISON, B., JOHNSTONE, J.S., POTTER, E.C.E. & WEIR, V. (1988). The effects of otter guards on the fishing efficiency of eel fyke nets. The Vincent Wildlife Trust, London.

JEFFERIES, D.J., FRENCH, M.C. & STEBBINGS, R.E. (1974). Pollution and Mammals. In: Monks Wood Experimental Station Report for 1972-73, 13-15. Natural Environment Research Council, Huntingdon.

JEFFERIES, D.J., GREEN, J., GREEN, R. & CRIPPS, S.J. (1993). Otter mortalities due to commercial fishing 1975-1992. In: P.A. Morris (ed.) Proceedings of the National Otter Conference, Cambridge, September 1992, 25-29. The Mammal Society, Bristol.

JEFFERIES, D.J., HANSON, H.M. & HARRIS, E.A. (1990). The prevalence of *Pseudoterranova decipiens* (Nematoda) and *Corynosoma strumosum* (Acanthocephala) in otters *Lutra lutra* from coastal sites in Britain. Journal of Zoology, London 221: 316-321.

JEFFERIES, D.J., JOHNSON A., GREEN, R. & HANSON, H.M. (1989). Entanglement with monofilament nylon fishing net: a hazard to otters. Otters, Journal of the Otter Trust, 1988, 2 (2): 11-16.

JEFFERIES, D.J., STAINSBY, B. & FRENCH, M.C. (1973). The ecology of small mammals in arable fields drilled with winter wheat and the increase in their dieldrin and mercury residues. Journal of Zoology, London 171: 513-539.

JEFFERIES, D.J., WAYRE, P., JESSOP, R.M., MITCHELL-JONES, A.J. & MEDD, R. (1985). The composition, age, size and pre-release treatment of the groups of otters *Lutra lutra* used in the first releases of captive-bred stock in England. Otters, Journal of the Otter Trust, 1984, 1 (8): 11-16.

JEFFERIES, D.J., WAYRE, P., JESSOP, R.M. & MITCHELL-JONES, A.J. (1986). Reinforcing the native otter *Lutra lutra* population in East Anglia: an analysis of the behaviour and range development of the first release group. Mammal Review 16: 65-79.

JEFFERIES, R. (1883). Nature near London. Chatto & Windus, London.

JENKINS, D. (1980). Ecology of otters in northern Scotland. I. Otter (*Lutra lutra*) breeding and dispersion in mid-Deeside, Aberdeenshire in 1974-79. Journal of Animal Ecology 49: 713-735.

JENKINS, D. & BURROWS, B.O. (1980). Ecology of otters in northern Scotland. III. The use of faeces as indicators of otter (*Lutra lutra*) density and distribution. Journal of Animal Ecology 49: 755-774.

JENKINSON, S. (1991). The influence of public access on the badger *Meles meles*: an initial study. Unpublished Post-graduate Diploma thesis, Manchester Polytechnic.

JENSEN, S. (1966). Report of a new chemical hazard. New Scientist 32: 612.

JESSOP, R.M. (1985). Status and conservation of the otter in Norfolk and Suffolk. Transactions of the Norfolk and Norwich Naturalists' Society 27 (2): 144-149.

JESSOP, R.M. & CHEYNE, D.L. (1993). The re-introduction of the European otter *Lutra lutra* into lowland England carried out by the Otter Trust 1983-92: A progress report. In: P.A. Morris (ed.) Proceedings of the National Otter Conference, Cambridge, September 1992, 12-16. The Mammal Society, Bristol.

JESSOP, R.M. & MACGUIRE, F. (1990). Norfolk Otter Survey 1988/89: Implications for the status of the otter. Otters, Journal of the Otter Trust, 1989, 2 (3): 9-12.

KENNEDY, A.H. (1951). The mink in health and disease. Fur Trade Journal of Canada, Toronto.

KEYMER, I.F. (1993a). Diseases of the otter (*Lutra lutra*). In: P.A. Morris (ed.) Proceedings of the National Otter Conference, Cambridge, September 1992, 30-33. The Mammal Society, Bristol.

KEYMER, I.F. (1993b). Blindness in otters. Mammal News 92: 8-9.

KEYMER, I.F., WELLS, G.A.H., MASON, C.F. & MACDONALD, S.M. (1988). Pathological changes and organochlorine residues in tissues of wild otters (*Lutra lutra*). Veterinary Record 122: 153-155.

KIHLSTRÖM, J.E., OLSSON, M., JENSEN, S., JOHANSSON, A., AHLBOM, J. & BERGMAN, A. (1992). Effects of PCB and different fractions of PCB on the reproduction of the mink (*Mustela vison*). Ambio 21: 563-569.

KING, A. & POTTER, A. (1980). A guide to otter conservation for Water Authorities. The Vincent Wildlife Trust, London.

KING, C. (1989). The natural history of weasels and stoats. Christopher Helm, London.

KITCHENER, A. (1989). Teenagers eat humble crab. BBC Wildlife 7: 359.

KRUUK, H. & CONROY, J.W.H. (1987). Surveying otter *Lutra lutra* populations: A discussion of problems with spraints. Biological Conservation 41: 179-183.

KRUUK, H. & MACDONALD, D. (1985). Group territories of carnivores: empires and enclaves. In: R.M. Sibley and R.H. Smith (eds) Behavioural Ecology - Ecological Consequences of Adaptive Behaviour, 521-536. Blackwell Scientific Publications, Oxford.

KRUUK, H., CARSS, D.N., CONROY, J.W.H. & DURBIN, L. (1993a). Otter (*Lutra lutra* L.) numbers and fish productivity in rivers in north east Scotland. Symposium of the Zoological Society, London 65: 171-191.

KRUUK, H., CONROY, J.W.H., GLIMMERVEEN, U. & OUWERKERK, E.J. (1986). The use of spraints to survey populations of otters *Lutra lutra*. Biological Conservation 35: 187-194.

KRUUK, H., CONROY, J.W.H. & MOORHOUSE, A. (1987). Seasonal reproduction, mortality and food of otters *Lutra lutra* in Shetland. Symposium of the Zoological Society, London 58: 263-278.

KRUUK, H., CONROY, J.W.H. & CARSS, D.N. (1993b). Otters, eels and contaminants. Institute of Terrestrial Ecology, Report to Scottish Natural Heritage.

LAGERWERFF, J,V. & SPECHT, A.W. (1970). Contamination of roadside soil and vegetation with cadmium, nickel, lead and zinc. Environmental Science & Technology 4: 583-586.

LENTON, E. (1982). Otters and the Otter Haven Project. Nature in Devon 3: 27-43.

LENTON, E.J., CHANIN, P.R.F. & JEFFERIES, D.J. (1980). Otter Survey of England 1977-79. Nature Conservancy Council, London.

LEONARDS, P.E.G., SMIT, M.D., De JONGH, A.W.J.J. & VAN HATTUM, B. (1994). Evaluation of dose-response relationships for the effects of PCBs on the reproduction of mink (*Mustela vison*). Institute for Environmental Studies, Vrije Universiteit, Amsterdam.

LINN, I.J. & STEVENSON, J.H.F. (1980). Feral mink in Devon. Nature in Devon 1: 7-27.

LINN, I.J., GREEN, B.H., ELLIOTT, H., LUCAS, G., MAITLAND, P.S., RAWLINS, C.G.C., SANDS, T. & STUTTARD, R.M. (1979). Wildlife Introductions to Great Britain: Report by the Working Group on Introductions to the U.K. Committee for International Nature Conservation. Nature Conservancy Council, London.

LLOYD, J.L. (1962). Where are the otters? Gamekeeper and Countryside 65: 299-300.

LOCKIE, J.D. (1966). Territory in small carnivores. Symposium of the Zoological Society, London 18: 143-165.

LYNCH, J.M., CONROY, J.W.H., KITCHENER, A.C., JEFFERIES, D.J. & HAYDEN, D.J. (1996). Variation in cranial form and sexual dimorphism among five European populations of the otter *Lutra lutra*. Journal of Zoology, London 238: 81-96.

MACDONALD, D. (1987). Running with the fox. Unwin Hyman, London.

MACDONALD, S. & MASON, C.F. (1976). The status of the otter (*Lutra lutra* L.) in Norfolk. Biological Conservation 9: 199-124.

MACDONALD, S.M. & MASON, C.F. (1983). Some factors affecting the distribution of otters (*Lutra lutra*). Mammal Review 13: 1-10.

MACDONALD, S.M. & MASON, C.F. (1987). Seasonal marking in an otter population. Acta Theriologica 32: 449-462.

MACDONALD, S.M., MASON, C.F. & COGHILL, I.S. (1978). The otter and its conservation in the River Teme catchment. Journal of Applied Ecology 15: 373-384.

MARSHALL, H. (1991). Recent otter records in Devon. Report and Transactions of the Devonshire Association for the Advancement of Science 123: 137-148.

MASON, C.F. (1988). Concentrations of organochlorine residues and metals in tissues of otters *Lutra lutra* from the British Isles 1985-1986. Lutra 31: 62-67.

MASON, C.F. & MACDONALD, S.M. (1980). The winter diet of otters (*Lutra lutra*) on a Scottish sea loch. Journal of Zoology, London 192: 558-561.

MASON, C.F. & MACDONALD, S.M. (1986). Otters: Ecology and Conservation. Cambridge University Press, Cambridge.

MASON, C.F. & MACDONALD, S.M. (1987). Acidification and otter (*Lutra lutra*) distribution on a British river. Mammalia 51 (1): 81-87.

MASON, C.F. & MACDONALD, S.M. (1993). Pollution and otter conservation in a European context. In: P.A. Morris (ed.) Proceedings of the National Otter Conference, Cambridge, September 1992, 17-20. The Mammal Society, Bristol.

MASON, C.F., FORD, T.C. & LAST, N.I. (1986a). Organochlorine residues in British otters. Bulletin of Environmental Contamination and Toxicology 36: 656-661.

MASON, C.F., LAST, N.I. & MACDONALD, S.M. (1986b). Mercury, cadmium and lead in British otters. Bulletin of Environmental Contamination and Toxicology 37: 844-849.

McKAY, J. (1993). Be on your guard. Shooting Times and Country Magazine, December 2-8, 1993: 24.

MELLANBY, K. (1980). The biology of pollution, 2nd edition. Studies in Biology, The Institute of Biology. Arnold, London.

MELLANBY, K. (1981). Farming and Wildlife. New Naturalist Series. Collins, London.

MILLAIS, J.G. (1904-06). The Mammals of Great Britain and Ireland. Longmans Green, London.

MILLER, S.H. & SKERTCHLY, S.B.J. (1878). The Fenland past and present. Longmans Green, London.

MILLS, D.H., GRIFFITH, D. & PARFITT, A. (1978). A survey of the freshwater fish fauna of the Tweed Basin. Nature Conservancy Council, London.

MITCHELL-JONES, A.J., JEFFERIES, D.J., TWELVES, J., GREEN, J. & GREEN, R. (1984). A practical system of tracking otters *Lutra lutra* using radio-telemetry and 65-Zn. Lutra 27: 71-84.

MOORE, N.W. (1957). The past and present status of the buzzard in the British Isles. British Birds 50: 173-197.

MOORE, N.W. (1965). Pesticides and birds: A review of the situation in Great Britain in 1965. Bird Study 12: 222-252.

MOORE, N.W. (1987). The bird of time: The science and politics of nature conservation, a personal account. Cambridge University Press, Cambridge.

MORIARTY, F. & FRENCH, M.C. (1977). Mercury in waterways that drain into the Wash in eastern England. Water Research 11: 367-372.

MULDER, J.L. (1990). The stoat *Mustela erminea* in the Dutch dune region, its local extinction, and a possible cause: the arrival of the fox *Vulpes vulpes*. Lutra 33: 1-21.

NATIONAL RIVERS AUTHORITY (1991). The Quality of Rivers, Canals and Estuaries in England and Wales: Report of the 1990 Survey. Water Quality Series No 4. National Rivers Authority, Almondsbury, Bristol.

NATIONAL WATER COUNCIL (1978). River water quality: the next stage. National Water Council, London.

NEAL, E. (1986). The Natural History of Badgers. Christopher Helm, London.

NEWTON, I. (1986). The sparrowhawk. T. & A.D. Poyser, Calton.

NEWTON, I. & HAAS, M.B. (1984). The return of the sparrowhawk. British Birds 77: 47-70.

NICHOLSON, J.K., KENDALL, M.D. & OSBORNE, D. (1983). Cadmium and mercury nephrotoxicity. Nature, London 304: 633-635.

NILSSON, S.I., MILLER, H.G. & MILLER, J.D. (1982). Forest growth as possible cause of soil and water acidification: an examination of the concepts. Oikos 39: 40-49.

NOVIKOV, G.A. (1956). Carnivorous mammals of the USSR. Translation 1962. Israel Programme for Scientific Translations, Jerusalem.

O'CONNOR, F.B., BARWICK, D., CHANIN, P.R.F., FRAZER, J.F.D., JEFFERIES, D.J., JENKINS, D., NEAL, E. & SANDS, T.S. (1977). Otters 1977: First report of the Joint Otter Group. Nature Conservancy Council and Society for the Promotion of Nature Conservation, London and Nettleham, Lincoln.

O'CONNOR, F.B., CHANIN, P.R.F., JEFFERIES, D.J., JENKINS, D., NEAL, E., RUDGE, J., SANDS, T.S., WEIR, V. & WOODS, M.S. (1979). Otters 1979: A second report of the Joint Otter Group. Society for the Promotion of Nature Conservation, Nettleham, Lincoln.

ÖSTMAN, J., HILLEGAART, V. & SANDEGREN, F. (1985). Behavioural changes in captive female otters (*Lutra lutra* L.) around parturition. Otters, Journal of the Otter Trust, 1984, 1 (8): 58-63.

PARSLOW, J.L.F. & JEFFERIES, D.J. (1973). Relationships between organochlorine residues in livers and whole bodies of guillemots. Environmental Pollution 5: 87-101.

PICKFORD, T. (1994). Death on the moors: The slaughter of Lancashire's raptors. Birdwatch 21: 7.

'POLESTAR' (1983). Mr Courtenay Tracy and his hounds, being a history of the C.T.O.H. Privately published, The C.T.O.H. Club.

PORTER, D.D. (1986). Aleutian Disease: A persistent parvovirus infection of mink with a maximal but ineffective host humoral immune response. Progress in Medical Virology 33: 42-60.

RESTT, I. (1965). An enquiry into the recent breeding status of some of the smaller birds f prey and crows in Britain. Bird Study 12: 196-221.

RESTT, I. (1970). Organochlorine pollution of rivers and the heron (*Ardea cinerea* L.). apers and Proceedings of the 11th Technical Meeting of the International Union for onservation of Nature and Natural Resources, 1969, New Delhi, 1: 95-102.

RESTT, I., JEFFERIES, D.J. & MOORE, N.W. (1970). Polychlorinated biphenyls in ild birds in Britain and their avian toxicity. Environmental Pollution 1: 3-26.

ATCLIFFE, D.A. (1970). Changes attributable to pesticides in egg breakage frequency nd eggshell thickness in some British birds. Journal of Applied Ecology 17: 67-107.

ATCLIFFE, D.A. (1972). The peregrine population in 1971. Bird Study 19: 117-156.

ATCLIFFE, D.A. (1984). The peregrine breeding population of the United Kingdom in 981. Bird Study 31: 1-18.

ATCLIFFE, D. (1993). The Peregrine Falcon, 2nd edition. T. & A.D. Poyser, Calton.

AUSCH, R.L. (1953). Studies on the helminth fauna of Alaska. XIII: Disease in the sea tter, with special reference to helminth parasites. Ecology 34: 584-604.

EASON, P., HARRIS, S. & CRESSWELL, P. (1993). Estimating the impact of past ersecution and habitat changes on the numbers of badgers *Meles meles* in Britain. Aammal Review 23: 1-15.

EYNOLDS, C.M. (1974). The census of heronries, 1969-73. Bird Study 21: 129-134.

IDDING, M.O. & SMITH, H.D. (1988). Post-release monitoring of the second group f otters. Otters, Journal of the Otter Trust, 1987, 2 (1): 30-36.

ISEBROUGH, R.W., RIECHE, P., PEAKALL, D.B., HERMAN, S.G. & IRVEN, M.N. (1968). Polychlorinated biphenyls in the global ecosystem. Nature, ondon 220: 1098-1102.

AFE, S. (1984). Polychlorinated biphenyls (PCBs) and polybrominated biphenyls (PBBs): iochemistry, toxicology and mechanisms of action. CRC Critical Reviews in Toxicology 3: 319-395.

AFE, S. (1990). Polychlorinated biphenyls (PCBs), dibenzo-p-dioxins (PCDDs) libenzofurans (PCDFs) and related compounds: Environmental and mechanistic onsiderations which support the development of toxic equivalency factors (TEFs). CRC Critical Reviews in Toxicology 21: 51-88.

AGAR, M. (1984). The History of the Border Counties (N.W.) Otterhounds. Privately ublished by M. Sagar & S.Maxfield, Welshpool.

SCHROEDER, H.A. & BALASSA, J.J. (1963). Cadmium: uptake by vegetables from superphosphate in soil. Science, New York 140: 819-820.

SCOTT, D.R. (1960). The badger in Essex. Essex Naturalist 30: 272-275.

SERPELL, J. (1991). It's the elephant by a nose. BBC Wildlife 9: 849-851.

SHEAIL, J. (1985). Pesticides and Nature Conservation: The British experience 1950-1975. Clarendon Press, Oxford.

SKINNER, C., SKINNER, P. & HARRIS, S. (1991a). An analysis of some of the factors affecting the current distribution of badgers *Meles meles* in Essex. Mammal Review 21: 51-65.

SKINNER, C., SKINNER, P. & HARRIS, S. (1991b). The past history and recent decline of badgers *Meles meles* in Essex: an analysis of some of the contributory factors. Mammal Review 21: 67-80.

SMAL, C. (1991). Feral American mink in Ireland. The Office of Public Works, Dublin.

SMIT, M.D., LEONARDS, P.E.G., VAN HATTUM, B. & De JONGH, A.W.J.J. (1994). PCBs in European otter (*Lutra lutra*) populations. Institute for Environmental Studies, Vrije Universiteit, Amsterdam.

SOKAL, R.R. & ROHLF, F.J. (1981). Biometry: The principles and practice of statistics in biological research, 2nd edition. W.H. Freeman, New York.

SPALTON, J.A. & CRIPPS, S.J. (1989). The report of the North Norfolk Otter Project, 1986-1987. The Vincent Wildlife Trust and Norfolk Naturalists' Trust, London and Norwich.

STEPHENS, M.N. (1957). The natural history of the otter. Universities Federation for Animal Welfare, London.

STRACHAN, R. & JEFFERIES, D.J. (1993). The water vole *Arvicola terrestris* in Britain 1989-1990: its distribution and changing status. The Vincent Wildlife Trust, London.

STRACHAN, R., BIRKS, J.D.S., CHANIN, P.R.F. & JEFFERIES, D.J. (1990). Otter Survey of England 1984-1986. Nature Conservancy Council, Peterborough.

STRACHAN, R., JEFFERIES, D.J. & CHANIN, P.R.F. (in press). Pine marten survey of England and Wales 1987-1988. Joint Nature Conservation Committee, Peterborough.

STROUD, D.A. & GLUE, D. (1991). Britain's birds in 1989/90: the conservation and monitoring review. British Trust for Ornithology and Nature Conservancy Council, Thetford.

STROUD, D.A., REED, T.M., PIENKOWSKI, M.W. & LINDSAY, R.A. (1987). Birds, Bogs and Forestry: the peatlands of Caithness and Sutherland. Nature Conservancy Council, Peterborough.

STUBBE, M. (1977). Der Fischotter *Lutra lutra* (L. 1758) in der DDR. Zoologischer Anzeiger 199: 265-285.

TAPPER, S. (1992). Game Heritage: An ecological review from shooting and gamekeeping records. Game Conservancy, Fordingbridge.

TAYLOR, J.C. & BLACKMORE, D.K. (1961). A short note on the heavy mortality in foxes during the winter 1959-60. Veterinary Record 73: 232-233.

THAPAR, V. (1986). Tiger, portrait of a predator. Collins, London.

THAPAR, V. (1989). Tigers, the secret life. Elm Tree Books, London.

THOMPSON, H.V. & SOUTHERN, H.N. (1964). Red fox *Vulpes vulpes* (L.). In: H.N. Southern (ed.) The Handbook of British Mammals, 1st edition, 352-357. Blackwell, Oxford.

THOMPSON, P.M. & HALL, A.J. (1993). Seals and epizootics - what factors might affect the severity of mass mortalities? Mammal Review 23: 149-154.

TURNER, T.W. (1954). Memoirs of a gamekeeper (Elveden, 1868-1953). Geoffrey Bles, London.

TWELVES, J. (1983). Otter (*Lutra lutra*) mortalities in lobster creels. Journal of Zoology, London 201: 585-588.

TYLER, C. (1995). The emasculators. BBC Wildlife 13 (6): 74-76.

WAYRE, P. (1979). The private life of the otter. Batsford, London.

WAYRE, P. (1985). Report of Council 1984. Otters, Journal of the Otter Trust, 1984, 1 (8): 4-10.

WAYRE, P. (1986). Report of Council 1985. Otters, Journal of the Otter Trust, 1985, 1 (9): 3-7.

WAYRE, P. (1988). Report of Council 1987. Otters, Journal of the Otter Trust, 1987, 2 (1): 1-9.

WAYRE, P. (1989). Operation Otter. Chatto & Windus, London.

WAYRE, P. (1993a). Report of Council 1992. Otters, Journal of the Otter Trust, 1992, 2 (6): 1-10.

WAYRE, P. (1993b). The first twenty-one years of the Otter Trust and its re-introduction programme. Otters, Journal of the Otter Trust, 1992, 2 (6): 17-22.

WAYRE, P. (1994). Conservation Report and Release update. Otters, Journal of the Otter Trust, 1993, 2 (7): 6-8.

WEBER, J.-M. (1991). Gastrointestinal helminths of the otter, *Lutra lutra*, in Shetland. Journal of Zoology, London 224: 341-346.

WEIR, V. (1984). The otter, 1st edition. The Vincent Wildlife Trust, London.

WEIR, V. & BANISTER, K.E. (1972). The food of the otter in the Blakeney area. Transactions of the Norfolk and Norwich Naturalists' Society 22: 377-382.

WEIR, V. & BANISTER, K.E. (1977). Additional notes on the food of the otter in the Blakeney area. Transactions of the Norfolk and Norwich Naturalists' Society 24: 85-88.

WELLS, G.A.H., KEYMER, I.F. & BARNETT, K.C. (1989). Suspected Aleutian Disease in a wild otter (*Lutra lutra*). Veterinary Record 125: 232-235.

WEST, R.B. (1975). The Suffolk otter survey. Suffolk Natural History 16 (6): 378-388.

WICKENS, J.D. (1991). Otter survey of Roaringwater Bay, Southwest Cork, Ireland, 1990. Bulletin of Sherkin Island Marine Station 12: 1-38.

WILLIAMS, J. (1989). Blindness in otters. Bulletin of I.U.C.N. Otter Specialist Group 4: 29-30.

WILLIAMS, J. (1993). Blindness in otters. Mammal News 92: 8.

WILSON, A. (1969). Further review of certain persistent organochlorine pesticides: Report of the Advisory Committee on Poisonous Substances used in Agriculture and Food Storage. H.M.S.O., London.

WISE, M.H., LINN, I.J. & KENNEDY, C.R. (1981). A comparison of the feeding biology of mink *Mustela vison* and otter *Lutra lutra*. Journal of Zoology, London 195: 181-213.

WOODROFFE, G.L. (1993). Mammals. In: A. Cleave (ed.) Wildlife Reports. British Wildlife 5 (2): 109-110.

WOODROFFE, G.L. (1994). The status and distribution of the otter (*Lutra lutra* L.) in North Yorkshire. Naturalist 119: 23-25.

WORSNIP, J.V. (1976). Otters in Gloucestershire. Unpublished report, Gloucestershire Trust for Nature Conservation.

YALDEN, D.W. (1982). When did the mammal fauna of the British Isles arrive? Mammal Review 12: 1-57.

YOUNG, P.C. & LOWE, D. (1969). Larval nematodes from fish of the subfamily Anisakinae and gastro-intestinal lesions in mammals. Journal of Comparative Pathology 79: 301-313.

Appendix 1. The standard survey form as completed for the Otter Survey of England 1991-94.

A — 1 Site No: E 1944 | 2 Site Name: R. Creedy Crediton | 3 Grid Reference: SS 846011

B — 1 Recorder: RS | 2 County: Devon | 3 Altitude: 38 m | 4 Date of Visit: 8.5.1993

C HABITAT TYPE — Sea Coast | Sea Loch | Estuary | Lowland Lake/Broad | Upland Loch/Tarn | Reservoir | Running Water ✓ | Bog/Marsh | Canal

D SHORE TYPE — Boulders | Stones ✓ | Gravel ✓ | Sand ✓ | Silt | Earth ✓ | Rock Cliffs | Earth Cliffs ✓

E CURRENT — Rapid | Fast | Slow ✓ | Sluggish | Static

F WIDTH — <1m | 1 - 2m | 2 - 5m ✓ | 5 - 10m | 10 - 20m | 20 - 40m | >40m

G MEAN DEPTH — <0.5m | 0.5 - 1m ✓ | 1 - 2m | >2m

H VEGETATION Bankside: — Trees D | Shrubs S | Herbs S | Emergent S | Floating Attached A | Free Floating A | Submerged S

J LAND USE BORDERING — Upland Grassland | Permanent/Temp. Grassland ✓ | Mixed/Broadleaf Woodland | Conifer Woodland | Acid Peat Bog | Arable | Salt Marsh | Heath | Urban/Industrial | Park/Garden ✓ | Fen

K BANK TREATMENT — Canalised | Maintained | Wild ✓

L WEED CONTROL — Mechanical | Chemical | None ✓

M WATER USE — Water Abstraction | Boating/Powered ✗ | Boat/Sail ✗ | Boat/Manpower ✗ | Bank/Angling ✓ | Bankside/Shooting | Keepered | None | Reserve ✗

N POLLUTION — UNPOLLUTED | DOMESTIC: organic, others | AGRICULTURE: organic, pesticide, fertilizer | INDUSTRIAL: organic, toxin, solid, temperature

Appendix 1. Reverse of standard survey form.

P MINK SIGNS	Present ✓ tracks Absent
Q OTTER HUNTING	Yes No
R FISH (Species present)	
S APPARENT DISTURBANCE FACTOR	1–2

Description or sketch of site

large silt bank under bridge

Pasture

A3072

Crediton

Playing fields

good tree line along both bank of mature oak, Alder + Sycamor Plenty of checke sprainting sites

Distance surveyed 600 downstream

Description or sketch of spraint site	Otter signs seen (and number)
Earth bank under bridge Leaning tree + tree root "saddles"	– Fresh padding (tracks), sandcastles and two large spraints – 5 spraints found in total.
	Salmonid ova in spraints

Appendix 2. Statistical comparison of the numbers of positive full survey sites found at the 1977-79, 1984-86 and 1991-94 surveys (data from Table 2).

Note 1: Chi-squared 2 x 2 test used, comparing numbers of positive and negative sites found at consecutive surveys.

Note 2: Chi-squared 2 x 2 test used, with Yates correction for small numbers.

Note 3: Fisher's Exact Probability Test used (Sokal & Rohlf, 1981).

Note 4: Little or no change in numbers of positive sites. Numbers too low for meaningful test.

NS: Difference not statistically significant.

*: Probably significant difference (at the 5% level).

**: Significant difference (at the 1% level).

***: Highly significant difference (at the 0.1% level).

Region	Comparison of 1977-79 & 1984-86 positive sites	Significance of Difference (Notes)	Comparison of 1984-86 & 1991-94 positive sites	Significance of Difference (Notes)
North West	9 to 31/324 $\chi^2 = 12.896$ $p<0.001$	* * * (1)	31 to 93/324 $\chi^2 = 38.336$ $p<0.001$	* * * (1)
Northumbrian	14 to 17/168 $\chi^2 = 0.320$ $0.50<p<0.90$	NS (1)	17 to 45/168 $\chi^2 = 15.506$ $p<0.001$	* * * (1)
Yorkshire	4 to 5/226 Little change	NS (4)	5 to 25/226 $\chi^2 = 12.889$ $p<0.001$	* * * (2)
Severn-Trent	13 to 22/567 $\chi^2 = 2.388$ $0.10<p<0.50$	NS (1)	22 to 120/567 $\chi^2 = 77.315$ $p<0.001$	* * * (1)
Anglian	20 to 8/623 $\chi^2 = 5.261$ $p<0.05$	* (1)	8 to 52/623 $\chi^2 = 33.899$ $p<0.001$	* * * (1)
Thames	0 to 0/170 No change	NS (4)	0 to 4/170 Little change	NS (4)
Wessex	2 to 1/154 Little change	NS (4)	1 to 29/154 $p = 0.0416$	* (3)
South West	91 to 169/386 $\chi^2 = 35.283$ $p<0.001$	* * * (1)	169 to 259/386 $\chi^2 = 42.472$ $p<0.001$	* * * (1)
Southern	5 to 7/241 $\chi^2 = 0.0855$ $0.50<p<0.90$	NS (2)	7 to 9/241 $\chi^2 = 0.0646$ $0.50<p<0.90$	NS (2)
Welsh	12 to 24/81 $\chi^2 = 5.143$ $p<0.05$	* (1)	24 to 51/81 $\chi^2 = 18.099$ $p<0.001$	* * * (1)
Totals	**170 to 284/2940** $\chi^2 = 77.315$ $p<0.001$	* * * **(1)**	**284 to 687/2940** $\chi^2 = 200.343$ $p<0.001$	* * * **(1)**

Appendix 3. River Quality Gradings.

The river quality classification used in this report (see Section 5) is that devised by the National Water Council (1978) and which was in use at the time of the 1991-94 survey.

Class	Description	Current potential use
1A	Good quality	Waters of high quality suitable for potable supply abstractions; game or other high-class fisheries; high amenity value.
1B	Good quality	Waters of less high quality than Class 1A, but usable for substantially the same purposes.
2	Fair quality	Waters suitable for potable supply after advanced treatment; supporting reasonably good coarse fisheries; moderate amenity value.
3	Poor quality	Waters which are polluted to the extent that fish are absent or only sporadically present; may be used for low-grade industrial abstraction purposes; considerable potential for further use if cleaned up.
4	Bad quality	Waters which are grossly polluted and are likely to cause nuisance.

Owing to problems experienced in the application of the above classification, two new schemes have been brought into use by the NRA for reporting and management of river water quality in 1994; the General Quality Assessment (GQA) and the statutory Water Quality Objectives (WQO).

The GQA provides a means of accurately assessing the general state of controlled waters that is nationally consistent and independent of the uses to which waters may be put. At present, the assessment is based on water chemistry, but future assessment will include biological data, nutrient input and aesthetic quality. The GQA class scheme is designed to be based upon three years of data.

The basic chemical grades of the GQA scheme are defined by the concentrations of biological oxygen demand, total ammonia and dissolved oxygen (see Table below).

GRADE	Description	Dissolved Oxygen (% sat.) 10 %tile	BOD (mg/1) 90 %tile	Total Ammonia (mg/1) 90 %tile
A	Good	80	2.5	0.25
B	Good	70	4.0	0.60
C	Fair	60	6.0	1.30
D	Fair	50	8.0	2.50
E	Poor	20	15.0	9.00
F	Bad	<20	–	–

Appendix 3 continued

Chemical quality objectives for watercourses were formerly set using the classifications of the NWC, known as the River Quality Objectives (RQO). These have been replaced by the WQO scheme, which addresses the chemical quality requirements to the needs of the uses to which the river can be put.

The WQO scheme establishes clear quality targets, to provide a planning framework for regulatory bodies and dischargers alike. The uses include River Ecosystem (RE), Special Ecosystem, Abstraction for potable supply, Agricultural/Industrial Abstraction and Water-sports. Standards defining a five-tiered RE classification were introduced by the Surface Water (River Ecosystem) (Classification) Regulations 1994 (see Table below).

River Ecosystem Class	**Description**
RE1	Very good; suitable for all fish species
RE2	Good; suitable for all fish species
RE3	Fair; suitable for high-class coarse fish
RE4	Fair; suitable for coarse fish populations
RE5	Poor; likely to limit coarse fish populations
Unclassed	Bad; fish unlikely to be present or insufficient data to classify water quality

The standards for the other uses are still under development.

In addition to the above classifications, the NRA also carries out biological monitoring to provide data on river water quality, assigning scores to particular invertebrate taxa that show varying degrees of sensitivity to pollution (Biotic Index) (see Table below).

Biotic Class	**Description**	**Biotic Index**
A	Good	150+
B	Fair	101–150
C	Moderate	51–100
D	Poor	16–50
E	Very poor	0–15

Appendix 4. The mean initial number of signs (spraints and sets of prints) per positive site in each Region surveyed in Scotland in 1977-79 and 1984-85, compared to the percentage occupation of sites in that Region (data from Green & Green, 1987).

Region of Scotland	Percentage occupation of sites surveyed		Mean initial number of signs per positive site	
	1977-79	1984-85	1977-79	1984-85
Shetland	98		5.14	
Western Isles	97		4.90	
Orkney	94		3.00	
Highland	92		4.23	
Dumfries & Galloway	82	87	3.05	4.25
Grampian	77	79	3.10	4.05
Tayside	61	76	3.06	3.29
Strathclyde	52	59	3.08	3.15
Central	49	66	2.94	3.00
Borders	31	39	1.86	2.33
Fife	5	23	1.50	2.65
Lothian	0	7	0.00	1.43

Appendix 5. The mean number of signs (largely spraints) per positive site in each Region surveyed in England in 1984-86 and 1991-94, compared to the percentage occupation of sites in that Region.

Region	Total Survey Sites	No. of positive sites		Percentage positive		Total signs		Mean signs/positive site	
		1984-86	1991-94	1984-86	1991-94	1984-86	1991-94	1984-86	1991-94
North West	333	31	93	9.31	27.93	69	513	2.23	5.52
Northumbrian	174	17	46	9.77	26.44	74	256	4.35	5.57
Yorkshire	270	6	28	2.22	10.37	11	89	1.83	3.18
Severn-Trent	610	22	126	3.61	20.66	43	433	1.95	3.44
Anglian	725	8	58	1.10	8.00	9	358	1.13	6.17
Thames	180	0	4	0.00	2.22	0	5	0.00	1.25
Wessex	154	1	29	0.65	18.83	1	104	1.00	3.59
South West	386	169	259	43.78	67.10	985	1,559	5.83	6.02
Southern	275	8	12	2.91	4.36	24	46	3.00	3.83
Welsh	81	24	51	29.63	62.96	77	267	3.21	5.24
Overall	**3,188**	**286**	**706**	**8.97**	**22.15**	**1,293**	**3,630**	**4.52**	**5.14**

Appendix 6. The influence of altitude on the percentage of sites found to be occupied and the mean number of signs (largely spraints) found per positive site for two Regions of England.

Altitude range in metres	Sites surveyed No. (% of total surveyed)	Found positive (with signs) No. (% of sites surveyed)	Total signs found	Mean no. of signs per positive site
SOUTH WEST REGION				
0 – 50	57 (14.77)	42 (73.68)	267	6.36
51 – 100	69 (17.88)	49 (71.01)	311	6.35
101 – 150	69 (17.88)	50 (72.46)	324	6.48
151 – 200	65 (16.84)	44 (67.69)	291	6.61
201 – 250	56 (14.51)	40 (71.43)	238	5.95
251 – 300	30 (7.77)	18 (60.00)	78	4.33
301 – 350	23 (5.96)	11 (47.83)	40	3.64
351 – 400	8 (2.07)	3 (37.50)	7	2.33
> 400	9 (2.33)	2 (22.22)	3	1.50
Totals	**386**	**259 (67.10)**	**1,559**	**6.02**
SEVERN-TRENT REGION				
0 – 50	103 (16.89)	28 (27.18)	99	3.54
51 – 100	112 (18.36)	34 (30.36)	127	3.74
101 – 150	135 (22.13)	25 (18.52)	94	3.76
151 – 200	108 (17.70)	23 (21.30)	71	3.09
201 – 250	61 (10.00)	8 (13.11)	25	3.12
251 – 300	31 (5.08)	4 (12.90)	10	2.50
301 – 350	27 (4.43)	2 (7.41)	3	1.50
351 – 400	17 (2.79)	1 (5.88)	2	2.00
> 400	16 (2.62)	1 (6.25)	2	2.00
Totals	**610**	**126 (20.66)**	**433**	**3.44**

Appendix 7. The percentage site occupation *(x)* for each Region of England, Wales and Scotland showing an increase at the second and third surveys. The percentage site occupation is shown at the start of the first and second surveys (column 3) and after a period of seven years (column 4). The increase in percentage site occupation in that time is the difference between the two (column 5). The percentage increase (column 6) is that difference calculated as a percentage of the starting figure. In column 7 this percentage is converted to the logarithm *(y)*. The relationship between *x* and *y* is shown graphically in Figure 23.

Country	Region	Percentage occupation at start *(x)*	Percentage occupation at finish	Difference	Percentage increase	Log percentage increase *(y)*
England 1977-79 to 1984-86	North West	2.78	9.57	6.79	244.24	2.3878
	Northumbrian	8.33	10.12	1.79	21.49	1.3322
	Yorkshire	1.77	2.21	0.44	24.86	1.3955
	Severn-Trent	2.29	3.88	1.59	69.43	1.8415
	South West	23.58	43.78	20.20	85.67	1.9328
	Southern	2.07	2.90	0.83	40.10	1.6031
	Welsh	14.81	29.63	14.82	100.07	2.0003
England 1984-86 to 1991-94	North West	9.57	28.70	19.13	199.90	2.3008
	Northumbrian	10.12	26.79	16.67	164.72	2.2167
	Yorkshire	2.21	11.06	8.85	400.45	2.6025
	Severn-Trent	3.88	21.16	17.28	445.36	2.6487
	Anglian	1.28	8.35	7.07	552.34	2.7422
	Wessex	0.65	18.83	18.18	2796.92	3.4467
	South West	43.78	67.10	23.32	53.27	1.7265
	Southern	2.90	3.73	0.83	28.62	1.4567
	Welsh	29.63	62.96	33.33	112.49	2.0511
Wales 1977-78 to 1984-85	Cleddau	41.07	53.57	12.50	30.44	1.4834
	Clwyd	3.70	33.33	29.63	800.81	2.9035
	Conwy	3.23	19.35	16.12	499.07	2.6982
	Dee	30.19	39.62	9.43	31.24	1.4947
	Dyfi	9.78	31.52	21.74	222.29	2.3469
	Severn	39.60	67.33	27.73	70.03	1.8453
	Teifi	38.36	39.73	1.37	3.57	0.5527
	Tywi	12.79	67.44	54.65	427.29	2.6307
	Usk	10.71	25.00	14.29	133.43	2.1253
	Wye	23.73	61.86	38.13	160.68	2.2060
	Ystwyth	30.16	60.32	30.16	100.00	2.0000
Wales 1984-85 to 1991	Cleddau	53.57	71.43	17.86	33.34	1.5230
	Clwyd	33.33	62.96	29.63	88.90	1.9489
	Conwy	19.35	35.48	16.13	83.36	1.9210
	Dee	39.62	49.06	9.44	23.83	1.3771
	Dyfi	31.52	52.17	20.65	65.51	1.8163
	Taff	4.76	28.57	23.81	500.21	2.6992
	Teifi	39.73	58.90	19.17	48.25	1.6835
	Tywi	67.44	68.60	1.16	1.72	0.2355
	Usk	25.00	50.00	25.00	100.00	2.0000
	Wye	61.86	83.05	21.19	34.25	1.5347
	Ystwyth	60.32	73.02	12.70	21.05	1.3233
Scotland 1977-79 to 1984-85	Grampian	77.13	78.95	1.82	2.36	0.3729
	Tayside	61.22	75.85	14.63	23.90	1.3784
	Central	48.63	66.44	17.81	36.62	1.5637
	Strathclyde	52.51	59.22	6.71	12.78	1.1065
	Fife	4.55	22.73	18.18	399.56	2.6016
	Borders	31.18	38.71	7.53	24.15	1.3829
	Dumfries & Galloway	82.37	86.96	4.59	5.57	0.7459

Appendix 8. The overall percentage site occupation ***(x)*** for England, Wales and Scotland at the start of the first and second surveys (column 2) and after a period of seven years (column 3). The increase in percentage site occupation in that time is the difference between the two (column 4). The percentage increase (column 5) is that difference calculated as a percentage of the starting figure. In column 6 this percentage increase is converted to the logarithm ***(y)***. The relationship between x and y is shown graphically in Figure 23.

Country	Percentage occupation at start *(x)*	Percentage occupation at finish	Difference	Percentage increase	Log percentage increase *(y)*
England 1977-79 to 1984-86	5.78	9.66	3.88	67.06	1.8265
England 1984-86 to 1991-94	9.66	23.37	13.71	141.90	2.1520
Wales 1977-78 to 1984-85	20.54	38.99	18.45	89.82	1.9534
Wales 1984-85 to 1991	38.99	52.48	13.49	34.60	1.5391
Scotland 1977-79 to 1984-85	57.02	64.79	7.77	13.63	1.1346

Appendix 9. The equation $y = 2.4328 - 0.02196x$ provides the overall relationship between percentage site occupation at the start (x) and the log of its percentage increase after a seven-year period (y) using British data. Using this equation and an initial figure of 1.0% site occupation at year 0 allows the calculation of a figure of 3.5% site occupation after seven years. A cumulative series of figures can then be calculated until a figure of 99.9% site occupation is reached after 112 years.

Years at start	Percentage site occupation at start (x)	Calculated log percentage increase (y)	Percentage increase	Difference in percentage site occupation after 7 years	Percentage site occupation at finish of period	Years at finish
0	1.0	2.4108	257.5135	2.5751	3.5	7
7	3.5	2.3543	226.0997	8.0833	11.6	14
14	11.6	2.1768	150.2450	17.5162	29.1	21
21	29.1	1.7921	61.9584	18.0761	47.2	28
28	47.2	1.3952	24.8428	11.7384	58.9	35
35	58.9	1.1374	13.7214	8.0941	67.0	42
42	67.0	0.9597	9.1138	6.1138	73.1	49
49	73.1	0.8254	6.6896	4.8966	78.0	56
56	78.0	0.7179	5.2228	4.0786	82.1	63
63	82.1	0.6283	4.2491	3.4916	85.6	70
70	85.6	0.5516	3.5612	3.0507	88.7	77
77	88.7	0.4846	3.0521	2.7077	91.4	84
84	91.4	0.4252	2.6620	2.4336	93.8	91
91	93.8	0.3717	2.3534	2.2088	96.0	98
98	96.0	0.3232	2.1047	2.0219	98.0	105
105	98.0	0.2788	1.9002	1.8638	99.9	112

Appendix 10. The survey results for the 386 sites examined in the South West Region are shown divided into groups of 25 sites numbered in order of survey. The numbers of mink and otter-occupied sites per 25 surveyed are shown for the surveys of 1984-86 and 1991-94. The 16th group was composed of only 11 survey sites, so the number in brackets is the calculated number of occupied sites per 25 surveyed, for comparison with the 15 groups above it. The mean (± standard error) occupation level per 25 sites for the two species in the two surveys is provided, together with the variance about the mean; indicative of the variability of the data. Overall percentage site occupation is also shown. Finally, columns 7 and 8 list the changes (gains and losses) in the numbers of sites occupied by mink and otters between the surveys of 1984-86 and 1991-94.

SOUTH WEST REGION							
Group	No. of survey sites	1984-86		1991-94		Change in no. of mink sites	Change in no. of otter sites
		Mink present	Otter present	Mink present	Otter present		
1	25	20	14	7	18	– 13	+ 4
2	25	16	5	7	15	– 9	+ 10
3	25	19	12	12	20	– 7	+ 8
4	25	16	4	11	14	– 5	+ 10
5	25	19	8	10	17	– 9	+ 9
6	25	16	5	9	13	– 7	+ 8
7	25	15	9	7	18	– 8	+ 9
8	25	18	11	11	19	– 7	+ 8
9	25	18	16	10	20	– 8	+ 4
10	25	16	16	9	21	– 7	+ 5
11	25	20	14	7	18	– 13	+ 4
12	25	19	16	10	18	– 9	+ 2
13	25	17	12	9	13	– 8	+ 1
14	25	14	12	7	13	– 7	+ 1
15	25	18	13	6	17	– 12	+ 4
16	11	8 (18.2)	2 (4.5)	3 (6.8)	5 (11.4)	– 5	+ 3
Totals	**386**	**269**	**169**	**135**	**259**	**– 134**	**+ 90**
Mean occupied sites per 25 surveyed ± S.E.		17.45 ± 0.45	10.72 ± 1.07	8.67 ± 0.47	16.59 ± 0.74		
Variance of above		3.2133	18.3323	3.4767	8.7758		
% overall site occupation		69.69%	43.78%	34.97%	67.10%		
Overall % change						49.81% loss	53.25% gain

Appendix 11. The survey results for the 610 sites examined in the Severn-Trent Region are shown divided into groups of 25 sites numbered in order of survey. The numbers of mink and otter-occupied sites per 25 surveyed are shown for the surveys of 1984-86 and 1991-94. The 25th group was composed of only 10 survey sites, so the number in brackets is the calculated number of occupied sites per 25 surveyed, for comparison with the 24 groups above it. The mean (± standard error) occupation level per 25 sites for the two species in the two surveys is provided, together with the variance about the mean. Overall percentage site occupation is also shown. Finally, columns 7 and 8 list the changes (gains and losses) in the numbers of sites occupied by mink and otters between the two surveys.

SEVERN-TRENT REGION

Group	No. of survey sites	1984-86		1991-94		Change in no. of mink sites	Change in no. of otter sites
		Mink present	Otter present	Mink present	Otter present		
1	25	8	0	10	3	+ 2	+ 3
2	25	13	0	12	3	− 1	+ 3
3	25	12	2	14	7	+ 2	+ 5
4	25	13	6	16	16	+ 3	+ 10
5	25	9	0	13	9	+ 4	+ 9
6	25	12	2	10	6	− 2	+ 4
7	25	9	3	11	8	+ 2	+ 5
8	25	10	2	15	7	+ 5	+ 5
9	25	16	7	14	16	− 2	+ 9
10	25	11	0	13	0	+ 2	0
11	25	18	0	10	19	− 8	+ 19
12	25	10	0	8	3	− 2	+ 3
13	25	10	0	15	2	+ 5	+ 2
14	25	8	0	7	0	− 1	0
15	25	16	0	21	8	+ 5	+ 8
16	25	10	0	13	3	+ 3	+ 3
17	25	17	0	15	9	− 2	+ 9
18	25	12	0	14	2	+ 2	+ 2
19	25	10	0	14	2	+ 4	+ 2
20	25	9	0	13	3	+ 4	+ 3
21	25	9	0	13	0	+ 4	0
22	25	3	0	10	0	+ 7	0
23	25	7	0	15	0	+ 8	0
24	25	3	0	4	0	+ 1	0
25	10	0	0	1 (2.5)	0	+ 1	0
Totals	**610**	**255**	**22**	**301**	**126**	**+ 46**	**+ 104**
Mean occupied sites per 25 surveyed ± S.E.		10.20 ± 0.85	0.88 ± 0.38	12.10 ± 0.78	5.04 ± 1.09		
Variance of above		18.0833	3.6100	15.2500	29.9567		
% overall site occupation		41.80%	3.61%	49.34%	20.66%		
Overall % change						18.04% gain	472.73% gain

Appendix 12. The survey results for the 725 sites examined in the Anglian Region are shown divided into groups of 25 sites numbered in order of survey. The numbers of mink and otter-occupied sites per 25 surveyed are shown for the surveys of 1984-86 and 1991-94. The mean (± standard error) occupation level per 25 sites for the two species in the two surveys is provided, together with the variance about the mean. Overall percentage site occupation is also shown. Finally, columns 7 and 8 list the changes (gains and losses) in the numbers of sites occupied by mink and otters between the two surveys.

ANGLIAN REGION

Group	No. of survey sites	1984-86		1991-94		Change in no. of mink sites	Change in no. of otter sites
		Mink present	Otter present	Mink present	Otter present		
1	25	9	0	16	0	+ 7	0
2	25	12	0	13	0	+ 1	0
3	25	10	0	17	0	+ 7	0
4	25	11	1	18	3	+ 7	+ 2
5	25	13	0	14	3	+ 1	+ 3
6	25	8	1	12	1	+ 4	0
7	25	9	0	15	4	+ 6	+ 4
8	25	4	0	6	4	+ 2	+ 4
9	25	5	0	10	7	+ 5	+ 7
10	25	1	1	7	3	+ 6	+ 2
11	25	0	2	7	3	+ 7	+ 1
12	25	0	0	0	1	0	+ 1
13	25	0	0	7	4	+ 7	+ 4
14	25	5	0	14	0	+ 9	0
15	25	3	0	8	4	+ 5	+ 4
16	25	0	0	2	2	+ 2	+ 2
17	25	1	0	18	2	+ 17	+ 2
18	25	4	0	15	0	+ 11	0
19	25	1	3	0	11	− 1	+ 8
20	25	0	0	3	1	+ 3	+ 1
21	25	1	0	2	1	+ 1	+ 1
22	25	0	0	1	1	+ 1	+ 1
23	25	2	0	1	0	− 1	0
24	25	4	0	1	0	− 3	0
25	25	5	0	1	1	− 4	+ 1
26	25	1	0	3	0	+ 2	0
27	25	7	0	6	2	− 1	+ 2
28	25	2	0	4	0	+ 2	0
29	25	0	0	4	0	+ 4	0
Totals	**725**	**118**	**8**	**225**	**58**	**+ 107**	**+ 50**
Mean occupied sites per 25 surveyed ± S.E.		4.07 ± 0.76	0.28 ± 0.13	7.76 ± 1.14	2.00 ± 0.46		
Variance of above		16.9236	0.4926	37.4039	6.1429		
% Overall site occupation		16.28%	1.10%	31.03%	8.00%		
Overall % change						90.68% gain	625.00% gain

Appendix 13. The numbers and release sites of captive-bred and rehabilitated otters released in England by the Otter Trust and The Vincent Wildlife Trust, respectively, from 1983 to 1993 (data from P. Wayre and J. Green, pers. comm.).

*(Note *: replacements released after the death of the original male due to road traffic accidents and other causes, see Section 7.2.5.3.)*

Release Number	Year	Number of otters ♂/♀	Date	River	County
(a) Releases of captive-bred otters by the Otter Trust					
01	1983	1/2	July	Black Bourn	Suffolk
02	1984	1/2	July	Thet	Norfolk
03	1984	1/1	October	Waveney	Norfolk
04	1985	1/2	July	Minsmere	Suffolk
05	1987	1/0*	June	Minsmere	Suffolk
06	1987	1/1	September	Glaven	Norfolk
07	1988	1/1	September	Catfield	Norfolk
08	1989	1/1	May	Stour	Dorset
09	1989	1/1	December	Wylye	Wiltshire
10	1989	1/1	December	Yare	Norfolk
11	1990	1/1	July	By Brook	Wiltshire
12	1990	1/1	October	Catfield	Norfolk
13	1991	1/2	May	Glaven	Norfolk
14	1991	1/2	May	Stour	Dorset
15	1991	1/2	October	Lea/Ash	Hertfordshire
16	1991	1/2	December	Stort	Hertfordshire
17	1992	1/1	March	Waveney	Norfolk
18	1992	1/2	July	Wissey	Norfolk
19	1992	1/0*	August	Glaven	Norfolk
20	1992	1/0*	August	Yare	Norfolk
21	1993	1/2	May	Tiffey	Norfolk
22	1993	1/2	June	Deben	Suffolk
23	1993	1/2	August	Itchen	Hampshire
24	1993	1/0*	October	Wissey	Norfolk
Total		**24/31 = 55**			
(b) Releases of rehabilitated otters by The Vincent Wildlife Trust					
				River Derwent catchment	
01	1990	2/2	April	Helmsley	N. Yorkshire
02	1991	0/1	June	Elleron Lo.	N. Yorkshire
03	1991	1/0	July	Elleron Lo.	N. Yorkshire
04	1991	1/1	July	Wykeham	N. Yorkshire
05	1991	1/1	September	Hilla Green	N. Yorkshire
06	1992	2/1	April	Thornton Ings	N. Yorkshire
07	1992	3/0	May	Buttercrambe	N. Yorkshire
08	1992	0/1	July	Thornton Ings	N. Yorkshire
09	1992	1/1	September	Thornton Ings	N. Yorkshire
10	1993	1/1	July	Buttercrambe	N. Yorkshire
				River Esk catchment	
11	1993	1/1	December	Grosmount	N. Yorkshire
12	1993	1/1	December	Egton Bridge	N. Yorkshire
Total		**14/11 = 25**			
Grand total to 1993		**38/42 = 80**			

Appendix 14. Organochlorine insecticide (plus metabolites) and polychlorinated biphenyl residues found in the livers of 12 English otters reported in the literature. These concentrations are expressed as mg/kg^{-1} of extractable lipid in liver samples. The amount of extractable lipid in the sample expressed as milligrams of lipid per gram of liver tissue and the multiplication factors for obtaining wet weight concentrations from lipid concentrations are given in column 4 for otters 9 to 12 (see Section 7.3.1.2.3). Analyses 1 to 8 were completed at the University of Essex and 9 to 12 at Monks Wood (Institute of Terrestrial Ecology). Otter 1 was found dead and emaciated, and the causes of death of otters 9 to 12 are as given in Appendix 15. Unfortunately Mason *et al* (1986a) and Mason (1988) only list road traffic accidents, drownings and found dead from unknown causes as the causes of death of all their samples without noting the cause of death of individual animals. (N = below detectable limit of 0.01 mg/kg^{-1}; tr = trace, present at very low concentrations and left unquantified.)

Reference No.	County	Date/Sex	Extractable lipid in mg/g tissue (lipid to wet weight factor)	Concentrations in liver expressed as mg/kg^{-1} of extractable lipid							Reference
				PCB	Lindane	Dieldrin	DDE	DDD	DDT	Heptachlor epoxide	
1	Norfolk	1984/F	?	232.0	8.3	59.0	100.0	N	N	N	Mason *et al* (1986a)
2	Norfolk	1982-85/M	?	14.2	N	7.9	tr	N	5.9	N	"
3	Norfolk/ Suffolk border	1982-85/M	?	147.3	N	27.8	43.1	N	N	N	"
4	Shropshire	1982-85/M	?	4.4	tr	14.7	32.1	22.0	9.1	N	"
5	Cambridge	1985-86/F	?	20.0	7.4	N	5.3	1.5	N	N	Mason (1988)
6	Devon	1985-86/F	?	tr	2.6	tr	1.5	1.6	N	N	"
7	Devon	1985-86/M	?	tr	1.2	N	4.7	4.4	N	N	"
8	Devon	1985-86/M	?	0.3	1.8	N	21.0	10.0	N	N	"
9	Norfolk	1986/F	23.67 (x 0.02356)	73.0	N	12.0	20.5	N	N	N	Spalton & Cripps (1989)
10	Norfolk	1984/M	51.84 (x 0.05115)	5.67	3.50	15.17	10.26	8.32	N	2.63	Jefferies (1985)
11	Suffolk	1987/M	42.93 (x 0.04267)	N	1.24	4.35	12.42	N	N	N	Jefferies & Hanson (1988a)
12	Suffolk	1987/M cub	53.91 (x 0.05389)	61.98	0.69	3.23	9.22	N	N	N	"

Appendix 15. Heavy metal (mercury, cadmium, copper, zinc) concentrations found in the livers and kidneys of four English otters reported in the literature. These concentrations are expressed as mg/kg^{-1} of dry weight of tissue and all analyses were completed at Monks Wood (Institute of Terrestrial Ecology). The reference numbers in column 1 relate to those for the same otters in Appendix 14 (organochlorines). The causes of death were drowning in fyke net (10); found blind, sick, emaciated and dying (9); road traffic accidents (11,12). (N = below detectable limit.)

Ref. No.	County	Date/Sex	Concentrations expressed as mg/kg^{-1} dry weight of tissue				Reference
			Hg	Cd	Cu	Zn	
Liver							
10	Norfolk	1984/M	0.90	N	9.44	53.70	Jefferies (1985)
9	Norfolk	1986/F	6.80	N	13.02	162.50	Spalton & Cripps (1989)
11	Suffolk	1987/M	7.23	0.14	44.80	123.00	Jefferies & Hanson (1988a)
12	Suffolk	1987/M cub	2.27	N	10.20	125.00	"
Kidney Cortex							
10	Norfolk	1984/M	0.75	N	11.46	57.46	Jefferies (1985)
9	Norfolk	1986/F	9.18	N	15.19	78.13	Jefferies (pers. comm.)
11	Suffolk	1987/M	6.38	0.44	16.30	97.00	Jefferies & Hanson (1988a)
12	Suffolk	1987/M cub	2.39	N	10.70	90.70	"

Appendix 16. Scientific names and taxonomy of species mentioned in the text by their English names.

Order (Family)	English name	Scientific name (Authority)
VERTEBRATES		
Mammalia		
Insectivora (Soricidae)	Water shrew	*Neomys fodiens* (Pennant)
Chiroptera (Vespertilionidae)	Pipistrelle	*Pipistrellus pipistrellus* (Schreber)
Lagomorpha (Leporidae)	Rabbit	*Oryctolagus cuniculus* (L.)
Rodentia (Muridae)	Bank vole	*Clethrionomys glareolus* (Schreber)
" "	Field vole	*Microtus agrestis* (L.)
" "	Water vole	*Arvicola terrestris* (L.)
" "	Field mouse	*Apodemus sylvaticus* (L.)
" "	Brown rat	*Rattus norvegicus* (Berkenhout)
" (Capromyidae)	Coypu	*Myocastor coypus* (Molina)
Cetacea (Phocoenidae)	Harbour porpoise	*Phocoena phocoena* (L.)
Carnivora (Canidae)	Domestic dog	*Canis familiaris* L.
" "	Red fox	*Vulpes vulpes* (L.)
" "	Wolf	*Canis lupus* L.
" (Mustelidae)	Pine marten	*Martes martes* (L.)
" "	Stoat	*Mustela erminea* L.
" "	Weasel	*Mustela nivalis* L.
" "	Polecat	*Mustela putorius* L.
" "	Ferret	*Mustela furo* L.
" "	American mink	*Mustela vison* Schreber
" "	Badger	*Meles meles* (L.)
" "	Eurasian otter	*Lutra lutra* (L.)
" "	Sea otter	*Enhydra lutris* (L.)
" "	Asian short-clawed otter	*Aonyx cinerea* (Illiger)
" "	Indian smooth-coated otter	*Lutrogale perspicillata* (Geoffroy)
" (Felidae)	Domestic cat	*Felis catus* L.
" "	Tiger	*Panthera tigris* (L.)
" "	Leopard	*Panthera pardus* (L.)
" "	Jaguar	*Panthera onca* (L.)
" "	Puma	*Felis concolor* L.
Pinnipedia (Phocidae)	Common seal	*Phoca vitulina* L.
" "	Grey seal	*Halichoerus grypus* (F.)
Aves		
Ciconiiformes (Ardeidae)	Grey heron	*Ardea cinerea* L.
Falconiformes (Falconidae)	Peregrine falcon	*Falco peregrinus* Tunstall
" (Accipitridae)	Sparrowhawk	*Accipiter nisus* (L.)
" "	Common buzzard	*Buteo buteo* (L.)
" "	White-tailed eagle	*Haliaeetus albicilla* (L.)

Appendix 16 continued

Order (Family)	English name	Scientific name (Authority)
VERTEBRATES (*cont.*)		
Aves (cont.)		
Galliformes (Phasianidae)	Pheasant	*Phasianus colchicus* L.
" "	Grey partridge	*Perdix perdix* (L.)
" "	Domestic chicken	*Gallus domesticus* L.
Charadriiformes (Alcidae)	Guillemot	*Uria aalge* (Pontoppidan)
" "	Razorbill	*Alca torda* L.
Columbiformes (Columbidae)	Woodpigeon	*Columba palumbus* L.
Coraciiformes (Alcedinidae)	Kingfisher	*Alcedo atthis* L.
Amphibia		
Anura (Bufonidae)	Natterjack toad	*Bufo calamita* Laurenti
Pisces		
Isospondyli (Clupeidae)	Allis shad	*Alosa alosa* (L.)
" (Salmonidae)	Salmon	*Salmo salar* L.
" "	Sea trout	*Salmo trutta* L.
" "	Brown trout	*Salmo trutta* L.
" "	Rainbow trout	*Salmo gairdneri* Richardson
" (Thymallidae)	Grayling	*Thymallus thymallus* (L.)
Haplomi (Esocidae)	Pike	*Esox lucius* L.
Ostariophysi (Cyprinidae)	Carp	*Cyprinus carpio* L.
" "	Minnow	*Phoxinus phoxinus* (L.)
" "	Chub	*Leuciscus cephalus* (L.)
" "	Dace	*Leuciscus leuciscus* (L.)
" "	Roach	*Rutilus rutilus* (L.)
" (Cobitidae)	Stone loach	*Noemacheilus barbatulus* (L.)
Apodes (Anguillidae)	Eel	*Anguilla anguilla* (L.)
Anacanthini (Gadidae)	Cod	*Gadus morrhua* Day
Perciformes (Percidae)	Perch	*Perca fluviatilis* L.
Scleroparei (Cottidae)	Bullhead	*Cottus gobio* L.
INVERTEBRATES		
Insecta		
Diptera (Chloropidae)	Frit fly	*Oscinella frit* L.
" (Muscidae)	Wheat bulb fly	*Leptohylemyia coarctata* (Fall.)
" (Tipulidae)	Leatherjacket [larva of]	*Tipula* sp.
" (Calliphoridae)	Greenbottle [causing fly strike]	*Lucilia caesar* L.
Coleoptera (Elateridae)	Wireworm [larva of]	*Agriotes lineatus* (L.)

Appendix 16 continued

Order (Family)	English name	Scientific name (Authority)
INVERTEBRATES ***(cont.)***		
Crustacea		
Decapoda (Portunidae)	Shore crab	*Carcinus maenas* (L.)
" (Cancridae)	Edible crab	*Cancer pagurus* L.
" (Nephropsidae)	Lobster	*Homarus vulgaris* Milne Edwards
" (Astacidae)	Crayfish	*Austropotamobius pallipes* (Lereboullet)
Nematoda		
(Ascaridata)	Sealworm	*Pseudoterranova decipiens* (Krabbe)
TREES		
(Fagaceae)	Beech	*Fagus sylvatica* L.
"	Oak	*Quercus robur* L.
(Oleaceae)	Ash	*Fraxinus excelsior* L.
(Aceraceae)	Sycamore	*Acer pseudoplatanus* L.

Notes

Notes